Handbook of Biodiesel

Handbook of Biodiesel

Edited by **Kurt Marcel**

LANRYE
INTERNATIONAL

New Jersey

Published by Clanrye International,
55 Van Reypen Street,
Jersey City, NJ 07306, USA
www.clanryeinternational.com

Handbook of Biodiesel
Edited by Kurt Marcel

International Standard Book Number: 978-1-63240-255-4 (Hardback)

Contents

Permissions

List of Contributors

Preface

This book covers nearly every aspect of biodiesel production and its applications. All facets from the development of feedstocks and its processing mechanisms are described in the book. This book would appeal to the students, scientists, researchers and professionals who are involved in this genre of science.

After months of intensive research and writing, this book is the end result of all who devoted their time and efforts in the initiation and progress of this book. It will surely be a source of reference in enhancing the required knowledge of the new developments in the area. During the course of developing this book, certain measures such as accuracy, authenticity and research focused analytical studies were given preference in order to produce a comprehensive book in the area of study.

This book would not have been possible without the efforts of the authors and the publisher. I extend my sincere thanks to them. Secondly, I express my gratitude to my family and well-wishers. And most importantly, I thank my students for constantly expressing their willingness and curiosity in enhancing their knowledge in the field, which encourages me to take up further research projects for the advancement of the area.

Editor

Feedstocks

Algal Biorefinery for Biodiesel Production

Didem Özçimen, M. Ömer Gülyurt and Benan İnan

Additional information is available at the end of the chapter

1. Introduction

In recent years, the rapid depletion of fossil fuels, increase in energy demand, global warming, increase in price of fossil fuels depends on economic and political behaviors increased orientation to alternative energy sources. In this context, biodiesel that is one of the renewable alternative energy sources draws attention because of its useful features such as easily biodegradable and environmentally friendly. However, biodiesel production from oil crops does not meet the required demand of vehicle fuel, and recently it is not economic and feasible. It needs to be improved to produce more economically to be able to compete with diesel in the market. Vegetable oils and crops which biodiesel produced from are a kind of human food sources and the shortage on food source cause to go up prices and make the biodiesel high-priced. To meet the requirements, the interest on algae is increased day by day since this technology has potential to meet global demand [1]. Microalgae have higher productivity per area and no need for farm field to grow as opposed to oil crops and animal fat. Microalgae use sunlight to reduce CO_2 to biofuels, foods, fertilizers, and valuable products. Furthermore, microalgae can be used to get different types of biofuels. Using microalgae as fuel source is not a novel idea but recently the prices of diesel and global warming hit this solution to the top [2].

Microalgae have lots of advantages for biodiesel production over other raw materials such as crops, waste cooking oils, and so on. Microalgae have short doubling time which is around 12-24 h since they have a simple structure and capable to high photosynthetic efficiency and they contain much more amount of oil than other oil crops that can be used as oil source for biodiesel production. Compared with the oil yields from various oil crops such as corn (172 L/ha), soybean (446 L/ha), canola (1190 L/ha), jatropha (1892 L/ha), coconut (2689 L/ha) and oil palm (5959 L/ha), oil yield from microalgae is very high as 136900 L/ha and 58700 L/ha for 70% oil in biomass and 30% oil in biomass, respectively [2-4].

The other significant feature is that algae can grow everywhere and every season in a year since there are thousands of algae species that have different adaptations and different properties. They can grow in saltwater, freshwater, lakes, deserts, marginal lands, etc. In addition to biodiesel production, algae can be also used as feedstock to produce different valuable products such as fertilizer, energy, neutraceuticals, protein, animal feed etc. The other significant property is that microalgae can remove some heavy metals, phosphorous, and nitrogen from water during its growth. Algae also clean up the water. Moreover, microalgae sequester lots of carbon by photosynthesis. Utilization of carbon dioxide by algae is significantly lowering the risk for greenhouse gas effects. Lastly, usage of microalgae for biodiesel almost cancels out the carbon dioxide and sulfur release to atmosphere [5]. These reasons mentioned above are enough to believe that microalgae can take the place of fossil fuels completely.

There are many of microalgae studies for biodiesel production. Because the most of the scientists believe that microalgae will take the place of the petroleum diesel, however, algal biodiesel production is not feasible yet since there is no much commercial or large scale production of microalgae for biodiesel. That is why most of the works are focused on decreasing the cost of biodiesel production or make it competitive versus petroleum diesel. Surely, until these improvements are achieved, algal biodiesel can not be an accurate alternative. The current problems making biodiesel expensive can be improved with some innovations. The first of all is about the algae strain which is also first step of algal biodiesel production. The algae strain should be better than recent ones. There are natural many kinds of algae strains and isolation of new natural algae strain may help procedure to be cost effective. The algae strain has to have high lipid productivity and adaptability to new environments. These features let it produce more and obtain more oil content [6, 7]. As an example, if the flue gas is used as carbon dioxide source, microalgae have to be adapted for this situation so that it can tolerate the high concentration of SO_x, NO_x, and other gases [8]. That will reduce the cost and increase the biomass growth rate. The other important innovation should focus on cultivation of algae. The large-scale production is one of the most cost-intense parts. The innovative thinking should show a tendency to lower the cost of operation and capital for cultivation systems. As it is explained below, open ponds are the cheapest way but the efficiency of them has to be worked on. Moreover, the closed photobioreactors (PBR) are also being improved for a cheaper way to control and lighten the system. Furthermore, microalgae can be fixed in a cultivation system with an immobilization technique to get higher biomass. The last way to lower the cost is to produce sub-products from microalgae beyond biodiesel. There are lots of high value products and sub-products produced from microalgae such as biogas [9, 10], biobutanol, acetone [11], Omega 3 oil [12], eicosapentaenoic acid [13], livestock feed [14], pharmaceuticals and cosmetics [15, 16]. Especially sub-products can be preferred for economic support of main process.

For example, recovery of methane from microalgae pulp after biodiesel production develops renewability of conversion of microalgae biomass to biodiesel process as much as it makes the cost of process and environmental effects less. The microalgae pulps after oil removed contain significant amounts of protein and carbohydrate that can convert to biogas by anaerobic fermentation. Conversion of algal waste to biogas by anaerobic fermentation will play a dual role for renewable energy production and also sustainable development of microalgal biodiesel industry [17, 18].

Algae can be also used in bioethanol production. Algae are more uniform and continuous than terrestrial plant, due to lack of functional parts such as root and leaf composition. Their cell walls made of polysaccharides that can hydrolyze to the sugar. For this reason, microalgae can be used as carbon source in fermentation process. Ethanol produced by fermentation can be purified for using as a fuel, CO_2 as a nutrient may also be recycled to algae culture to grow microalgae [19, 20].

In this chapter, algae production methods that cover the algae strain and location selection, algae cultivation, harvesting, oil extraction, and algal biodiesel production processes are presented in detail with alternatives. New progresses in this area are also explained.

2. Algae strains and properties

Algae are simple organisms including chlorophyll. They can be found in seas, soils and lakes wherever they can use the light for their photosynthesis. There are two types of main algae groups. The first group is macro algae, which includes green, brown and red algae. The second group is microalgae as phytoplankton in the coasts, lakes and oceans, which includes diatoms, dynoflagellates, green and brownish flagellate, and blue-green algae [21].

The classification of algae can be done in many ways since there is a millions of kind. Also there is no standard on classification so you can see different types of classification. The taxonomic group of algae can be given as follow: *Archaeplastida, Chlorophyta*(green algae), *Rhodophyta*(red algae), *Glaucophyta, Chlorarachniophytes, Euglenids, Heterokonts, Bacillariophyceae*(diatoms), *Axodine, Bolidomonas, Eustigmatophyceae, Phaeophyceae*(brown algae), *Chrysophyceae*(golden algae), *Raphidophyceae, Synurophyceae, Xanthophyceae*(yellow-green algae), *Cryptophyta, Dinoflagellates, Haptophyta*[22].

Algae are the most common wide photosynthetic bacteria ecologically. To grow algae some parameters such as amount and quality of ingredients, light, pH, turbulence, salinity, and temperature become prominent. Macro (nitrate, phosphate, silicate) and micro (some metals, B1, B12 and biotin vitamins) elements are required in the growth of algae. Light intensity has also an important role, the light demand changes up to microalgae density and type of microalgae. The other parameter pH is mostly between 7 and 9 for most of algae strains and mostly the optimum range is 8. 2-8. 7. The last parameter salinity should be between 20-24 ppt. Moreover, nitrogen also affects the growth of some algae strains as such as green algae [22-25].

2.1. Macroalgae

Macroalgae are adapted to life in ocean and it is a plant mostly seen on the costal strips. There are plenty of macro algae types. Algae can be classified as brown, red, and green based on type of pigments. Recently, several brown algae types have been used in the industry and energy production as an alternative source to fossil fuels, and green algae is also studied to produce biodiesel [26].

Brown algae have xanthophyll pigments and fucoxanthin, which results the colour of brown algae. These substances mask the other pigments [27]. Polysaccharides and higher alcohols are nutrition reserves of brown algae but the main carbohydrate reserve is laminarin. The cell walls of brown algae are made of cellulose and alginic acid. Brown algae have a lot of features such as: Cytotoxic and antitumor activity, Antifungal activity, Anti-inflammatory activity, Antiviral activity, Protection against herbivorous animals (fish, sea urchins), Anti-oxidant activity [21, 28, 29]. Composition of brown algae can vary according to species, their location, salinity and season. According to analysis, brown algae contain about 85% high moisture and 25 % high sodium carbonate [26].

Green algae contain chlorophyll a and b. Presence of these pigments makes green color of the green algae. There are a few reports about second metabolites of green algae. [21]. Moisture content of green algae is higher than brown algae but they have similar sodium carbonate content. Green algae species can access higher sugar levels and this makes them useful energy sources. They also have high cellulose content [26]. Green algae have a lot of features such as: Anti-inflammatory substances, Cytotoxic and immunosuppressive activities, Anti-bacterial activity, Antiviral activity, Antifungal activity [30].

Red Algae have phycoerythrin and phycothcyanin pigments that make red color of these algae. These pigments mask the other pigments. The cell walls of red algae made of cellulose, agar and carrageenan [27]. There are approximately 8000 red algae species. In comparison of the other algae species, red algae are considered as the most important active metabolite resource. They have a lot of features such as: Cytotoxic activities, Antiviral activity, Anti-inflammatory activity, Antimicrobial activity, Free radical scavenger activity [21, 31].

2.2. Microalgae

There are at least 30000 microalgae species in the world. Microalgae are mostly defined as unicellular photosynthetic cells but some complex associations create larger colonies. This is a heterogenic group, which contains prokaryotic organisms similar to bacteria and eukaryotic cells [26, 32]. Microalgae production is concentrated on particular species, which have special tolerance for extreme conditions in their growth. This situation enables the production in open ponds and canals. In future, microalgae production will focus on more advanced species for the demand of energy and pure monocultures which have specific capabilities like production of carbohydrate, lipid or hydrogen will be cultivated [33]. According to use of algae, biomass of microalgae has variable chemical composition. They can be rich or balanced composition of protein, lipid and sugar. Microalgae selection should be made according to desired biofuels. Microalgae have important lipid content even in the extreme conditions they reach higher lipid content [26].

Green algae or diatoms are the most used microalgae species for production of alternative energy derives. Just a handful of these species has commercial importance. This group contains *Chlorella, Spirulina, Dunaliella* and *Haematococcus*. Only *Dunaliella* is a dominant sea species. These are usually cultivated for extraction of high value component like pigments or proteins [26].

Blue-green algae (cyanobacteria) have a lot of common structural features with bacteria. They are classified as algae because they contain chlorophyll and other components. They have also nitrogenic components because all of the prokaryote species convert atmospheric nitrogen to ammonium [21, 34]. Morphologically blue green algae can have filamentous, conical or unicellular shape. They have a lot of features such as: anticancer and cytotoxic activities, antibacterial activity, antifungal activity, immunosuppressive activity [21, 35, 36].

Pyrrhophyta (Dinoflagellates) are unicellular organisms, which are classified as primitive algae. Large amount concentrations of these organisms exist in ocean surface and they cause fish deaths. Also because of their pigments, dinoflagellates give the water brown to red coloration in the sea [34, 37]. Particular dinoflagellate species produce toxin in case of consumed by species such as shellfish. Consumption of contaminated shellfish by humans can cause a lot of health problems including death [21].

Bacillariophyceae (Diatoms) are the most versatile and frequent family. They are more feasible for large-scale productions due to short doubling time and easy to grow. Unlike *Dinoflagellates* they create less second metabolites [38].

Microalgae are investigated as biodiesel feedstock because of their high photosynthetic efficiency, their ability to produce lipids. Macroalgae usually don't contain lipids too much and they are taken into consideration for the natural sugars and other carbohydrates that they contain. These contents can be fermented to produce alcohol-based fuels or biogas.

2.3. Lipid content of microalgae species

As the structure of many microalgae species can accumulate significant amounts of lipid and provide high oil yield. Their average lipid contents can be reached to 77% of dry biomass under some certain conditions [39]. Table 1 shows lipid content of some microalgae species.

Microalgae	Oil content (dry weight %)
Botryococcus braunii	25-75
Chlorella protothecoides	14-57
Crypthecodinium cohnii	20-51
Dunaliella tertiolecta	16-71
Nannochloris sp.	20-56
Neochloris oleoabundans	29-65
Phaeodactylum tricornutum	18-57
Schizochytrium sp.	50-77
Skeletonema coastatum	13-51

Table 1. Lipid content of some microalgae species [15, 39, 40-45].

Also high productivity is very important beside high oil content. As shown in table 1, microalgal lipid content can reach 77% by weight of dry biomass but it is observed that there can be low productivity of *Botryococcusbraunii*, however, *Chlorella* appears to be a good choice in biodiesel production, since it has high productivity though lower oil content [39].

Lipid content can be affected by several parameters such as nutrition, environment, cultivation phases and conditions growth can affect fatty acid composition [32], Fatty acid composition is important in microalgae selection because it has a significant effect on biodiesel properties. For example, if unsaturated fatty acid content is high in algal oils and their presence reduces the efficiency of esterification to produce biodiesel [39].

Value chain stages of biodiesel production from microalgae can be given as algae and site selection, algae cultivation, harvesting, filtration, dewatering, oil extraction and biodiesel production [39].

3. Biodiesel production from microalgae

The selection of species depends on some factors like ability to usage of nutrition or grow under specific environment conditions. All these parameters should be evaluated for biodiesel production.

3.1. Selection of algae strain and location

To make algal biodiesel cost effective lots of researchers keep going on algae culturing. The criteria to select location and sources are mentioned below [46]:

• Water sources and demand, salinity, content

• The region information such as topography, geology

• Weather conditions, isolation, evaporation

• Availability of carbon and food resources

The next decision should be on the algae culturing process type. It can be either batch or continuous process. Depending on microalgae strain, environmental conditions, availability of nutrition and moreover industrial pollutions the process type has to be selected. The devices and apparatuses also have to be adjusted for these conditions and nutrients [39].

Algae strains have different contents, different doubling time (the total biomass per time and volume) and resistance to change in environmental conditions. Biodiesel production directly depends on the oil content of microalgae and its efficiency. So that, even the process and culturing systems are selected perfectly, time and other related factors plays an important role [39].

3.2. Methods used for algae growth

Not only the microalgae strain is important for efficiency of oil but also growing conditions are important. There are different ways to grow algae. Each type of microalgae has a different mechanism which let them to respond different weather and environmental conditions [39, 47]. Different growing conditions affect the microalgae doubling time. There are 4 growing type basically: phototrophic, heterotrophic, mixotrophic, and photo heterotrophic. All of them will be explained in detail.

3.2.1. Phototrophic growth

Microalgae are mostly thought to be phototrophic since it requires light [48]. Phototrophic growing method is based on using light and carbon dioxide to produce chemical energy during photosynthesis. This is the most common way used to grow microalgae. The best advantage of the process is using carbon dioxide as a carbon source to grow or produce fatty acid. Since carbon dioxide is only the carbon source, locations close to fabrics and companies could be selected to procure carbon dioxide. If it is compared to other growing types, phototrophic method has the lowest contamination risk [49].

3.2.2. Heterototrophic growth

Some microalgae are not able to grow phototrophic conditions but they can grow in dark using organic carbon as a carbon source like bacteria. If microalgae is using organic carbon these microalgae are heterotrophic growing algae. Heterotrophic growth has advantages over phototrophic growth because light is not required. The biggest problem with the phototrophic is the light penetration when the density of the culture gets higher. In that way one of the biggest problems is solved with heterotrophic growth. Heterotrophic growth will be more cost effective compared to phototrophic growth [48]. And this method is said the most practical and promising way to increase the productivity [50-52]. Also higher oil rates and efficiency can be obtained when the algae grow heterotrophic, but the contamination risk is much higher compared to phototrophic [49].

Microalgae uses different organic carbon sources such as glucose, acetate, glycerol, fructose, sucrose, lactose, galactose, and mannose, especially growth with sugar is more efficient [49].

Mostly the organism growing heterotrophic should have adaptation property to new habitat as soon as possible since when culturing to new media the lag phase should be too short, and durability during processing in fermenters and other machines [48].

3.2.3. Mixotrophic growth

Mixotrophic growth is a combination of phototrophic and heterotrophic growth. Mixotrophic growth is using organic and inorganic carbon and the process requires light because of photosynthesis. Thus the microalgae have ability to live in both conditions. Microalgae uses organic compounds and carbon dioxide as a carbon source and the released carbon dioxide are also captured with the photosynthesis. Although mixotrophic-growing method mostly is not preferred compared to heterotrophic and phototrophic growth [49], because of other advantages even so mixotrophic method is applied in some studies. For example; Park *et al.* found that biomass and lipid productivities were boosted by mixotrophic cultivation [53]. Bhatnagar et al. found the mixotrophic growth of some microalgae strains resulted in 3–10 times more biomass production compared to that obtained under phototrophic growth conditions [54].

3.2.4. Photoheterototrophic growth

When microalgae use organic compounds as carbon sources, sometimes it requires light. The main difference between mixothrophic and photoheterotrophic is that mixotrophic growth using organic compounds as energy sources, as photoheterotrophic growth requires light as energy source. This method is mostly used for production of some beneficial metabolites; however, it is rarely used for biodiesel production [49]. Metabolisms can split into groups due to pH changes. *Chlorella vulgaris, Haematococcus pluvialis, Arthrospira (Spirulina) platensis* strains are the examples for the growth by mixotrophic, phototrophic and heterotrophic methods. *Selenastrum capricornutum* and *Scenedesmus acutus* are able to grow in phototrophic, heterotrophic, photoheterotrophic conditions [47].

Algae require more than organic carbon, sugar, protein, oil or any carbon sources. Algae cannot grow without vitamins, salts, or some other nutrients (nitrogen and phosphor). Moreover, there are lots of parameters has to be controlled during algae growth to maximize and stabilize the production. Some of these parameters are oxygen rate, carbon dioxide rate, pH, heat, light intensity and so on. When appropriate weather conditions and enough nutrients are provided microalgae grow faster. Mostly doubling time is between 3. 5 h and 24 h [39].

As a result, if we compare different methods mentioned above for microalgae growth;Heterotrophic growth is much better than the others for the application of biodiesel. These methods can produce more oil than other growing types. However, heterotrophic cultures may contaminate especially in open pond systems and result in big problems in large-scale production. Moreover, organic carbon as a carbon source is an expensive raw material and makes the process cost higher. Phototrophic growth is an easily scalable and mostly uses the carbon dioxide from exhaust gas for the production of oil. However, the efficiency of the oil is lower than heterotrophic growth because the biomass doubling time is higher and total biomass rate is lower at the end. Phototrophic method mostly preferred to set a cost effective system [49].

3.2.5. Conditions for growth of algae

3.2.5.1. Light

The microalgae growing photosynthetically needs light and the light intensity is the most significant limiting factor. Algae culture systems mostly use both sun and lamp light. Mostly lamp-lightened algae culture systems uses wider screens to be able to absorb more light from the system. For photosynthetic production, at least 50 % of the volume of PBR has to get enough light [55]. Open raceway ponds, plate, plate PBR, Vertical-column PBRs, Internally-illuminated PBRs, inclined tubular type, horizontal/continuous type, bubble column and air-lift PBRs are the systems used for photosynthetic algae growth. Plate photo bioreactor is more efficient than tubular photo bioreactor because the light can penetrate to bottom more in plate design. Recent works are on closed system photobioreactors to improve the capacity. Some works are done to increase the capacity; however the light penetration becomes a major problem. Light source for open ponds is only Sun. That is why the alteration is not possible for raceway ponds. The depth of the pond that the only thing can be changed. Thus mostly researches are going on closed systems to optimize light emission. Mostly photobioreactors in lab scale are lightened by fluorescence lights from inside and outside [56]. The light wavelength should be between 600-700 nm to maximize the photosynthesis. Light intensity depends on microalgae density. Higher algae density requires higher light intensity. Light also affects the lipid content. Yeesang and Cheirsilp reported that the lipid contents in all strains increased with increasing light intensity in their study [57].

Changes in light intensity and quality can alter biofuel quality [58]. Each type of microalgae has its own optimal light absorbing point. If this point exceeds the optimum point, microalgae light absorption ratio decreases. After a specific point, light decreases the biomass production and this is called photoinhibition. Photoinhibition processes depend on time and after stress of light for a few minute biomass loss starts. 10-20 min later more than 50 % damage can be seen. Cheirsilp and Torpee investigated the effect of light intensity on growth and lipid content of marine *Chlorella sp.* and *Nannochloropsis sp.* The growth of marine *Chlorella sp.* increased when the light intensity was increased from 2000 to 8000 lux. But up to 10000 lux its growth decreased. They reported that this could be some extent of effect from photoinhibition. The growth of *Nannochloropsis sp.* continuously increased up to the maximum level when increasing light intensity up to a maximum light intensity of 10000 lux. [59]. High light intensity limited algal growth, but gave the benefit of higher lipid content and yield. It can be seen in Ruangsomboon's study whose cultures exposed to low light intensity showed a higher biomass compared to others [60].

To increase the microalgae production, photoinhibition should be cut off or exceed to high light intense. In addition, photorespiration decreases the photosynthetic efficiency. Therefore the process has to avoid photorespiration. Photorespiration occurs when the oxygen concentration increases depending on carbon dioxide [56].

Sara et al. investigated the light effects on microalgae. The research was done by using red and blue lasers as light source for photosynthetic growth of green algae. The results showed that the both blue and red lasers increased the algae cell count [61].

Allen and Arnon tested the effect of light on green algae growth. The light intensity was around 16000 lux. There were two samples. One of the samples was analyzed under 11 h darkness and 13 h light. The other sample was analyzed under light for 24 h and the results showed that the growth rate was same. However after 5 days the growth rate for the sample with 24 h light was declined [62].

The effects of light on *Parietochloris incisa* was analyzed by Solovchenko et al. The results showed that best growth was seen on high light (400 μmol photons m^{-2} s^{-1}). With high light condition, total fatty acid and arachidonic amount was increased due to increase in biomass [63].

Another study (Yeh et al.) was focused on effects of different light sources on microalgae (*C. vulgaris*) growth. In the sutdy, three different light sources was used which are tungsten lamp, fluorescent lamp (TL5), fluorescent lamp (helix lamp). The results showed that fluorescence lamps were much better for algae growth. In an other study by Floreto et al., it was mentioned that high light intensity increased the palmitic acid and most fatty acids ratio [64].

3.2.5.2. Carbon dioxide

Carbon dioxide is the natural carbon source of the microalgae culture. Oxygen is releasing depending on decreasing carbon amount and it is delivered to the medium. Carbon dioxide is an general carbon source for photosynthetic microalgae. When the carbon amounts get low, oxygen is produced by photolysis of water and released to media. Since algae lives in high carbon dioxide concentration, greenhouse gases, nitrogen dioxide and atmospheric pollutants came from different sources became a food for algae. The exhausted gases can feed algae production facilities from fossil fuels and also its efficiency would be increased. Works on usage of stack gases as carbon source were done but the toxicity of the stack gas components couldn't be documented well. The amount of carbon dioxide required for the growth relates to type of microalgae and photo bioreactor. Some types of algae strains are able to keep growing in high carbon dioxide conditions, in contrast for faster growth lower carbon dioxide concentration is required [56]. Widjaja studied the effect of CO_2 on growth and it was seen that this effect correlates directly to the lipid productivity since growth was enhanced tremendously by increasing the CO_2 concentration [65]. CO_2 requirement can change up to strains. VirthieBhola et al. reported in their studies that at 15% CO_2 concentration there is a 3-fold decline in biomass yield when compared to the yield produced at a 4% CO_2 concentration. This suggests that the strain under study could not endure CO_2 concentrations greater than 4% [66]. Also Ebrahimzadeh et al. reported that increasing CO_2 injection had a significant effect on microalgae growth [67]. CO_2 input is also important. Sonnekus reported that the CO_2 should make up 0. 2 -5% of the total gas flow and being careful about the CO_2 input does not lower the pH of the culture [68].

3.2.5.3. Heat

Algal growth is also dependent on temperature. For maximum growth there is a need to know the optimal temperature. The temperature changes also lipid production and composition [69]. The degree of unsaturation of algal membrane lipids increases if cultures are maintained at temperatures below their optimum [70]. Other than this temperature is significant for dissolubility of carbon particles, which helps carbon to be used for photosynthesis. Heat effects respiration and photorespiration more than photosynthesis. However, if carbon dioxide and light are the limiting factor, the effect of heat is not significant anymore. Optimal temperature for microalgae cultures is between 20-24 °C. This can be different according to media composition, type of culture and strain. The most general cultured microalgae can tolerate the temperature between 16- 27 °C. The temperatures lower than 16 °C will increase the duplication time and higher than 35 °C will have a fatal effect on algae [56]. However, these ranges can be changed by environmental factors such as salinity, pH, carbon dioxide etc.

In the study of Floreta etal., the factors affecting algae growth were determined. Temperature effect was determined with salinity simultaneously. The results showed that low temperature (15 °C) with high salinity is the best choice. Low temperature increases the level of oleic and linoleic fatty acids. Moreover, high salinity increases the amount of C_{16} and C_{18} poly-unsaturated fatty acids [71].

3.2.5.4. pH

Microalgae require different pH values according to the media. During high pH concentration, the carbon dioxide might be limiting factor for growth and photosynthesis. The most used pH range for algal growth is around 7-9. The optimal pH for algae is between 8. 2- 8. 7. But it can change with different strains. For example, Weissel and Stadler studied with *Cryptomonas sp.* which showed positive population growth rates over a wide pH range, from 4. 4 to 9. 65 [72]. Appropriate pH can be adjusted by ventilation or gassing. There is a complex relationship between CO_2 concentration and pH in microalgal bioreactor systems, owing to the underlying chemical equilibrium among such chemical species as CO_2, H_2CO_3, HCO_3 and CO_3. Increasing CO_2 concentrations can increase biomass productivity, but will also decrease pH and this causes important effect upon microalgal physiology [73]. Water contaminated with a high pH has negative effects on algal abundance [74]. If there is not enough CO_2 gas supply, algae will utilize carbonate to maintain its growth [75].

Although high concentration of carbon dioxide provides high biomass efficiency, on the other side higher contamination risk and effect of low pH on microalgae physiology occurs [56].

Except the parameters mentioned above; there are also some parameters which affect on algal growth or lipid accumulation. Nitrogen, phosphorus and salinity can be examples for these parameters [76]. Widjaja et al. studied about nitrogen starvation effect on lipid accumulation. They reported that longer time of nitrogen starvation obviously resulted in higher accumulation of lipid inside the cells. Under all CO_2 concentrations, the lipid content tend to increase when the algae was exposed to nitrogen starvation condition that total lipid content was higher than lipid obtained during normal nutrition [75]. Ruangsomboon found the highest biomass concentration was found under the highest phosphorus concentration [60]. Li Xinet all. have reported in their study that lipid productivity was not at its highest when the lipid content was highest under nitrogen or phosphorus limitation [77]. Yeesang and Cheirsilp also studied about nitrogen and salinity effect. They found an increase in algal biomass under nitrogen-rich condition for all strains and in the absence of a nitrogen source, no growth was observed. They reported that although some loss in algal biomass was found, the lipid contents of four strains increased. They also noticed that growth and lipid accumulation by these microalgae could be affected by salinity. Under nitrogen-rich condition, all strains survived at high salinity but growth of some strains decreased [57, 78].

3.3. Microalgae cultivation systems

Cultivating microalgae can be achieved in open systems like lakes and ponds and in high controlled closed systems called photobioreactor. A bioreactor is defined as a system, which carries out biological conversion. Photobioreactors are reactors, which used for prototroph to grow inside or photo biological reactions to occur [79].

3.3.1. Open ponds

Generally open ponds are used in microalgae cultivation. Open ponds have various shapes and forms and certain advantages and disadvantages. In the scientific investigations and industrial applications, raceway ponds, shallow big ponds, circular ponds tanks and closed ponds are used [80]. Area where pool exist is critical factor for selection of pond type. Ponds become local climate function due to lack of control in open ponds [80, 81]. Therefore, area contributes to the success. Open ponds are limited by key growth parameters, which include light intensity, temperature, pH and dissolved O_2 concentration. Another problem seen in open ponds is contamination. It limits cultivation system of algae, which can grow under certain conditions [79].

Cost of cultivation systems is an important factor for comparison of open and closed systems. Construction, operation and maintaining costs are less than photobioreactors in ponds and these systems are simpler than the others [79, 82].

3.3.2. Photobioreactors

Nowadays researches are made for designing photobioreactors due to cultivating microalgae. Photobioreactors offer better control than open systems [2]. Their controlled environment allows high yield for cultivating.

Productivity is the most important indicator for bioreactor technology. It is very difficult to compare productivity of bioreactors due to various strains and scale of microalgae [80].

Photobioreactors basically can be tubular and flat type. When it is compared with the other bioreactors, tubular reactors considered as more suitable for open cultivating. Large illumination surface of reactor, which made of transparent tubes, is the main factor to being suitable for cultivation. Tubes can be adjusted in various types, adjustments convenience is depend to the specification of system.

A general configuration includes straight line and coiling tubes [83]. Reactor geometry is also important, tubular reactors can be vertical, horizontal or inclined shape. There are important differences between configurations of vertical and horizontal. Vertical designs provide more mass transfer and reduce energy consumption; horizontal designs can be scaled but needs more space. There are more studies about tubular photobioreactors but usually flat type photobioreactors is preferred because it can offer high cell density [84]. In addition, this type of reactors is advantageous due to low energy consumption and high mass transfer capacity, reduction of oxygen increases, high photosynthetic efficiency, no dark volumes when compared with the other photobioreactors. Suitable reactor design should be provided with maximum cell mass. Various flat-plate photobioreactor designs are made of glass, thick transparent PVC materials and V-shape and inclined. Although the other designs are cheap and easy to construct, glass and PVC is more transparent for maximum light penetration [80, 84-86].

3.3.2.1. Flat-plate photobioreactors

These systems have large illuminated surfaces. Generally these photobioreactors are made of transparent materials to utilize the solar light with maximum degree. Dissolved oxygen concentration is low compared to the horizontal tubular photobioreactors. In this system high photosynthetic activity can achieve. Although it is very suitable for culturing algae but it has some limitations [83].

3.3.2.2. Tubular photobioreactors

Most of tubular photobioreactors are made of glass or plastic tubes. They can be horizontal, serpentine, vertical, near horizontal, conical and inclined photobioreactors. Ventilation and mixing is generally performed by pump or ventilation systems. Tubular photobioreactor is suitable with their illuminated surfaces. But one of the important limitations of this system is poor mass transfer. It is a problem when photobioreactor is scaled. Also photoinhibition is seen in photobioreactors [83, 87].

If there is not sufficient mixing system cells don't have enough light for their growth. Developing mixing systems can provide effective light distribution.

Also controlling culture temperature is very difficult in these systems. Thermostat can be used but it is expensive and hard to control. Also cells can attach the walls of tubes. Long tubular photobioreactors are characterized with transfer of oxygen and CO_2 [83, 88].

Vertical column photobioreactors are low cost, easily constructed and compact systems. They are promising for large scale of algae production. Bubble column and airlift photobioreactors can reach specific growth rate [56].

3.3.2.3. Internally illuminated photobioreactors

Florescent lamps can illuminate some photobioreactors internally. Photobioreactor is equipped with wheels for mixing algal cultures. Sprayer provides air and CO2 to culture. This type of photobioreactors can utilize solar light and artificial light [90]. When solar light intensity is low (night or cloudy day) artificial light is used. Also in some researches, it is told that solar light can be collected and distributed with optic fibers in cylindrical photobioreactors [91]. Another advantages of this system are can be sterilized with heat under pressure and minimizing the contamination [56, 83].

3.3.2.4. Pyramid photobioreactor

The Pyramid photobioreactor is using fully controlled and automatic system that increases the production rate. With this system, it is easy to grow any microalgae at any climate conditions. The design is in pyramid shape to absorb light more effectively. As mentioned above, light is one of the significant parameters affecting algae growth rate and with this recent system algae can be supplied with optimal light intensity. That is why the shape of the system is the last innovation for production step. So, having optimal light intensity during high microalgae production decreases the energy consumption. The body design is angled to reduce to pump costs by using air-lifting method and decrease the deformation on cell walls. Thermo-isolated and high technologic materials are used to avoid energy lost and over heating [92].

3.4. Biocoil microalgae production system

Biocoil is a holozoic tubular photobioreactor which made of plastic tubes with small diameter (between 2.4-5 cm), centrifuges, diaphragm pumps or peristaltic pumping are utilized in this system. Biocoil design provides equal mixing and reduces the attachment of algae to the walls. It automates the production process. It is not suitable for all algae species. Some of algae species damages by circulation system and some of them attach to the internal surface of tubes and affects algae production negative. In this system, when the level of algae increases maximum degree, because of the light limitation photosynthesis can slow. Biocoil systems with utilizing solar light in or outsides can executable. Light is given with an angle so algal cell can utilize better and photosynthesis occurs easily [89, 93, 94].

3.4.1. Design of culture growth systems

Depends of local conditions and suitable materials various culture systems can be designed by various sizes, shapes of construction material, slope and mixing type. These factors affect performance, cost and resistance. To construct suitable photobioreactor material has main importance. Materials like plastic or glass relax and rigid shouldn't be toxic, they should have mechanical power, resistance, chemical stability and low cost. Tubular photobioreactors are the most suitable ones for open culture systems. They have big illumination surface, good biomass productivity and they aren't expensive because they are made of glass or plastic tubes. Flat-type photobioreactors are made of transparent materials to utilize solar light energy in maximum degree. This type of photobioreactors allows good immobilization of algae and they are cleaned easily [56]. Pond walls and deep side can made of simple sand, clay, brick or cement even PVC, glass fiber or polyurethane. For coating mostly long lasting plastic membrane is used. (e. g., 1-2 mm thick, UV-resistant, PVC or polyethylene sheets) sometimes to lower the cost uncoating ponds are used but that time some problems occur like contamination, a layer of mud and sand [39].

3.4.2. Mixing

Mixing is a process for increasing the productivity of biomass in photobioreactors. Mixing provides distribution of light intensity, sufficient CO_2 transfer and maintains uniform pH. Mixing is necessary for preventing algae sedimentation and avoiding cell attachment to the reactor wall. Mixing is also provides equal light and nutrients to all cells and increases the gas transfer between culture medium and air [95]. The second of priority measures is carbon supply for using in photosynthesis. In very dense cultures, CO_2 from air (includes 0.035 % of CO_2) and bubbles during the culture can be limited for algal growth. CO_2 addition creates a buffer for the result of changing pH in the water [56].

Poor mixing allows cells to clumping like different size of aggregates; therefore it leads 3 phase (solid-liquid-gas) system in reactor. This situation tends to reduce the mass transfer. But all algae cannot tolerate agitation. Because they are sensitive to hydrodynamic stress. High mixing rate can cause the damaging of cells. Mixing in bubble column and air lift reactors can characterize with axial dispersion coefficient, mixing time, circulation time and Bodenstein number [96]. Analysis of mixing in bubble column shows it has shorter time than airlift reactors. Bubbles beyond the suction pipe provide less blurry area and causes better exposure to the light. In addition, existence of suction pipe in airlift reactors causes more effective mixing because internal loop provides a circulation. Airlift reactor gives information about fluid flow and high gas-liquid mass transfer rate. Bubble column causes unbalance cell density and these causes to death of algae [56, 97].

3.4.3. Light penetration

Another key of successfully scale up is light penetration. Ilumination in the photobioreactor affects biomass composition, growth rate and products. Microalgae need light for their photosynthesis [98]. Photosynthetic active radiation wave changes about 400-700 nm and this is equal to the visible light [99]. In intense cultures, light gradient changes over the photobioreactor radius due to the weakening of the light. Reduction of light intensity related to wave length, cell concentration, photobioreactor geometry and distance of the light transmittance. Light intensity in photobioreactor related to light way, cell concentration and light which emits by microalgae [56].

3.4.4. Gas injection

Supplement of CO_2 by bubbles is an important factor to be considered in designs. Injection of CO_2 bases on giving CO_2 to photobioreactor artificially. Researches show that rich ventilation of CO_2 provides CO_2 to algae, supports deooxygenation of suspension, to improve cycling provides mixing and limits the light inhibition [100]. But high ventilation rate leads to higher cost that is why in large scale of microalgae production it is not recommended. These researches results for microalgae production necessary optimum aeration rate of CO_2 gas. Includes about 5% or 10% of CO_2 (v/v), rate of 0.025-1 vvm [100]. Volume of air/medium/time is found cost effective for air mass culture [56].

3.4.5. Comparison of open and closed culture systems

Open and closed culture systems have advantages and disadvantages. Construction and operation of open culture systems are cheaper and they are more resistant than closed reactors and have large production capacity [101]. Ponds use more energy to homogenize to nutrients and to utilize the solar energy for growth their water level cannot be less than 15 cm [41]. Ponds are exposing to air conditions because water temperature evaporation and illumination cannot be controlled. They produce large amounts of microalgae but they need larger areas than closed systems and they are open to other contaminations from the other microalgae and bacteria. Also when atmosphere has only 0. 03-0.06 % of CO_2 mass transfer limitation slows the growth of microalgae cell.

Photobioreactors are flexible systems, which can operate for biological and physiological characteristics of cultured microalgae. It can be possible to produce microalgae, which cannot produce in ponds. Exchange of gas and contaminants between atmosphere and cultured cells in photobioreactor is limited or blocked by reactor walls [39]. Depends on the shape and design, photobioreactors have more advantages than open ponds. Culture conditions and growth parameters can be controlled better, it prevents evaporation, reduces loss of CO_2, provides high microalgae density or cell concentration, high yield, creates more safe and preserved environment, prevents contamination. Despite the advantages, photobioreactors have problems to be solved. Over heating, biological pollution, accumulation of oxygen, difficulty of scale-up, high cost of construction and operation and cell damage because of shear stress and degradation of material in photo phase are main problems in photobioreactors [39].

Comparing photobioreactors and open ponds is not easy because growth of algae related to al lot of different factors. Three parameters are considered in algae production units for yield [41]:

- Volumetric productivity (VP): productivity per unit reactor volume (expressed as g/L. d).

- Areal productivity (AP): productivity per unit of ground area occupied by the reactor (expressed as g/m^2d).

- Illuminated surface productivity (ISP): productivity per unit of reactor illuminated surface area (expressed as g/m^2d).

According to researches closed systems don't provide advantage for areal productivity but provide volumetric productivity (8 times) and cell concentration (16 times) more than open ponds [39, 41].

3.4.6. Comparison of batch and continuous process

Photobioreactors can be operated in batch or continuous process. There are a lot of advantages for using continuous bioreactors than batch bioreactors. Continuous bioreactors provide more control than batch bioreactors. Growth rates can be regulating in long time periods, can be saved and with variable dilution rates biomass concentration can be controlled. With steady state continuous bioreactors results is more dependable, products can be easily produced and can be reached desired product quality. Continuous reactions offer many opportunities for system research and analysis [102].

But some type of bioreactors is not suitable for continuous process. For some productions, cell aggregation and wall growth can inhibit the steady state growth. Another problem is loss of original product strain in time. Mixtures viscosity and heterogenic nature make difficult for maintaining filamentous organisms. Long growth periods increase the contamination risks [83].

3.5. Harvesting alternatives

There are several ways to harvest microalgae and dry them. Some main harvesting methods are sedimentation, flocculation and filtration.

Sedimentation: When a particle moves continuously in a phase, the velocity is affected by two factors. First of them is increasing the velocity because the density gradient between particle and fluid create buoyant force. At the end, buoyant force gets equal to dragging force and particle starts moving with a constant velocity. The same idea is applied to collect microalgae from the ponds. Gravity force is used for settling of suspended particles in fluid. This method is cheap and easy. However, the particles suspended in the fluid have to be incompressible. The problem with the *Scenedesmus sp.* and *Chlorella sp.* is that they are compressible. That is why sedimentation cannot be used for these types [103]. For low value products, sedimentation might be used if it is improved with flocculation [104].

Flocculation: is also used for harvesting microalgae. The general idea is microalgae carries negative charge on it and if the flocculants disappear the negative charge, algae starts coagulation. Some used flocculants are $Al_2(SO_4)_3$, $FeCl_3$, $Fe_2(SO_4)_3$ [105].

Filtration: This is one of the most competitive methods for the collection of algae. There are different types of filtrations, for example, dead end, microfiltration, ultrafiltration, pressure filter and vacuum filter. Mostly filtrations require the liquid media with algae to come through filtration. Filter can be fed until a thick layer of microalgae is collected on the screen. This method is very expensive for especially microalgae. The pore sizes of the filters are the most important part. If the pore size is bigger than algae you cannot collect it. In contrast, if the pore size is too small it might result in decrease of the flow rate and block the pores [106].

3.6. Extraction of lipid from microalgae

There are a lot of methods for extraction of lipid from microalgae but the most common techniques are oil presses, liquid-liquid extraction (solvent extraction), supercritical fluid extraction (SFE) and ultrasonic techniques. Oil presses are usually used for extracting of lipids from nuts and seeds. The same process and devices can be used for lipid extraction from microalgae. For the purpose of this process to be effective, firstly microalgae must be dried. Presses use pressure for breaking cells and removing oil [107]. This method can extract 75% of oil but in longer extraction times it is less effective [80].

Solvent extraction is more successful for extracting lipids from microalgae. In this method organic solvents such as hexane, acetone, and chloroform are added in the algae paste. Solubility of oil is higher in organic solvents than water. Therefore solvent breaks the cell wall and extracts oil easily. Solvent extraction continues with distillation process for separating oil from the solvent [108]. Hexane is cheap and has high extraction capacity. For this reason it is reported to be the most effective solvent in extractions.

In addition to this studies, 2 stage process using ethanol improves lipid extraction. The yield of recovery of oil reaches about 80%. Butanol is also effective in extraction of lysophospholipids. But evaporation of butanol is difficult and there are some impurities because of its high polarity [80].

Supercritical extraction uses high pressure and temperature for breaking cells. This method is widely used and efficient for extraction time. Studies reported that temperature and pressure don't affect the yield of components but it affects extraction rate. Similar effects are seen in SFE system and solvent extraction [109].

Another method is using ultrasonic techniques. In this method microalgae is exposed to high intensity ultrasonic waves and these waves creates bubbles around the cell. Shock waves are emitted by collapsing bubbles. It breaks cell wall and desired components release to the solution. This method is also improves the extraction rate with the same way. This technique is widely used in laboratory scale but in commercially scale there is not enough information about cost and applicability [110, 80].

3.7. Biodiesel Production from Oil

After extraction there are 4 main methods for producing biodiesel: direct used and mixing with raw oils; microemulsion; pyrolysis and transesterification.

3.7.1. Dilution

This is a dilution method that certain proportion of vegetable and waste oils blended with diesel fuel and another solvent. The most used oils for producing biodiesel with this way are waste oils and vegetable oils like sunflower and rapeseed.

Direct use or blending generally considered being unsatisfactory and impractical for both direct and indirect diesel engines. There are specific problems such as high viscosity, acid composition, free fatty-acid content, gum formation because of oxidation, polymerization during storage and combustion, carbon deposits and also lubricating-oil thickening [111].

Dilution of vegetable oils with solvents lowers the viscosity. The viscosity of oil can be lowered by blending with pure ethanol [112]. The low viscosity is good for better performance of engine, which decreases with increasing the percentage of diesel [33]. In this method there is no chemical process and viscosity can be lower but there are also carbon deposits and lube pollution problems to be solved. To solve problems caused by high viscosity, micro-emulsion, pyrolysis and transesterification methods are used [113].

3.7.2. Micro-emulsion

It is defined that the size of 1-150 nm, the two immiscible liquid organic mixtures with ionic or non-ionic, self-formed stable colloidal distribution. With this method it is possible to form alternative diesel fuels except petroleum [28]. In this method vegetable oils with an ester and dispersant (co-solvent), or of vegetable oils, an alcohol and a surfactant, with or without diesel fuels can be used to make a microemulsion. Due to their alcohol contents, microemulsions have lower volumetric heating values than diesel fuels. But these alcohols have high latent heats of vaporization and also tend to cool the combustion chamber, which cause a reduction of nozzle coking. A microemulsion made of methanol and vegetable oils can perform like diesel fuels [111]. To solve the problem of the high viscosity of vegetable oils, microemulsions with solvents and immiscible liquids, such as methanol, ethanol, 1-butanol and ionic or non-ionic amphiphiles have been studied [114].

3.7.3. Pyrolysis

Pyrolysis is the conversion of organic substance into another by means of heat or by heat in the presence of a catalyst. Vegetable oil, animal fat, algae oil, natural fatty acids or methyl esters of fatty acids can be pyrolyzed [111]. Although this method is not very cheap, however, fuel can be produced without extraction of lipids or hydrocarbons. More uniform product can be obtained and ideally increases yields over transesterification with this method [115]. Products are chemically similar derived from petroleum products, which are to gasoline and diesel fuel derived [28]. Also with pyrolysis some low value materials and sometimes more gasoline than diesel fuel are produced [116]. In comparison between pyrolysis and the other cracking processes, pyrolysis is seen more simple, pollution free and effective [33]. Sharma et al. reported that pyrolysis of the vegetable oil can produce a product which has high cetane number, low viscosity, acceptable amounts of sulfur, water and sediments contents, acceptable copper corrosion values [117].

3.7 4. Transesterification

Transesterification of the oil is the most promising solution to the high viscosity problem [114]. In this process, triglycerides are converted to diglycerides, then the diglycerides are converted to monoglycerides, and the monoglycerides are converted to esters (biodiesel) and glycerol (by-products) [118]. There are three common kinds of catalysts used in transesterification process such as lipase catalysts, acid catalysts and alkali catalysts. Each catalyst has advantages and disadvantages [113].

In the acid-catalytic transesterification, the reaction can be catalyzed by sulfuric, phosphoric, hydrochloric and organic sulfonic acids. Very high yields can be obtained by using this catalyst. These reactions need the use of high alcohol-to-oil molar ratios in order to obtain good product yields in practical reaction times. But ester yields do not proportionally increase with molar ratio and the reaction time is very long (3–48 h) [114, 119, 120]. Xu et al. studied the acidic transesterification of microalgae (Heterotrophic C. Protothecoides) oil. They used methanol for alcohol and they achieved 80% of FAME yield [121].

Johnson made a study on *Schizochytrium limacinum* microalgae species. He converted this algal oil to biodiesel with acidic transesterification and he achieved 82.6% of biodiesel yield [122].

In the alkali-catalytic transesterification, the reaction can be catalyzed by alkaline metal alkoxides, and hydroxides, as well as sodium or potassium carbonates. Sodium methoxide is the most widely used biodiesel catalyst. This reaction is faster than acid-catalytic transesterification and reactions can occur in low temperatures with a small amount for catalyst and with little or no darkening of colour of the oil [114]. High quality can be obtained however this process is very sensitive to the presence of water and free fatty acids and needs lots of methanol. If the raw materials have a high percentage of free fatty acids or water, the alkali catalyst reacts with the free fatty acids to form soaps [113]. There are some studies on micro-algae oil to produce biodiesel by using alkali transesterification. Velasquez-Orta et al. studied on biodiesel production from *Chlorella vulgaris*. In that study, alkali transesterification was used for conversion and they achieved 71% of FAME yield [123]. Ferrentino et al. studied on biodiesel production from microalgae too. They used Chlorella sp. oil and their production method was alkali transesterification. They have obtained high yield from their experiment [124]. In another study, Carvalho et al. used alkali transesterification for biodiesel production from algae oil. In their study, they used *Chlorella emersonii* oil and they have obtained 93% conversion yield [124].

It can be seen that there are some problems such as recovery of glycerol or removing catalysts from product and need of wastewater treatment in acid or alkali-catalytic transesterification. Enzymatic catalysts like lipases are able to catalyze the transesterification of triglycerides effectively. With this process glycerol can be easily recovered however enzymatic catalysts are often more expensive than chemical catalysts. The high cost of enzyme production is the main obstacle to the commercialization of enzyme-catalyzed processes. But using solvent-tolerant lipases and immobilized lipases can be a solution for this. Lipase-catalyzed transesterification is considered to be one of the most effective reactions for production of biodiesel [114]. In another study Tran et al. used microalgae oil (*Chlorella vulgaris* ESP-31) for producing biodiesel. Their method was enzyme-catalyzed transesterification and they used lipase in this process. In the result, they reported that they achieved 94.78 % of FAME yield [126]. Table 2 presents the transesterification studies for biodiesel production from microalgae oil.

Supercritical process, microwave-assisted method and ultrasonic-assisted process are novel methods used in biodiesel production area. Since these methods are novel methods and also algae are new materials for biofuel area, there is a few studies biodiesel production from algae oil with these novel methods, these studies were reviewed and presented below.

With *supercritical process* biodiesel production can be easily achieved without catalysts. Supercritical fluid is a substance whose temperature and pressure is above the critical point. These fluids are environmentally friendly and economic. Usually water, carbon dioxide and alcohol is used for supercritical fluid. In biodiesel production generally supercritical methanol and supercritical ethanol is used. Advantages of this process are being easier for purification, shorter the reaction time and more effective reaction [130]. In the study of Patil et al., using supercritical methanol produced biodiesel. The wet algae were used and the ratio of alcohol/ oil was chosen as 9:1. The temperature of the reaction occurred at 255 and 1200 psi and resulted in 90% of FAME yield [131].

Microwaves activate differences in small degrees of polar molecules and ions, because the molecular friction and chemical reactions start. Molecules have not the enough time to relax and heat generation occurs in a short time because energy interacts with molecules very quickly. Transesterification reaction is carried out with microwave in a short time and microwave results in an efficient manner. As a result in a short time separation and pure products with high yield is obtained. Thus, production costs and the formation of by-product are reduced [130]. Patil et al., made a study on biodiesel production from dry microalgae by using microwave-assisted process. KOH was used as catalyst in the study and microwave condition is set to 800 W. The performance of the study is around 80% [132]. The other study with macroalgae for microwave-assisted algal biodiesel was showed that methanol to macroalgae ratio of 1:15 was the best condition. In the study, sodium hydroxide concentration was 2 wt % and reaction time of 3 min for the best condition [133]. Koberg et al. was reported the study used *Nannochloropsis* for algal biodiesel production with microwave-assisted method. The higher biodiesel yield was observed which was around 37.1% with microwave technique. The same conditions for sonication technique resulted in lower yield [134].

Algae strain	Method	Alcohol	Alcohol / oil molar ratio	Temp.	Time	Results	Ref.
Heterotrophic C. Protothecoides (microalga)	Acidic trasesterification	Methanol	56:1	30 °C	4 h	80% (FAME Yield	[121]
Chlorella vulgaris ESP-31 (microalga)	Enzymatic transesterification (Lipase)	Methanol	98.81	25-40 °C	48 h	94.78% (FAME Yield)	[126]
Chlorella vulgaris (microalga)	in situ alkaline transesterification	Methanol	600:1	60 °C	75 min	71% (FAME Yield)	[123]
Nannochloropsis oculata (microalga)	heterogeneous transesterification	Methanol	30:1	50 °C	4 h	97.5% (FAME Yield)	[127]
Chlorella (microalga)	In-situ acidic transesterification	Methanol	315:1	23 and 30 °C	15 min-2 h	70-92% (FAME Yield)	[17]
Chlorella sp. (microalga)	Alkali Transesterification	Methanol	-	100 °C	25 h	90 (Fluorometric Reading)	[124]
Schizochytrium limacinum (microalga)	Acidic Transesterification	Methanol	-	90 °C	40 min.	82.6% (biodiesel Yield)	[122]
Chlorella emersonii	Alkali trasesterification	Methanol	5:1	60 °C	2 h	93% conversion	[125]
Fucus spiralis (macroalga)	Alkali Transesterification	Methanol	6:1	60 °C	4 h	1.6-11.5% (Process Yield)	[128]
Commercially refined macroalga (Kelp)	McGyan process	Methanol	32:1	360 °C	30 s	94.7% (FAME Yield)	[129]

Table 2. The transesterification studies for biodiesel production from microalgae oil

Recent years, *ultrasonic-assisted process* is widely used in biodiesel production. Mixing is very important factor for biodiesel yield in transesterification reactions. It is an effective mixing method in liquid-liquid mass transfer to provide better mixing. Powerful mixing creates smaller droplets than the conventional mixing and increases the contact areas between the oil phases. Also it provides the activation energy, which needs for initiating transesterification reactions [130]. In the study of Eihaze et al., they are focused on the in situ transesterification of microalgae by ultrasound technique. The reaction takes 1 h with the use of methanol/oil ratio to 315:1. The result was 0.295 ± 0.003 g biodiesel/g dry *Chlorella* which shows that this is higher than mechanically stirred in situ technique [135].

3.8. Design of algae and biodiesel production

In this section of study, algae production stages that cover the algae strain and location selection, algae cultivation, harvesting, oil extraction, and biodiesel production process from microalgae are presented by using ChemCad design program. All stages are given in this process flow diagram (pfd) and equipment table in detail. As it is seen in a process flow diagram (pfd), the streams between 1-8 are the area of the process where algae growth occurs. The algae bodies contain a lipid, which can be extracted and converted into a type of biofuel. The area where between stream1-8 has several large ponds to grow algae containing large amounts of lipid in preparation for lipid extraction. Once a pond is harvested, it is re-inoculated for another crop of algae (stream 11-13). Once the algae reach maturity in the growth ponds and have the desired lipid content, the cells are harvested in the area where stream 9-10. This area at a concentration of 1g-algae/L water. The algae collected will be dewatered, and the usable lipid is extracted for the reaction process where stream 9,10,14-16. The remaining algal biomass will be sent to algal pulp tank, it may be evaluated for biogas production in digesters. Lipids, catalysts and alcohol are sent for fuel conversion to heat-jacketed transesterification reactor. Once the lipid is harvested from the algae cells, the usable triglycerides are converted to biofuel in streams 16-18. Then products sent to the separator to separate biodiesel and byproduct glycerol in stream 21-27. The byproduct of this reaction is glycerol, which is removed and treated as waste. The biofuel is then ready to be used in modern farm equipment, or as a fuel supplement for diesel. All the equipments, tanks and ponds are labeled in the Figure 1.

4. Conclusion

Nowadays, demands on energy are caused to reduction of sources and environmental problems let the world to use alternative fuels. Microalgae have important potential as an alternative energy source. A lot of valuable products can be produced from microalgae such as biodiesel, biogas, bioethanol, medicines and nutraceuticals. Biodiesel is one of the most important alternative fuels. Microalgal biodiesel production is very new technology. In this study, microalgae and their classifications, important steps of biodiesel production from microalgae have been mentioned. In production sections, steps are explained briefly and easily understandable. Also advantages and disadvantages in the production are mainly dis-

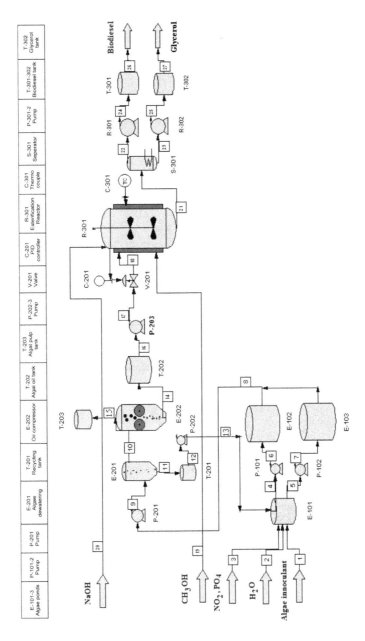

Figure 1. The process flow diagram of biodiesel production process from microalgae by ChemCAD.

cussed. At the end of this chapter, a biodiesel production from microalgae is designed by ChemCadprogram, which shows a simple process flow diagram for who desires to produce biodiesel from microalgae. Recently, microalgae are not economically viable. The main problems are the cost of capital cost. The rate of return is not short as it is expected. The operation cost is also affecting the total cost significantly. The main part, which makes the process expensive due to operation and capital costs, are algae growth, harvesting, dewatering, and fuel conversion. Beyond these, oil extraction step significantly increases the cost. If the oil could be extracted easily and at higher rates, the cost would be much lower. However, there are needs to innovate new ways to make the process economically feasible. Regardless, microalgae are seen as important resources for the future and there will be a lot of improvements on recent technology.

Author details

Didem Özçimen, M. Ömer Gülyurt and Benan İnan

*Address all correspondence to: ozcimen@yildiz.edu.tr

YıldızTechnical University, Faculty of Chemical and Metallurgical Engineering, Bioengineering Department, Istanbul, Turkey

References

[1] Szulczyk KR, McCarl BA. Market penetration of biodiesel. Renewable and Sustainable Energy Reviews 2010;14: 2426–2433.

[2] Chisti Y. Biodiesel from microalgae. Biotechnology Advances 2007; 25 (3): 294-306.

[3] Khan SA, Rashmi, Hussain MZ, Prasad S, Banerjee UC. Prospects of biodiesel production from microalgae in India. Renewable Sustainable Energy Reviews 2009; 13: 2361–2372.

[4] Teixeira CM, Morales ME. Microalga como mate' ria-prima para a produc‚aˇo de biodiesel. Revista: Biodiesel o Novo combustı'vel do Brasil 2007; 91–96.

[5] Wu X, Ruan R, Du Z, Liu Y. Current Status and Prospects of Biodiesel Production from Microalgae. Energies 2012; 5: 2667-2682.

[6] Yoo C, Jun SY, Lee JY, Ahn CY, Oh HM. Selection of microalgae for lipid production under high levels carbon dioxide. Bioresource Technology 2010; 101: 571–574.

[7] Jiang LL, Luo SJ, Fan XL, Yang ZM, Guo RB. Biomass and lipid production of marine microalgae using municipal wastewater and high concentration of CO_2. Applied Energy 2011; 88: 3336–3341.

[8] Lee JN, Lee JS, Shin CS, Park S. C, Kim SW. Methods to enhance tolerances of *Chlorella* KR-1 to toxic compounds in flue gas. Applied Biochemistry and Biotechnology 2000; 84–86: 329–342.

[9] Vergara-Fernandez A, Vargas G, Alarcon N, Velasco A. Evaluation of marine algae as a source of biogas in a two-stage anaerobic reactor system. Biomass and Bioenergy 2008; 32:338–44.

[10] Yen HW, Brune DE. Anaerobic co-digestion of algal sludge and waste paper to produce methane. Bioresource Technology 2007; 98: 130–134.

[11] Nakas JPS, Parkinson CM, Coonley CE, Tanenbaum SW. System development for linked-fermentation production of solvents from algal biomass. Applied and Environmental Microbiology 1983; 46:1017–23.

[12] Belarbi EH, Molina EY. A process for high yield and scalable recovery of high purity eicosapentaenoic acid esters from microalgae and fish oil. Process Biochemistry 2000; 35:951–69.

[13] Cheng-Wu Z, Zmora O, Kopel R, Richmond A. An industrial-size flat plate glass reactor for mass production of *Nannochloropsis* sp. (Eustigmatophyceae). Aquaculture 2001; 195:35–49.

[14] Besada V, Andrade JM, Schultze F, Gonzalez JJ. Heavy metals in edible seaweeds commercialised for human consumption. Journal of Marine Systems 2009; 75:305–13.

[15] Spolaore P, Joannis-Cassan C, Duran E, Isambert A. Commercial applications of microalgae. Journal of Bioscience and Bioengineering 2006; 101: 87-96.

[16] Olaizola M. Commercial development of microalgal biotechnology: from the test tube to the marketplace. Biomolecular Engineering 2003; 20: 459-466.

[17] Ehimen EA, Sun ZF, Carrington CG. Variables affecting the in situ transesterification of microalgae lipids. Fuel 2010; 89(3): 677–684.

[18] Yang Z, Guo R, Xu X, Fan X, Luo S. Hydrogen and methane production from lipid-extracted microalgal biomass residues. İnternational Journal of Hydrogen Energy 2011;36: 3465-3470.

[19] Hirayama S, Ueda R, Ogushi Y, Hirano A, Samejima Y, Hon-Nami K, Kunito S. Ethanol production from carbon dioxide by fermentative microalgae. Studies in Surface Science and Catalysis 1998; 114:657–60.

[20] Bush RA, Hall KM. Process for the production of ethanol from algae. U. S. Patent 7,135,308; 2006.

[21] Gamal A. Biological importance of marine algae. Saudi Pharmaceutical Journal 2010; 18: 1–25.

[22] Hasan MR. Use of algae and aquatic macrophytes as feed insmall-scale aquaculture. FAO Fisheries and aquaculture technical paper, 2009; 531.

[23] Hall J, Payne G. Factors controlling the growth of field population of Hydrodictyon reticulatum in New Zealand. Journal of Applied Phycology 1997; 9: 229-236.

[24] Rafiqul IM, Jalal KCA, Alam MZ. Environmental factors for optimisation of *Spirulina* biomass in laboratory culture. Biotechnology 2005; 4: 19-22.

[25] McHugh DJ. A guide to the seaweed industry. Fisheries Technical Paper 2003; No. 441. Rome, FAO, 118.

[26] Bruton T, Lyons H, Lerat Y, Stanley M. A Review of the Potential of Marine Algae as a Source of Biofuel in Ireland. Sustainable Energy Ireland 2009; 1-88.

[27] Bold HC, Wynne MJ. Introduction to the algae structure and reproduction, second ed., Prentice-Hall Inc., Englewood Cliffs, NJ, 07632,1985; 1–33.

[28] Guardia SD, Valls R, Mesguiche V, Brunei JM, Gulioli G. Enantioselective synthesis of (-)-bifuracadiol: a natural antitumor marine product. Tetrahedron Letters 1999; 40: 8359–8360.

[29] Bennamara A, Abourrichi A, Berrada M, Charrouf M, Chaib N, Boudouma M, Garneau XF. Methoxybifur-carenone: an antifungal and antibacterial meroditerpenoid from the brown alga *Cystoseira tamariscifolia*. Phytochemistry 1999; 52: 37–40.

[30] Awad NE. Biologically active steroid from the green alga *Ulva lactuca*. Phytotherapy Research 2000; 14: 641–643.

[31] Choi JS, Park HJ, Jung HA, Chung HY, Jung JH, Choi WC. A cyclohexanonyl bromophenol from the red alga *Symphyocladia latiuscula*. Journal of Natural Products 2000; 63: 1705–1706.

[32] Sheehan J, Dunahay T, Benemann J, Roessler P. A Look Back at the U. S. Department of Energy's Aquatic Species Program: Biodiesel from Algae 1998.

[33] Singh SP and Singh D. Biodiesel production through the use of different sources and characterization of oils and their esters as the substitute of diesel: A review. Renewable and Sustainable Energy Reviews 2010; 14: 200–216.

[34] Garson J. Marine natural products. Natural Product Reports 1989; 6, 143–170.

[35] Koehn FE, Longley RE, Reed JK. Microcolins A and B, new immunosuppressive peptides from the blue–green alga *Lyngbya majuscule*. Journal of Natural Products 1992; 55: 613–619.

[36] Gerwick WH, Roberts MA, Proteau PJ, Chen JL. Screening cultured marine microalgae for anticancer-type activity. Journal of Applied. Phycology 1994; 6: 143–149.

[37] Trease GE, Evanes WC. Pharmacognosy. 14th ed., W. B Saunders Company Ltd., London, Philadelphia, Toronto, Sydney, Tokyo, 1996. 18–27.

[38] Shimizu Y. Dinoflagellates as sources of bioactive molecules. Attaway, D. H., Zaborsky, O. R. (Eds.), Marine Biotechnology, vol. 1: Pharmaceutical and Bioactive Natural Products, New York, Plenum, 1993; 391–310.

[39] Mata MT, Martins AA, Caetano SN. Microalgae for biodiesel production and other applications: A review. Renewable and Sustainable Energy Reviews 2010; 14: 217–232.

[40] Li Y, Horsman M, Wu N, Lan CQ, Dubois-Calero N. Biofuels from microalgae. Biotechnology Progress 2008; 24(4): 815–20.

[41] Richmond A. Handbook of microalgal culture: biotechnology and applied phycology. Blackwell Science Ltd; 2004.

[42] Renaud SM, Thinh LV, Parry DL. The gross chemical composition and fatty acid composition of 18 species of tropical Australian microalgae for possible use in mariculture. Aquaculture 1999; 170:147–59.

[43] Rodolfi L, Zittelli GC, Bassi N, Padovani G, Biondi N, Bonini G, et al. Microalgae for oil: strain selection, induction of lipid synthesis and outdoor mass cultivation in a low-cost photobioreactor. Biotechnology and Bioengineering 2009; 102(1):100–12.

[44] Barclay B. Microalgae Culture Collection 1984–1985. Microalgal Technology Research Group (MTRG) 1984; SERI/SP-231-2486.

[45] Zhu CJ, Lee YK. Determination of biomass dry weight of marine microalgae. Journal of Applied Phycology 1997; 9:189–94.

[46] Maxwell EL, Folger AG, Hogg SE. Resource evaluation and site selection for microalgae production systems May 1985; SERI/TR-215-2484.

[47] Chojnacka K, Marquez-Rocha FJ, Kinetic. Stoichiometric relationships of the energy and carbon metabolism in the culture of microalgae. Biotechnology 2004; 3(1): 21–34.

[48] Wen Z, Chen F. Heterotrophic production of eicosapentaenoic acid by microalgae. Biotechnology Advances 2003; 21: 273–294.

[49] Chen C, Yeh K, Aisyah R. Cultivation, photobioreactor design and harvesting of microalgae for biodiesel production: A critical review. Bioresource Technology 2011; 102: 71–81.

[50] Chen F. High cell density culture of microalgae in heterotrophic growth. Trends in Biotechnology 1996; 14: 421–426.

[51] Li XF, Xu H, Wu QY. Large-scale biodiesel production from microalga *Chlorella protothecoides* through heterotrophic cultivation in bioreactors. Biotechnology and Bioengineering 2007; 98: 764–771.

[52] Doucha J, Lívansky K. Production of high-density *Chlorella* culture grown in fermenters. Journal of Applied Phycology 2012; 24: 35–43.

[53] Park KC., Whitney C, McNichol, JC, Dickinson KE, MacQuarrie S, Skrupski BP. Zou, J. ; Wilson, K. E. ; O'Leary, S. J. B. ; McGinn, P. J. Mixotrophic and Photoautotrophic cultivation of 14 microalgae isolates from Saskatchewan, Canada: Potential applica-

tions for wastewater remediation for biofuel production. Journal of Applied Phycology 2012; 24: 339–348.

[54] Bhatnagar A, Chinnasamy S, Singh M, Das KC. Renewable biomass production by mixotrophic algae in the presence of various carbon sources and wastewaters. Applied Energy 2011; 88: 3425–3431.

[55] Ogbonna JC, Yada H, Tanaka H. Light supply coefficient: A new engineering parameter for photobioreactor design, Journal of Fermentation and Bioengineering 1995; 80 (4): 369-376.

[56] Bitoga J. P, Lee I. -B, Lee C. -G, Kim K. -S, Hwang H-S, Hong S. -W, Seo I. -H, Kwon K. -S, Mostafa E. Application of computational fluid dynamics for modeling and designing photobioreactors for microalgae production, Computers and Electronics in Agriculture 2011; 76: 131–147.

[57] Yeesang C, Cheirsilp B. Effect of nitrogen, salt, and iron content in the growth medium and light intensity on lipid production by microalgae isolated from freshwater sources in Thailand. Bioresource Technology 2011; 102 (3): 3034-3040.

[58] Converti A, Casazza AA, Ortiz EY, Perego P, Del Borghi M. Effect of temperature and nitrogen concentration on the growth and lipid content of *Nannochloropsis oculata* and *Chlorella vulgaris* for biodiesel production. Chemical Engineering and Processing: Process Intensification 2009; 48 (6): 1146-1151.

[59] Cheirsilp B, Torpee S. Enhanced growth and lipid production of microalgae under mixotrophic culture condition: Effect of light intensity, glucose concentration and fed-batch cultivation, Bioresource Technology, 2012; 110: 510-516.

[60] Ruangsomboon S. Effect of light, nutrient, cultivation time and salinity on lipid production of newly isolated strain of the green microalga, *Botryococcus braunii KMITL 2*. Bioresource Technology 2012; 109: 261-265.

[61] Sara S. Kuwaharaa, Joel L. Cuelloa, Graham Myhreb, Stanley Pau. Growth of the green algae Chlamydomonas reinhardtii under red and blue lasers. Optics and Lasers in Engineering, 2010; 49: 434-438.

[62] Allan MB, Arnon DI. Studies on nitrogen-fixing blue-green algae. I. Growth and nitrogen fixation by Anabaena cylindrica Lemm. Plant Physiology 1955; 30:366–372.

[63] Solovchenko AE, Khozin-Goldberg I, Didi-Cohen S, Cohen Z, Merzlyak MN Effects of light intensity and nitrogen starvation on growth, total fatty acids and arachidonic acid in the green microalga *Parietochloris incise*. Journal of applied Phycology 2008; 20:245–251.

[64] Yeh KL, Chang JS. and Chen WM. Effect of light supply and carbon source on cell growth and cellular composition of a newly isolated microalga *Chlorella vulgaris* ESP-31. Engineering in Life Scince 2010; 10: 201–208.

[65] Widjaja A, Lipid production from microalgae as a promising candidate for biodiesel production. Makara Teknologi 2009, 13(1): 47-51.

[66] Bhola V, Desikan R, Santosh SK, Subburamu K, Sanniyasi E, Bux F. Effects of parameters affecting biomass yield and thermal behaviour of *Chlorella vulgaris*, Journal of Bioscience and Bioengineering 2011; 111 (3): 377-382.

[67] Ebrahimzadeh MM. Torkian Boldaji SA. Hosseini Boldaji, Sh. Ettehad. Effects of varying sodium nitrate and carbon dioxide quantities on microalgae growth performance of a designed tubular photobioreactor and a channel photobioreactor. Iranian Journal of Plant Physiology 2011; 1(4): 257-263.

[68] Sonnekus MJ, Effects of Salinity on the Growth and Lipid Production of Ten Species of Microalgae from the Swartkops Saltworks: A Biodiesel Perspective Nelson Mandela Metropolitan University, 2010.

[69] Sheng J, Kim H. W, Badalamenti J. P, Zhou C, Sridharakrishnan S, Krajmalnik-Brown R, Rittmann BE, Vannela R. Effects of temperature shifts on growth rate and lipid characteristics of *Synechocystis* sp. PCC6803 in a bench-top photobioreactor. Bioresource Technology 2011; 102 (24): 11218-11225.

[70] Sivakumara G, Xua J, Thompsonb RW, Yangc Y, Randol-Smithd P, Weathersc PJ. Integrated green algal technology for bioremediation and biofuel, Bioresource Technology 2012; 107: 1-9.

[71] Floreto EAT, Hirata H, Ando S, Yamasaki S. Effects of Temperature, Light Intensity, Salinity and Source of Nitrogen on the Growth, Total Lipid and Fatty Acid Composition of Ulva pertusa Kjellman (*Chlorophyta*)" Botanica Marina, 2012; 36 (2): 149-158.

[72] Weisse T, Stadler P. Effect of pH on growth, cell volume, and production of freshwater ciliates, and implications for their distribution. Limnology and Oceanography 2006; 51: 1708-1715.

[73] Kumar A, Ergas S, Yuan X, Sahu A, Zhang Q, Dewulf J, Malcata F. X, Langenhove H. Enhanced CO_2 fixation and biofuel production via microalgae: Recent developments and future directions. Trends in Biotechnology 2010; 28(7): 371-380.

[74] Bergstrom C, McKeel C, Patel S. Effects of pH on algal abundance: a model of Bay Harbor, Michigan, Biological Station of University of Michigan, 2007.

[75] Widjaja A, Chien CC, Ju YH. Study of increasing lipid production from fresh water microalgae Chlorella vulgaris, Journal of the Taiwan Institute of Chemical Engineers 2009; 40(1):13-20.

[76] Moisander PH, McClinton E. and Paerl HW. Salinity Effects on Growth, Photosynthetic Parameters, and Nitrogenase Activity in Estuarine Planktonic Cyanobacteria 2002; 43(4): 432-442.

[77] Xin L, Hong-ying H, Ke G, Ying-Xue S. Effects of different nitrogen and phosphorus concentrations on the growth, nutrient uptake, and lipid accumulation of a freshwater microalga *Scenedesmus* sp. Bioresource Technology 2010; 101 (14): 5494-5500.

[78] Lin Q, Lin J. Effects of nitrogen source and concentration on biomass and oil production of a *Scenedesmus rubescens* like microalga, Bioresource Technology 2011; 102 (2): 1615-1621.

[79] Harun R, Singh M, Forde GM, Danquah MK. Bioprocess engineering of microalgae to produce a variety of consumer products, Renewable and Sustainable Energy Reviews 2010; 14: 1037–1047.

[80] Borowitzka MA. Commercial production of microalgae: ponds, tanks, and fermenters, Progress in Industrial Microbiology 1999; 35 (C): 313-321.

[81] Masojidek J, Torzillo G. Mass cultivation of freshwater microalgae. Ecological Engineering. Encyclopedia of Ecolology 2008; 3: 2226. 2235.

[82] "Biodiesel from Algae Oil", www. oilgae. com", Access date: May 10, 2012.

[83] Ugwu CU, Aoyagi H, Uchiyama H. Photobioreactors for mass cultivation of algae. Bioresource Technology 2008; 99: 4021–4028.

[84] Hu Q, Guterman H, Richmond A. A flat inclined modular photobioreactor for outdoor mass cultivation of photoautotrophs. Biotechnology and Bioengineering 1996; 51:51–60.

[85] Ramos de Ortega A, Roux JC. Production of Chlorella biomass in different types of flat bioreactors in temperate zones. Biomass 1986; 10:141–56.

[86] Iqbal M, Grey D, Stepan-Sarkissian F, Fowler MW. A flat-sided photobioreactor for culturing microalgae. Aquacultural Engineering 1993 ;12: 183–90.

[87] Vonshak A, Torzillo G. Environmental stress physiology. In: Richmond, A. (Ed.), Handbook of Microalgal Culture. Blackwell Publishers, Oxford, 2004; 57–82.

[88] Camacho Rubio F, Acie ´n Ferna ´ndez F. G, Sa ´nchez Pe ´rez J. A, Garcı ´a Camacho F, Molina Grima, E. Prediction of dissolved oxygen and carbon dioxide concentration profiles in tubular photobioreactors for microalgal culture. Biotechnology and Bioengineering 1999; 62: 71–86.

[89] Yılmaz HK. Mikroalg Üretimi İçin Fotobiyoreaktör Tasarımları, MEU. Su Ürünleri Fakültesi, (2006).

[90] Ogbonna JC, Soejima T, Tanaka H. An integrated solar and artificial light system for internal illumination of photobioreactors. Journal of Biotechnology 1999; 70: 289–297.

[91] Mori K. Photoautotrophic bioreactor using visible solar rays condensed by fresnel lenses and transmitted through optical fibers. Biotechnology and Bioengineering Symposium 1985; 15: 331–345.

[92] "Microalgae production", SOLEY INSTITUTE, www. algaeinstitute. com

[93] Harun R, Davidson M, Doyle M, Technoeconomic analysis of an integrated microalgae photobioreactor, biodiesel and biogas production facility. Biomass and Bioenergy 2011; 35: 741-747.

[94] Kennedy CA, In Home Photosynthetic Bioreactor. Advanced Biology 1998.

[95] Carvalho AP, Meireles LA, Malcata FX. Microalgal reactors: a review of enclosed system designs and performances. Biotechnology Progress 2006; 22: 1490–1506.

[96] Miron SA, Garcia MC, Camacho GF, Molina GEY. Mixing in bubble column and airlift reactors. Chemical Engineering Research and Design 2004; 82 (A10): 1367–1374.

[97] Fan L, Zhang Y, Cheng L, Zhang, L, Tang D, Chen H. Optimization of carbon. 2007.

[98] Fernandes BD, Dragoner GM, Teixiera JA, Vicente AA. Light regime characterization in an airlift. Photobioreactor for production of microalgae with high starch content. Applied Biochemical Biotechnology 2010; 161: 218–226.

[99] Kommareddy AR, Anderson GA. Study of Light as a Parameter in the Growth of Algae in a Photo-Bio Reactor (PBR). ASAE Paper No. 034057. ASAE, St. Joseph, Michigan. 2003.

[100] Zhang K, Kurano N, Miyachi S. Optimized aeration by carbon dioxide gas for microalgal production and mass transfer characterization in a vertical flat-plate photobioreactor. Bioprocess Biosystems Bioengineering 2002; 25: 97–101.

[101] Scott AS, Davey MP, Dennis SJ. Biodiesel from algae: challenges and prospects. Current Opinion in Biotechnology 2010; 21: 277–288.

[102] Williams JA. Keys to Bioreactor selection. CEP Magazine 2002; 34–41.

[103] Becker EW. Microalgae: Biotechnology and Microbiology. ISBN 0521350204. 1995.

[104] Demirel Z. Eğirdir Gölünden İzole Edilen Yeşil Mikroalg: *Chlorophyta Scenedesmus protuberans fris.* 'in Antimikrobiyal ve Antioksidan Özelliğinin Araştırılması, MSc. Thesis, Ege University, İzmir. 2006.

[105] Molina Grima E, Belarbi E, Acie ́n Ferna ́ndez F, Robles Medina A, Chisti Y. Recovery of microalgal biomass and metabolites: process options and economics. Biotechnology Advances 2003; 20:491–515.

[106] Danquah M, Ang L, Uduman N, Moheimani N, Forde G. Dewatering of microalgal culture for biodiesel production: exploring polymer flocculation and tangential flow filtration. Journal of Chemical Technology and Biotechnology 2009; 84:1078–83.

[107] Popoola TOS, Yangomodou OD. Extraction, properties and utilization potentials of cassava seed oil. Biotechnology 2006; 5:38–41.

[108] Serrato AG. Extraction of oil from soybeans. Journal of the American Oil Chemistry 1981; 58-3:157-159.

[109] Macias-Sanchez MD, Mantell C, Rodriguez M, Martinez De La Ossa E, Lubian LM, Montero O. Supercritical fluid extraction of carotenoids and chlorophyll a from *Nannochloropsis gaditana*. Journal of Food Engineering 2005; 66:245–51.

[110] Wiltshire KH, Boersma M, Moller A, Buhtz H. Extraction of pigments and fatty acids from the green alga *Scenedesmus obliquus* (*Chlorophyceae*). Aquatic Ecology 2000; 34:119–26.

[111] Yusuf N, Kamarudin SK, Yaakub Z. Overview on the current trends in biodiesel production. Energy Conversion and Management 2011; 52: 2741–2751.

[112] Demirbas A. Progress and recent trends in biodiesel fuels. Energy Conversion and Management 2009; 50: 14–34.

[113] Lin L, Cunshan Zhou, Vittayapadung S, Xiangqian S, Mingdong D. Opportunities and challenges for biodiesel fuel. Applied Energy 2011; 88: 1020–1031.

[114] Balat M, Balat H. Progress in biodiesel processing. Applied Energy 2010; 87: 1815–1835.

[115] D'Elia L, Keyser A, Young C. Algae Biodiesel, An Interactive Qualifying Project Report. 2010.

[116] Maa F, Hanna M. Biodiesel production: a review. Bioresource Technology 1999; 70: 1-15.

[117] Sharma YC, Singh B, Upadhyay SN. Advancements in development and characterization of biodiesel: A review. Fuel 2008; 87(12): 2355–73.

[118] Ahmad AL, Yasin NH, Derek JK. Microalgae as a sustainable energy source for biodiesel production: A review. Renewable and Sustainable Energy Reviews 2011; 15: 584–593.

[119] Atabani AE, Silitongaa AS, Badruddina IA, Mahliaa TMI, Masjukia HH, Mekhilef S. A comprehensive review on biodiesel as an alternative energy resource and its characteristics. Renewable and Sustainable Energy Reviews 2012; 16: 2070– 2093.

[120] Rachmaniah O, Ju YH, Vali SR, Tjondronegoro I, Musfil AS. A study on acid-catalyzed transesterification of crude rice bran oil for biodiesel production. In: youth energy symposium, 19th World energy congress and exhibition, Sydney (Australia), September 5–9; 2004.

[121] Xu H, Miao X, Wu Q. High quality biodiesel production from a microalga *Chlorella protothecoides* by heterotrophic growth in fermenters. Journal of Biotechnology 2006; 126: 499–507.

[122] Johnson MB. Microalgal Biodiesel Production through a Novel Attached Culture System and Conversion Parameters. MS thesis Virginia Polytechnic Institute and State University; 2009.

[123] Velasquez-Orta SB, Lee JGM, Harvey A. Alkaline in situ transesterification of *Chlorella vulgaris*. Fuel 2012; 94: 544–550.

[124] Ferrentino JM, Farag IH, Jahnke LS. Microalgal Oil Extraction and In-situ Transesterification. Chemical Engineering, University of New Hampshire (UNH) Durham, NH http://www. nt. ntnu. no/users/skoge/prost/proceedings/aiche-2006/data/papers/ P69332. pdf (accessed 02. 08. 12).

[125] Carvalho J, Ribiero A, Castro J, Vilarinho C, Castro F. Biodiesel Production by Microalgae and Macroalgae from North Littoral Portuguese Coast. WASTES: Solutions, Treatments and Opportunities, 1st International Conference September 12th – 14th 2011.

[126] Tran D, Yeh K, Chen C, Chang J. Enzymatic transesterification of microalgal oil from *Chlorella vulgaris* ESP-31 for biodiesel synthesis using immobilized Burkholderia lipase. Bioresource Technology 2012; 108: 119–127.

[127] Umdu ES, Tuncer M, Seker E. Transesterification of Nannochloropsis oculata microalga's lipid to biodiesel on Al_2O_3 supported CaO and MgO catalysts. Bioresource Technology 2009; 100: 2828–2831.

[128] Maceiras R, Rodrı´guez M, Cancela A, Urréjola S, Sánchez A. Macroalgae: Raw material for biodiesel production. Applied Energy 2011; 88: 3318–3323.

[129] Krohn JB, McNeff VC, Yan B, Nowlan D. Production of algae-based biodiesel using the continuous catalytic Mcgyan process. Bioresource Technology 2011; 102: 94–100.

[130] Özçimen D, Yücel S. Novel Methods in Biodiesel Production, In: Marco Aurélio dos Santos Bernardes (Ed.) Biofuel's Engineering Process Technology. InTech; 2011. p353-384.

[131] Patil P, Deng S, Isaac Rhodes J, Lammers PJ. Conversion of waste cooking oil to biodiesel using ferric sulfate and supercritical methanol processes. Fuel 2009; 89 (2): 360-364.

[132] Patil PD, Gude VG, Mannarswamy A, Cooke P, Munson-McGee S, Nirmalakhandan N, Lammers P, Deng S. Optimization of microwave-assisted transesterification of dry algal biomass using response surface methodology. Bioresource Technology 2011;102: 1399-1405.

[133] Cancela A, Maceiras R, Urrejola S, Sanchez A. Microwave-Assisted Transesterification of Macroalgae. Energies 2012, 5: 862-871.

[134] Koberg M, Cohen M, Ben-Amotz A, Gedanken A. Bio-diesel production directly from the microalgae biomass of *Nannochloropsis* by microwave and ultrasound radiation. Bioresource Technology, 2011; 5 (102): 4265-4269.

[135] Ehimen EA, Sun ZF, Carrington CG. Use of ultrasound and co-solvents to improve the in-situ transesterification of microalgae biomass. Procedia Environmental Sciences 2012; 15: 47-55.

Chapter 2

Biodiesel Feedstock and Production Technologies: Successes, Challenges and Prospects

Y.M. Sani, W.M.A.W. Daud and A.R. Abdul Aziz

Additional information is available at the end of the chapter

1. Introduction

In order to achieve the biodiesel central policy of protecting the environment, replacing petroleum diesel and protecting and/or creating jobs, a good understanding of biodiesel history is essential. This is because consumers always tend to buy cheap rather than "green" fuels. Moreover, it is more difficult for a new technology to dislodge one that has reached societal standard. The more the popular technology is used, the more it improves; becoming less expensive due to wider market potentials. Petrodiesel has become the "life-blood" of our economy. It would be almost impossible to find a commercial product today that does not consume diesel fuel during its production and distribution [1-4]. Therefore, the aim of this chapter is to provide an overview on the history and motivation, successes, challenges and prospects of biodiesel as source of energy. This will provide a global outlook in making biodiesel an economical and eco-friendly alternative to petroleum diesel.

The historical developments of the biofuel industry in general and biodiesel in particular, is unlike many industries. This is because the driving factors for its advances are more of economics and politics than technological [5]. As early as 1853, transesterification was conducted on vegetable oil in the search for a cheap method to produce glycerine for producing explosives during World War II by E. Duffy and J. Patrick [6-8]. In 1937, G. Chavanne, Belgian scientist patented the "Procedure for the transformation of vegetable oils for their uses as fuels". "Biodiesel" as a concept was thus established [9]. It is a simple process where alkoxy group of an ester compound (oil or fat) is exchanged with an alcohol. However, it was not until 1977 that first patent on commercial biodiesel production process was applied for by Expedito Parente; a Brazilian scientist [10].

Prior to the discovery of and boom in fossil fuels, power was mainly generated from steam. However, the use of hydro-energy consumes large resources coupled with the inefficiencies of the steam engine where only about 10 to 12% efficiency is derived from new power generation plant. A patent for an efficient thermal engine which was to be operated on peanut oil was filed in 1892 by Rudolph Diesel in Germany. By 1893, Diesel's invention was demonstrated in an exhibition in Paris. Within five years of its invention, Diesel's engine ran on its own power with 75% efficiency against its initial 26% efficiency [11]. In 1912, Diesel published two articles [12,13] in which he reflected:

"The fact that fat oils from vegetable sources can be used may seem insignificant to-day, but such oils may perhaps become in course

of time of the same importance as some natural mineral oils and the tar products are now. (...) In any case, they make it certain that

motor power can still be produced from the heat of the sun, which is always available for agricultural purposes, even when all our

natural stores of solid and liquid fuels are exhausted."

The demand for biofuels began to increase in America from the 1890's to 1920's. These were attributed to the pioneering efforts on the diesel engine by Adolphus Busch and Clessie L. Cummins along with other engine manufacturers. However, the biofuel industry was faced with a major challenge of cheap and readily available feedstock. Unfortunately for the biofuel industry, at this same period, the petroleum industries found out more advanced technologies for improving the properties of the "black gold". The discoveries of large reservoirs and developments created new markets for this "black gold". Therefore, by 1940, diesel engines were altered to enable them use petroleum-based fuels which have lower viscosities. Thereafter, the sales of biodiesel were weakened and the production structure was pushed to the background. Therefore, no significant efforts were made to increase the public awareness on its potentials. This period witnessed increased demands for automobiles which were propelled by petroleum fuels. The availability of public funds, and new transportation infrastructure such as interstate and highway systems helped in this regard [14].

The early post-WWII fossil fuel demand and supply was influenced by the commencement of offshore oil and gas production in 1945 at the Gulf of Mexico and the invention of jet aircraft [14]. However in the 1970s, speculations regarding the finite nature of the fossil oil reserves became an issue worth pondering over. In 1973 and 1978, OPEC reduced oil supplies and increased the prices to meet with the shortages of the petroleum crisis of that time. This marked the reemergence of the potentials of biofuels in the public consciousness. Thus in 1979, South Africa started the commercial development of biodiesel. Sunflower oil was transesterified and refined to a standard similar to petroleum diesel fuel [15]. The outcome was the discovery of several sources and technologies that improved engine performance with reduced environmental impacts. Experiences from past were used in achieving improved efficiencies, while reducing costs by developing the renewable energy marketing advantage.

The procedure for the production, quality and engine-testing for biodiesel was finalized and published internationally in 1983. The South African technology was obtained by Gaskoks; an Austrian company. Gaskos established the first pilot plant for biodiesel production in 1987.

By April of 1989, the firm set up the first commercial-scale plant producing 20 million gallon per year (MGPY). However during this period, biodiesel was only being produced on a noncommercial scale in the United States. The growth in producing biodiesel in Europe began in 1991 because of the need to reduce environmental impacts from emissions of greenhouse gases (GHG). Three years later, the first commercial biodiesel production was started in America. By 2000, the Commodity Credit Corporation started subsidizing value-added agriculture towards biodiesel production. The past decade (2002 to 2012) witnessed an unprecedented production of biodiesel. Incentives from policy makers such as tax exemptions, tax credits and renewable fuel standards aided the biodiesel growth. However, some proper-ties of biodiesel also contributed to the unprecedented growth we are witnessing in the biodiesel industry [16-18].

The increasing interests on biodiesel is fueled by the need to find a sustainable diesel fuel alternative. This is mainly because of environmental issues, apprehensions over energy independence and skyrocketing prices. Several processing options are available for the biodiesel production. The various feedstocks and processing conditions provide several processing technologies. The choice of a particular technology is dependent on catalyst and the source, type and quality of feedstock. Others include postproduction steps such as product separation and purification and catalyst and alcohol recovery. The dominant factor in the production process is the cost of feedstock while capital costs contribute only about 7%. It is therefore essential to utilize cheap feedstock to reduce the overall production costs. In the same regards, some technologies are designed to handle variety of feedstocks.

2. Past achievments

Non-fossil fuel alternatives are favored because of their common availability, renewability, sustainability, biodegrablability, job creation, regional development and reduced environ-mental impacts. Table 1 summarizes some of the major successes of biodiesel.

2.1. Feedstocks

Numerous feedstocks have been experimented in biodiesel production. Advancements from such experimentations led to establishment of waste-to-wealth biodiesel production. Cheap and readily available raw materials such as used cooking oil and yellow grease are used for producing biodiesel. These efforts helped in reducing the environmental impacts associated with dumping in landfills as well as saves the cost of paying for such dumping. Another notable success is the use of Jatropha or the "miracle plant" in many developing countries. The fact that it can be cultivated almost anywhere with minimal irrigation and less intensive care, made it suitable for peasant farmers. Sustained high yields were obtained throughout its average life cycle of 30–50 years. Castor plantation are also intercropped with jatropha to improve the econmic viability of jatropha within the first 2 to 3 years [19]. Another oil crop that is used to improve soil quality is the nitrogen-fixing *Pongamia pinnata*. It produces seeds with significant oil contents.

2.2. Technologies

Biodiesel is one of the most thoroughly tested alternative fuel in the market today. Studies by many researchers have confirmed similar engine performance of biodiesel to petroleum diesel. Transesterification produce oil with similar brake power as obtained with diesel fuel. Minimal carbon deposits were noticed inside the engine except the intake valve deposits which were slightly higher. The level of injector coking was also reduced significantly lower than that observed with D2 fuel [7,17]. An important breakthrough in transesterification is the Mcgyan Process®, which can utilize various inexpensive, non-food-grade and free fatty acids (FFAs) containing feedstocks (Figure 1). The process can be small in physical size and it utilizes heterogeneous catalysts to produce biodiesel within 4 s [20,21]. The easy fatty acid removal or EFAR system ensures that no wastes are produced from the process. It eliminates post production costs such as the washing and neutralization steps. To achieve 100% conversion, it recycles all unreacted feedstock and excess alcohol back into the reactor. Energy efficiency is also achieved through heat transfer mechanism; in-coming cold reactants are preheated by the out-going hot products [20,21].

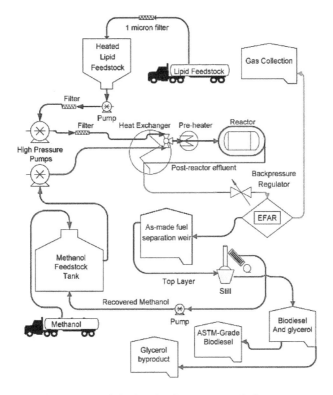

Figure 1. Process flow diafram of a biodiesel plant based on the Mcgyan proces[21]

Economic & social impact	Environment impact	Energy security
Sustainability; made from agricultural or waste resources	Reduced 78% GHG emissions	Reduced dependence on fossil fuels
Fuel diversity & improved fuel efficiency & economy	Reduced air pollution	Domestic targets
Improved rural economy	Biodegradability	Supply reliability
Increased income tax & trade balances	Improved land & water use	Readily available
International competitiveness	Carbon sequestration	Renewability
Increased investments on feedstocks & equipment	Lower sulfur content	Domestic distribution
Technological developments (R & D)	Lower aromatic content	Improved fuel economy
Higher cetane number (52 vs. 48), lubricity & flash point	Lesser toxicity	Comparable energy content (92.19%)
Knowledge development & diffusion	Safer handling & storage	Strict quality requirements are met
Strong growth in demand & market formation		Viscosity 1.3 to 1.6 times that of D2 fuel
Improved engine performance		Good energy balance (3.24:1 vs. 0.88:1)
Reduces the need for maintenance & prolongs engine life		
Compatible with all conventional diesel engines		
Offers the same engine durability & performance		
Has the potential of displacing petroleum diesel fuel		
Comparable start-up, torque range & haulage rates		

Table 1. Major achievements of biodiesel [16,23-27]

2.3. Environmental impacts and health effects

A 78% reduction in GHG emission was reported by the U.S. Departments of Agriculture and Energy with biodiesel usage. Essentially, biodiesel is non-aromatic and sulphur-free as compared with petrodiesel which contains 20 to 40 wt.% aromatic compounds and 500 ppm

SO_2 [7]. The potential of pure biodiesel to form ozone (smog) from hydrocarbons is 50% less. Also, sulfates and oxides of sulfur (major constituents of acid rain) are essentially eliminated from the exhaust emissions compared to petrodiesel. These help in curbing the increasing global warming problems. Average decrease of 22.5% for smoke density, 17.1% for CO and 14% for CO_2 have been reported when biodiesel was used [22]. Human life expectancy is thereby enhanced because of improved air quality.

2.3.1. Energy independence

Biodiesel reduces the excessive reliance on fossil fuels. This enhances the global energy security [17]. It also has the potential to replace oil importation since it is produced domestically, thereby providing additional market for agricultural products. It supports the rural communities where it is cultivated by protecting and generating jobs. Producing biofuels equivalent to 1% of automobile fuel consumption in the EU protected and/or created approximately 75,000 jobs [16]. Approximately, for every unit of fossil energy used in biodiesel production, 4.5 units of energy is gained. Moreover, lesser energy is required for biodiesel production than the energy derived from the final product [16].

3. Different feedstocks used in the production of biodiesel

More than 350 oil-bearing crops have been identified as potential sources for producing biodiesel. However, only palm, jatropha, rapeseed, soybean, sunflower, cottonseed, safflower, and peanut oils are considered as viable feedstocks for commercial production [28].

3.1. Edible feedstocks

Depending on availability, different edible oils are utilized as feedstocks for biodiesel production by different countries. Palm oil and coconut oil are commonly used in Malaysia and Indonesia. Soybean oil is majorly used in U.S. [30].

3.2. Non-edible feedstocks

In order to reduce production costs and to avoid the *food-for-fuel* conflict, inedible oils are used as the major sources for biodiesel production. Compared to edible oils, inedible oils are affordable and readily available. They are obtained from *Jatropha curcas* (jatropha or ratanjyote or seemaikattamankku), *Pongamia pinnata* (karanja or honge), *Calophyllum inophyllum* (nagchampa), *Hevca brasiliensis* (rubber seed tree), *Azadirachta indica* (neem), *Madhuca indica and Madhuca longifolia* (mahua), *Ceiba pentandra* (silk cotton tree), *Simmondsia chinensis* (jojoba), *Euphorbia tirucalli,* babassu tree, microalgae, *etc.* [31]. Among the 75 plant species which have more than 29% oil in their seed/kernel; palm, *Jatropha curcas*, and *Pongamia pinnata* (Karanja) were found to be the most suitable for biodiesel production [32]. Many European countries utilize rapeseed [29]. During World War II, oil from *Jatropha* seeds was used as blends with and substituted for diesel [33,34]. It has been reported that biodiesel produced from palm and

Jatropha have physical properties in the right balance; conferring it with adequate oxidation stability and cold performance [35]. Most of the strict requirements set by the American and European biodiesel standards for biodiesel have been achieved [36]. The major oils used for producing biodiesel are presented in Table 2.

Group	Source of oil
Major oils	Coconut (copra), corn (maize), cottonseed, canola (a variety of rapeseed), olive, peanut (groundnut), safflower, sesame, soybean, and sunflower.
Nut oils	Almond, cashew, hazelnut, macadamia, pecan, pistachio and walnut.
Other edible oils	Amaranth, apricot, argan, artichoke, avocado, babassu, bay laurel, beech nut, ben, Borneo tallow nut, carob pod (algaroba), cohune, coriander seed, false flax, grape seed, hemp, kapok seed, lallemantia, lemon seed, macauba fruit (Acrocomia sclerocarpa), meadowfoam seed, mustard, okra seed (hibiscus seed), perilla seed, pequi, (Caryocar brasiliensis seed), pine nut, poppy seed, prune kernel, quinoa, ramtil (Guizotia abyssinica seed or Niger pea), rice bran, tallow, tea (camellia), thistle (Silybum marianum seed), and wheat germ.
Inedible oils	Algae, babassu tree, copaiba, honge, jatropha or ratanjyote, jojoba, karanja or honge, mahua, milk bush, nagchampa, neem, petroleum nut, rubber seed tree, silk cotton tree, and tall.
Other oils	Castor, radish, and tung.

Table 2. Major oil species for biodiesel production [37]

3.2.1. Algae oil

Currently, algae-based biodiesel is the focus of many research interests because they have the potential to provide sufficient oil for global consumption. It has the potential to produce biodiesel yields >100 times those attainable per hectare from plant feedstock (Table 3). Besides their high lipid contents and fast growth rate, microalgae have the potential to mitigate the competitions for land-use and food-for-fuel conflicts. They are also able to reduce the GHG effect via CO_2 sequestration [38]. Microalgae can be cultivated in habitats which are not favorable for energy crops. Compared with oilseeds, the harvesting and transportation costs of microalgae are relatively low. *Nannochloropsis*, members of the marine green algae are considered the most suitable candidates for biodiesel production. These strains have shown high lipid content and biomass productivity. However, research in this area especially algal oil extraction is still limited and in early stages.

3.2.2. Other feedstocks

Used vegetable oils (UCO), yellow grease (8-12 wt% FFA), brown grease (>35 wt% FFA), and soapstock (by-product of refining vegetable oils) are potential feedstocks for biodiesel production. Their low costs and availability make them suitable for reducing the production costs of biodiesel. To achieve this however, the problems associated with high FFA which are common to these feedstocks, particularly when alkaline catalysts are employed need attention. Solid acid catalysts are currently receiving great attention because they are suitable for feedstocks containing FFAs [39-41]. Another process that has the potential of processing these feedstocks is supercritical transesterification. The pretreatment step, soap and catalyst removal common to alkaline catalysis are eliminated since the process requires no catalyst [42,43]. The process has fast reaction rate which significantly reduces the reaction time [44]. The process is insensitive to water and FFAs [43,45]. However, this method is not economical because it requires high reaction temperature, pressure and higher molar ratio of alcohol to feedstock [42,43,46]. Another interesting feedstock is *Salicornia bigelovii* (Halophytessuch). It can produce equal biodiesel yields obtained from soybeans and other oilseeds. They grow in saltwater of coastal areas unsuitable for energy crops.

Microalgae/Plant	Oil yield (L/ha/year)	Oil content (% wt in biomass)	Required land (M ha $^{-1a}$)	Biodiesel productivity (kg biodiesel/ha/year)
Microalgae[b] (high oil content)	136 900	70	2	121 104
Microalgae[c] (low to low oil content)	58 700 to 97 800	30 to 50	4.5	51 927-85 515
Oil palm (*Elaeis guineensis*)	5 950	30 to 60	45	4747
Jatropha (*Jatropha curcas* L.)	1 892	Kernel: 50 to 60 Seed: 35 to 40	140	656
Canola/Rapeseed (*Brassica napus* L.)	1 190	38 to 46	223	862
Soybean (*Glycine max* L.)	446	15 to 20	594	562
Corn/Maize (Germ) (*Zea mays* L.)	172	44 to 48	1540	152

Table 3. Estimated oil content, yields and land requirement for various biodiesel feedstocks.[36,47,48]

4. Methods of oil extraction

The three common methods used in extracting oil are: (i) Mechanical extraction, (ii) solvent extraction and (iii) enzymatic extraction.

4.1. Mechanical extraction method

This method is used by smaller production firms for processing less than 100,000 kg/day. Usually, an engine driven screw press or a manual ram press is used to extract 68–80% or 60–65% of the available oil respectively. Pretreatment such as dehulling and cooking increase oil yields to 89% and 91% after single and dual pass respectively [48,49]. However, most of the mechanical presses are designed for particular seeds which affect yields with other seeds. Also, extra treatments such as degumming and filtration are required for oil extracted by this technique.

4.2. Chemical (solvent) extraction method

The commonly used chemical methods are: (1) soxhlet extraction, (2) Ultrasonication technique and (3) hot water extraction [48,49]. Solvent extraction (or leaching) is typically used for processing more than 300,000 kg/day [50]. Yields are affected by particle size, solvent type and concentration, temperature and agitation. To increase the exposure of the oil to the solvent, the oilseeds are usually flaked. After extraction, the oil-solvent mixture or *miscella*, is filtered while heat is used to vaporize the solvent from the miscella. Steam is injected to remove any solvent remaining from the oil. The immiscibility of the solvent and steam vapors is used to separate them in a settling tank after condensation. The highest oil yields are obtained with n-hexane. However, the process requires higher energy and longer time compared to other methods. Furthermore, the human health and environmental impacts associated with toxic solvents, waste water generation and emissions of volatile organic compounds are challenges facing this method.

4.3. Enzymatic extraction method

Oilseeds are reduced to small particles and the oil is extracted by suitable enzymes. Volatile organic compounds are not produced by this method which makes it environmentally friendly when compared to the other methods. However, it has the disadvantage of long processing time and high cost of purchasing enzymes [51].

5. Technologies used in biodiesel production

Several researches were carried out to overcome or minimize the problems associated with producing biodiesel. The methods that have been used for minimizing the viscosity of vegetable oils for practical application in internal combustion engines include: pyrolysis, microemulsification, blending (diluting) and transesterification. Dilution and microemulsification are not production processes and are therefore not discussed in this chapter. A summary of vegetable oils and animal fats and the major biodiesel production technologies are presented in Table 4.

5.1. Pyrolysis or catalytic cracking

Pyrolysis is the heating of organic matter in the absence of air to produce gas, a liquid and a solid [52]. Heat or a combination of heat and catalyst is used to break vegetable oils or animal fats into smaller constituents. Olefins and paraffins are thus obtained with similar properties to petrodiesel where such products derived the name *"diesel-like-fuel"* [53]. Studies on effects of rapeseed particle size showed that the product yield is independent of the oilseed particle size [52]. The maximum temperature range for conversion of bio-oil is 400°C to 450°C [54]. Rapid devolatilization of cellulose and hemicellulose occur at this temperature. Heating rate and temperature have significant effects on bio-oil yields, char and gas released from olive [55]. The viscosity, flash and pour points and equivalent calorific values of the oil are lower than diesel fuel. Though the pyrolyzate has increased cetane number, it is however lower than that of diesel oil. Apart from reducing the viscosity of the vegetable oil, pyrolysis enables decoupling of the unit operation equipment in shorter time, place and scale. It produces clean liquids which needs no additional washing, drying or filtering. Product of pyrolysis consists of heterogeneous molecules such as water, particulate matter, sulfur, alkanes, alkenes and carboxylic acids [39,56]. Consequently, it is difficult to characterize fuel obtained from pyrolysis [52]. This process is energy consuming and needs expensive distillation unit. Moreover, the sulfur and ash contents make it less eco-friendly [57].

5.2. Transesterification (alcoholysis)

Transesterification is the most widely employed process for commercial production of biodiesel. It involves heating the oil to a designated temperature with alcohol and a catalyst, thereby restructuring its chemical structure. This conversion reduces the high viscosity of the oils and fats. For the transesterification of triglyceride (TG) molecule, three consecutive reactions are needed. In these reactions, FFA is neutralized by the TG from the alcohol. One mole of glycerol and three moles of alkyl esters are produced (for each mole of TG converted) at the completion of the net reaction. These separate into three layers, with glycerol at the bottom, a middle layer of soapy substance, and biodiesel on top [57]. Transesterification is a reversible reaction. To obtain reasonable conversion rates therefore, it requires a catalyst. The reaction conditions, feedstock compositional limits and post-separation requirements are predetermined by the nature of the catalyst. Table 5 presents a genaral overview of the several transesterification techniques for biodiesel production.

5.2.1. Homogeneous alkali-catalyzed transesterification

Alkali catalysts such as NaOH and KOH were preferred over other catalysts because of their ability to enhance faster reaction rates [63]. This is because they are readily available at affordable prices and enable fast reaction rates [24]. Detailed review on base-catalyzed transesterification of vegetable oils can be found in ref [64]. However, homogeneous catalysis has been faced with the been faced with the problems saponification, highly sensitive to FFAs, expensive separation requirement, waste water generation and high energy consumption.

5.2.2. Homogeneous acid-catalyzed transesterification

Though the performance of this method is not strongly affected by FFAs in the feedstock, the process is not as popular as the base-catalyzed process. This is because the use of strong acids such as H_2SO_4 [65,66], HCl, BF_3, H_3PO_4, and organic sulfonic acids [67], is associated with higher costs and environmental impacts. Moreover, the technique is about 4000 times slower than the homogeneous base-catalyzed reaction. The mechanism of the acid-catalyzed transesterification can be found in ref [68].

5.2.3. Heterogeneous acid and base-catalyzed transesterification

Solid acid can simultaneously catalyze the esterification and transesterification without the need for pretreating feedstocks with high FFAs. Thus, this technique has the potential of reducing the high cost of biodiesel production by directly producing biodiesel from readily available and low-cost feedstocks [67].
Solid basic catalysts also have the potential of reducing the cost of biodiesel production because of lesser catalyst consumption, reuse and regeneration. However, these catalysts have some disadvantages which hinder their wide acceptability. These include mass transfer (diffusion) problem which reduces the rate of reaction as a result of the formation of three phases with alcohol and oil. Other problems associated with base catalyzed transesterification are loss of catalyst activity in the presence of water and post-production costs such as product separation, purification and polishing.

5.2.4. Enzymatic transesterification

Some of the problems associated with homogeneous catalysts such as expensive product separation, wastewater generation, and the presence of side reactions are avoided with enzymatic transesterification [69]. Enzyme immobilization is usually done to enhance the product quality, increase the number of times the catalyst is reused and to reduce cost [28,70]. However, several technical difficulties such as high cost of purchasing enzymes, product contamination, and residual enzymatic activity are limiting the applicability of this technique.

5.2.5. Supercritical alcohol transesterification

Unlike the conventional transesterification of two heterogeneous liquid phases involving alcohol (polar molecule) and non-polar molecules (TGs), supercritical transesterification is done in single homogeneous phase. Subjecting solvents containing hydroxyl groups (such as water and alcohol) to conditions in excess of their critical points make them to act as superacids. Under supercritical conditions, alcohol serves a dual purpose of acid catalyst and a reactant [46,71]. The absence of interphase solves the mass transfer limitations which gives the possibility of completing the reaction in minutes rather than several hours. In fact, the Mcgyan Process® was used to produce biodiesel under 4 s [19,20]. However, this process is not economical especially for commercial production as it requires expensive reacting equipment due to high temperature and pressure [72]. Studies are currently being undertaken in order to reduce these high reacting conditions.

Direct use	Dilution with vegetable oils	Microemulsion of oils	Pyrolysis and catalytic cracking	Transesterification of oils and fats	
Advantages	Advantages	Advantages	Advantages	Catalytic	Non-catalytic
Simple process	Simple process and non-polluting	Simple process and non-polluting	Simple process & non-polluting no additional washing, drying or filtering required		
Disadvantages	Disadvantages	Disadvantages	Disadvantages	Acid-catalyzed	BIOX cosolvent process
Highly viscous	Highly viscous	Incomplete combustion	Contains heterogeneous molecules	Alkali catalytic	Supercritical alcohol
Highly unstable	Highly unstable	Injector needle sticking	Low purity	Enzyme-catalyzed	Microwave and ultrasound assisted
Low volatility	Low volatility	Carbon deposits	Requires high temperature	Catalytic supercritical alcohol	
Not suitable for commercial production	Not suitable for commercial production	Not suitable for commercial production	Requires expensive equipment	See Table 5 for advantages and disadvantages	

Table 4. Use of vegetable oils and animal fats and major biodiesel production processes.

5.3. Technologies

5.3.1. Microwave assisted transesterification

The microwave irradiation as energy stimulant has been attracting the attention of many researchers. This is because the reaction process fast (within minutes), it employs a lower alcohol-oil ratio and it reduces by-products quantities. It uses a continuously changing electrical and magnetic fields to activate the smallest degree of variance of the reacting molecules. These rapidly rotating charged ions interact easily with minimal diffusion limitation [73]. However, this process also has commercial scale-up problem because of high operating conditions and safety aspects [74]. An even more daunting challenge is in increasing the irradiation penetration depth beyond a few centimeters into the reacting molecules.

5.3.2. Ultrasound assisted transesterification

This process utilizes sound energy at a frequency beyond human hearing. It stretches and compresses the reacting molecules in an alternating manner. Application of high negative pressure gradient beyond the critical molecular distance forms cavitation bubbles. Some of the bubbles expand suddenly to unstable sizes and collapse violently. This causes emulsification and fast reaction rates with high yields since the phase boundary has been disrupted [75-77].

Chemical catalysed				Chemical catalysed (Modified)			Biochemical catalysed	Noncatalysed
Homogeneous acid	Homogeneous base	Heterogeneous acid	Heterogeneous base	Microwave irradiation	Ultrasound (sonication)	Oscillatory flow reactor	Enzyme	Supercritical methanol
Merits	Merits	Merits	Merits	Merits	Merits	Merits	Merits	Merits
Employs feedstocks with high FFAs (*/>2 wt %)	Reaction is 4000 times faster than homogeneous acid catalysed reactions	High possibility of reusing and regenerating catalyst many times	High possibility of reusing and regenerating catalyst many times	Speeds up rate of reaction (from hours to minutes)	Increases FAME production from seedcakes	Increases mixing of reactants	Operates at milder reaction conditions	Simultaneous transesterification of TGs and esterification of FA
No pretreatment required	Operates at mild temperature (50 to 80 °C)	Simultaneous transesterification of TGs and esterification of FA	Saves cost of purchasing catalyst	Improves catalyst activity and selectivity	*In situ* extraction transesterification	Efficient heat and mass transfer	Cleaner biodiesel and glycerol are produced	High biodiesel yield
	Lower alcohol-to-oil (5:1) molar ratio	Simpler and less energy intensive	Simpler and less energy intensive	Minimizes energy consumption	High FAME yields	Higher yield in shorter time compared to batch-type	Energy consumption is minimized	Simultaneous transesterification of TGs and esterification of FA
	High biodiesel yield	Does not require feedstock pretreatment	Easy separation of products		Eliminates of saponification	Reduces reactor length-to-diameter ratio	Waste generation is minimized	Requires no catalyst
	Catalysts are cheap and readily available	Mild reaction conditions & less prone to leaching	Mild reaction conditions		Low reaction time	Reduces costs		Relatively fast reaction rate
		Waste generation is minimized			Mild reaction conditions			Short reaction time (<30 minutes)
		Relatively fast reaction rates			Enhances mass transfer			
		Easy product separation						
		Saves cost of purchasing catalysts						
		Minimizes solvation of active sites by action of water						
		Eliminates of saponification						

	Chemical catalysed			Chemical catalysed (Modified)			Biochemical catalysed	Noncatalysed
Homogeneous acid	Homogeneous base	Heterogeneous acid	Heterogeneous base	Microwave irradiation	Ultrasound (sonication)	Oscillatory flow reactor	Enzyme	Supercritical methanol
		Reduces size & cost of reaction vessel						
		Very attractive commercially						
Challenges	*Challenges*	*Challenges*	*Challenges*	*Challenges*	*Challenges*	*Challenges*	*Challenges*	*Challenges*
Very slow reaction rate & mineral acids used are corrosive to the equipment	Highly sensitive to water and FFAs in the oil	Availability of specific catalysts at low cost. Researches are ongoing to find low cost precursors.	Requires feedstock pretreatment & catalyst get poisoned with prolong exposure to ambient air	Difficulties in process scale-up from laboratory scale to large-scale	Difficulties in process scale-up from laboratory scale to large-scale	Difficulties in process scale-up from laboratory scale to large-scale	High cost of enzymes	Energy intensive
Catalyst required in large quantities	Requires refined feedstock (0.5 % FFA; 0.06% H$_2$O)	Limitation due to diffusion problems. This is solved by designing catalysts with large interconnected pores with high concentration of acid sites	High cost of reacting vessels	Depth of radiation is limited to a few cm.	Requires advanced technology	Requires advanced technology	High production cost	Very expensive
Requires high alcohol-to-oil molar ratio	Water saponifies the esters and FFAs reacts with the catalyst		Two-step reaction of esterification and transesterification	Requires advanced technology	Safety issues in equipment handling	Safety issues in equipment handling	Enzymes easily denatured	Not commercially profitable
Higher temperature	Requires methanol-to-oil of 6:1 (or higher) molar ratio instead of the stoichiometric 3:1 ratio		Product contamination from leaching of active catalytic sites	Safety issues in equipment handling.			Not commercially profitable	Safety issues
Undesirable etherification reaction (dialkyl or glycerol ethers)	Soap formation (FFA"/>2%)		Water saponifies the esters and FFAs reacts with the catalyst making purification difficult				Very slow reaction rates (slower than homogeneous acid catalysed)	High temperature and pressure

Chemical catalysed			Chemical catalysed (Modified)				Biochemical catalysed	Noncatalysed
Homogene- ous acid	Homogeneous base	Heterogeneous acid	Heterogene- ous base	Microwave ir- radiation	Ultrasound (sonication)	Oscillatory flow reactor	Enzyme	Supercritical methanol
Separation and purifica- tion of glyc- erol	Loss of catalyst		Catalyst leach- ing leads to product con- tamination				Sensitive to methanol	Energy inten- sive
Not commer- cially profita- ble	Reduces biodie- sel yield & gen- erates wastewater		Purification de- creases biodie- sel yield					

Table 5. Merits and challenges surrounding transesterification processes[78]

6. Current challenges and future prospects

In order to make biodiesel profitable, several technical challenges need to be resolved. The most important challenge is in reducing the high cost of feedstock. Low-cost feedstocks such as algal oils, used cooking oils and animal fats are utilized to increase biodiesel profitability. However, presence of higher amounts of water and FFAs in these feedstocks poses the problems of saponification and extra pretreatment and purification costs with alkali catalysts. The challenge facing researchers currently is developing efficient heterogeneous acid catalysts that would alleviate these problems. Also, diversifying the by-product of biodiesel production processes is critical to ensuring its economic, social and environmental sustainability.

6.1. Vegatable oil as feedstock for biodiesel

Currently, biodiesel production costs are higher than those of petroleum diesel. Subsidies such as tax exempt and excise duty reductions are essential to make biodiesel price-competitive. It is not certain whether these political supports will be sustained in the future. It is therefore crucial for the biofuel industry to establish readily available and affordable feedstocks and efficient production systems to sustain its market growth.

6.2. Non-food crops

Early studies have indicated relative differences in the cultivation patterns and oil production management of the non-food feedstocks compared to food crops. These are still under investigation [79]. Therefore, more data is needed to evaluate the sustainability index to estimate the real global impact of these feedstocks. Microalgae are promising in solving most of the problems associated with energy crops. However, the cultivation and extraction technologies are still at their infancy and need major advancements for sustainable commercial production [1]. The oil extraction methods currently in use for algal oil are expensive. Efficient mixing from pumps or motionless mixers is required to en-sure homogeneity and to reduce mass transfer limitations. However, this increases the dispersion of glycerol into the FAME phase and the time required for separation. Techni-ques that utilize motionless mixing requires higher temperature and pressure to achieve shorter residence time. This increases energy consumption and cost implications. This as-pects of biodiesel production technology is still being developed.

6.3. Effects of moisture and FFA

The key parameters that determine the viability of most feedstocks is FFA and moisture content. Pretreatments to less than 0.05% FFA is required for homogeneous alkali catalysts [80,81]. Prolonged storage in the presence of water and air leads to microbial growth and fuel degradation. This contributes to deposit formation on fuel injectors and engine damage. Heterogeneous acid catalysts are utilized to avoid the pretreatment and post production costs and storage problems.

6.4. Pyrolysis

Pyrolysis generates aromatic toxins. The bio-oil produced is corrosive due to high acidity, water content and other impurities such as solids and salts. These and other problems such as variable viscosity make it unstable and unsuitable for direct use [82]. It has 40% less energy density compared to diesel fuel because of the high oxygen content [83]. Depending on the feedstock and reacting conditions used, bio-oil is 10 to 100% more expensive than petroleum diesel. There is also the need to establish standards for product quality, use and distribution [1]. In order to stabilize the composition of the bio-oil and reduce water and oxygen content, processes such as steam reforming, hydro-treatment, hydro-cracking and emulsification with mineral diesel for direct use are employed [84-87].

6.5. Alcohol

Methanol is toxic, highly flammable and contributes to global warming. Gaskets and rubber seals made from natural rubber get easily deteriorated when biodiesel containing a high level of alcohol is used [80]. Therefore, control or replacement of the alcohol content is required. The biodiesel produced with methanol from fossil sources has approximately 94 to 96% biogenic content. In order to produce a 100% renewable biodiesel (fatty acid ethyl ester; FAEE), bioethanol is currently experimented as a substitute for methanol [88]. However, it is expensive to purify and recover ethanol because it forms an *azeotrope* with water. Additionally, chemical grade ethanol is usually denatured with poisonous substances to prevent it from being abused. Therefore, it is difficult to obtain pure chemical grade ethanol.

6.6. Supercritical alcohol process

The residence time for this process is within 4 s to 10 min because of efficient mixing [71,72]. However, due to higher reacting conditions of temperature and pressure, the process is faced with some limitations. Process scale-up for commercial production is the major one amongst them. The process requires more energy at extra cost and higher molar ratio of alcohol-to-oil (42:1). Also, there is the need to quench the reaction in a rapid manner. This prevents the biodiesel from decomposing as a result of the high temperature and pressure. To reduce the high operating conditions and increase product yield, some researchers employ co-solvents, such as hexane, CO_2, and CaO [72]. Oil and alcohol are sparingly soluble in each other. However, small amount of hexane (2.5 wt%) added increased the biodiesel yield from 67.7% to 85.5% under supercritical conditions [72]. This was made possible because the co-solvent increased the homogeneity of the reactants. Supercritical CO_2 is a facile substance that can obtained at affordable cost. It is also environmentally friendly and can effectively used in the reaction and safely recovered via depressurization. A process that combines co-solvents in superctical conditions is promising in increasing product yield, reducing process time and overall production costs.

6.7. Biodiesel/glycerol separation and FAME quality

The slightly soluble nature of FAMEs and glycerol makes product separation a necessary step. The product is usually allowed to settle for some hours into the different phases. However, the solubility of glycerol in ester and vice versa is increased in the presence of excess unreacted methanol which acts as solvent. This solvent action by the methanol increases the post production costs. Besides, it is also essential to remove all traces of TGs which form emulsion layer between the two phases. The presence of this layer further makes separation difficult and expensive. On the other hand, the storage, transportation, distribution and retail infrastructure used for petroleum diesel can be used for biodiesel even in its neat form. This will reduce construction costs for establishing new insfrasctures for biodiesel. However, biodiesel degrades after long period of storage. In order to prevent this from occuring, advances in storage and distribution logistics have to be developed. Also, similar logistics employed by the petroleum industry could be adapted.

6.8. Use of cosolvents

A technique developed to overcome mass transfer limitations and to increase the rate of reaction is the use of cosolvents such as methyl tert-butyl ether (MTBE) and tetrahydrofuran (THF). High quality FAMEs is obtained at moderate conditions (30 °C) within 10 minutes. However, the process requires larger and special "leak proof" reacting vessels and complete removal of the cosolvent from the product.

6.9. NO_x emissions

Despite the favorable environmental impacts in terms of overall reduced GHG emissions, biodiesel has the potential to increase NO_x emissions. Approximately 3 to 4%, 4 to 6% and 6 to 9% over petroleum diesel is emitted from B20, B40 and B100 respectively [89]. Adjustments in combustion temperatures and injection timing [90], use of antioxidants [91] and catalytic conversion techniques were successful in reducing these emissions [90].

6.10. Economic analysis

As discussed in the introductory section, vegetable oils have other important uses. Recently, dielectric oils and synthetic lubricants used for electric transformers have joined the market competition for these raw materials. This will impact negatively towards the cost of raw materials for the biodiesel industry [92]. About 15% of lubricants used in vehicles in some European countries are from vegetable oil derivatives [93]. Additionally, the heating value of biodiesel is 10% lower than that of petroleum diesel. This is because of the substantial amount of oxygen in the fuel. Moreover, it also has a higher specific gravity of 0.88 when compared to 0.85 of petroleum diesel. Therefore, its overall energy content per unit volume is having an impact which is approximately 5% lower than that obtained from petroluem diesel [94]. This results in higher specific fuel consumption values of the biodiesel. Another problem encountered when switching from petroleum diesel to biodiesel in the same fuel system is the clogging of the fuel filters. This is because biodiesel acts as a solvent which dissolves sediments in diesel fuel tanks [95]. On a positive note, sales of purified glycerol (glycerine) saved 6.5% of the operational cost [97-99] while 25% saving was reported in ref [96] from the utilization of waste soapstock with respect to virgin soybean oil. However, it is necessary to compensate the negative cost implications from commercial production of biodiesel from such low value feedstocks before valid conclusions can be derived.

7. Conclusions

Some of the major challenges faced by the biodiesel industry include readily available and affordable feedstocks, competition from a popular and cheaper energy source, technological advancements and acceptability. Those challenges requiring immediate attention are product stability under long storage, lower energy content, cold flow properties, catalyst leaching, microalgal oil extraction and NOx emissions. Despite these challenges however, the historical development of biodiesel is intriguing. Biodiesel has successfully remained an energy source to be reckoned with even after being relegated to the background for so many years. Concerns over diminishing oil reserves, increasing crude oil prices and associated environmental impacts aided the reemergence of biodiesel; making it the fastest growing industry worldwide. Several technologies were developed while more advances are in the process of being established. Other successes associated with the biodiesel industry include reduction in environmental impacts, job creation, energy security and waste-utilization. Biodiesel is regarded as a viable alternative or additive to petrodiesel because of its good properties such as nontoxicity, clean-burning, renewability and acceptability. Consequently, the prospects of the biodiesel industry are numerous. The biodiesel production process is shifting from other sources to algal oil and heterogeneous acid catalysts. Algal oil is a more reliable and efficient source. It has the potential of producing yields of more than 100 times those attainable per hectare from oilseeds. Affordable and readily available non-food feedstocks such as microalgae have been produced in commercial scale without competing with arable land or causing deforestation. Additionally, the use of heterogeneous acid catalyst produces cleaner and higher yields.

It employs cheaper and readily available feedstocks and minimizes pre- and post-product costs. These and other factors such as waste-utilization and cleaner emissions will help ensure biodiesel as a cheaper energy source with greater economic benefits and healthier environments.

Author details

Y.M. Sani[1,2], W.M.A.W. Daud[1*] and A.R. Abdul Aziz[1]

*Address all correspondence to: ashri@um.edu.my

1 Department of Chemical Engineering, Faculty of Engineering, University Malaya, Kuala Lumpur, Malaysia

2 Department of Chemical Engineering, Ahmadu Bello University, Zaria, Nigeria

References

[1] Haung, D, Zhou, H, & Lin, L. Biodiesel: an alternative to conventional fuel. Energy Procedia (2012). , 1-1874.

[2] James, M. Utterback. Mastering the Dynamics of Innovation. Boston: Harvard Business School Press, (1994).

[3] Arthur, W. Competing Technologies, Increasing Returns, and Lock-In By Historical Events. Economic Journal (1989). , 99(394), 116-131.

[4] Hughes, T. P. The evolution of large technological systems. In: Bijker W.E., Hughes TP and Pinch T. The social construction of technological systems. Cambridge, Massachusetts: MIT Press. (1989). , 51-82.

[5] Luque, R, Herrero-davila, L, Campelo, J. M, Clark, J. H, Hidalgo, J. M, Luna, D, Marinas, J. M, & Romero, A. A. Biofuels: a technological perspective. Energy Environ. Sci. (2008). , 1-542.

[6] Lin, L, Cunshan, Z, Vittayapadung, S, Xiangqian, S, & Mingdong, D. Opportunities and challenges for biodiesel fuel. Applied Energy (2011). , 88-1020.

[7] Demirbas, A. Biofuels, in: Green Energy and Technology, (2009). Springer London, , 103-230.

[8] Shay, E. G. Diesel fuel from vegetable oils: status and opportunities. Biomass Bioenergy (1993). , 4-227.

[9] Knothe, G. Historical perspectives on vegetable oil-based diesel fuels. Inform (2001). , 11-1103.

[10] Parente, E. Lipofuels: biodiesel and biokerosene. National Institute of Standards and Technology. (2007).

[11] Lappe, P. The Biodiesel Industry: Past, Present and Future, GEAPS: Proceedings Sessions. http://www.geaps.com/knowledge/proceedings/article.cfm?id=15.Assessed 07 May (2012).

[12] Diesel, R. The Diesel oil-engine. Engineering (1912). , 93-395.

[13] Diesel, R. The Diesel oil-engine and its industrial importance particularly for Great Britain. Proc Inst Mech Eng. (1912). , 179-280.

[14] Jones, J. C. Energy Resources for the Past, Present and Future. The Open Thermodynamics Journal (2010). , 4-72.

[15] SAE technical paper series noSAE international of highway meeting, Milwaukee, Wisconsin, USA; (1983).

[16] Demirbas, A. Biofuels securing the planet's future energy needs. Energy Conversion and Management (2009). , 50-2239.

[17] Soy Biodiesel Supplier and User Guidehttp://www.mdsoy.com/consumer/biodiesel-guide.pdf,accessed 08 May (2012).

[18] Yusuf NNANKamarudin SK and Yaakob Z. Overview on the production of biodiesel from Jatropha curcas L. by using heterogeneous catalysts. Biofuels, Bioprod. Bioref. (2012). , 6-319.

[19] Gui, M. M, Lee, K. T, & Bhatia, S. Feasibility of edible oil vs. non-edible oil vs. waste edible oil as biodiesel feedstock. Energy (2008). , 33-1646.

[20] Mcneff, C. V, Mcneff, L. C, Yan, B, Nowlan, D. T, Rasmussen, M, Gyberg, A. E, Krohn, B. J, Fedie, R. L, & Hoye, T. R. A continuous catalytic system for biodiesel production. Appl. Catal. A: Gen. (2008). , 343, 39-48.

[21] Krohn, B. J, Mcneff, C. V, Yan, B, & Nowlan, D. Bioresourc. Production of algae-based biodiesel using the continuous catalytic Mcgyan® process. Technol. (2011). , 102, 94-100.

[22] Utlu, Z. Evaluation of biodiesel fuel obtained from waste cooking oil. Energy Sources Part A (2007). , 29-1295.

[23] Ban-weiss, G. A, Chen, J. Y, Buchholz, B. A, & Dibble, R. W. A numerical investigation into the anomalous slight NOx increase when burning biodiesel: A new (old) theory. Fuel Process. Technol. (2007). , 88-659.

[24] Vicente, G, Martinez, M, & Aracil, J. Integrated biodiesel production: a comparison of different homogeneous catalysts systems. Bioresourc. Technol. (2004). , 92-297.

[25] Nas, B, & Berktay, A. Energy potential of biodiesel generated from waste cooking oil: an environmental approach. Energy Sources Part B (2007). , 2-63.

[26] Wardle, D. A. Global sale of green air travel supported using biodiesel. Renew. Sustain. Energy Rev (2003). , 7-1.

[27] Chand, N. Plant oils-fuel of the future. J Sci Ind Res (2002). , 61-7.

[28] Al-zuhair, S. Production of biodiesel: possibilities and challenges, Biofuels, Bioprod. Bioref. (2007). , 1-57.

[29] Goering, E, Schwab, W, Daugherty, J, Pryde, H, & Heakin, J. Fuel properties of eleven vegetable oils. Trans ASAE (1982). , 25-1472.

[30] Demirbas, A. (2003). Fuel conversional aspects of palm oil and sunflower oil. Energy Sources , 25, 457-466.

[31] Karmee, S. K, & Chadha, A. Preparation of biodiesel from crude oil of Pongamia pinnata. Biores Technol (2005). , 96-1425.

[32] Azam, M. M, Waris, A, & Nahar, N. M. Prospects and potential of fatty acid methyl esters of some non-traditional seed oils for use as biodiesel in India. Biomass Bioenergy (2005). , 29-293.

[33] Foidl, N, Foidl, G, Sanchez, M, Mittelbach, M, & Hackel, S. Jatropha curcas L. As a source for the production of biofuel in nicaragua. Biores Technol (1996). , 58-77.

[34] Gubitz, G. M, Mittelbach, M, & Trabi, M. Exploitation of the tropical seed plant Jatropha Curcas L. Biores Technol (1999). , 67-73.

[35] Sarin, R, Sharma, M, Sinharay, S, & Malhotra, R. K. Jatropha-palm biodiesel blends: An optimum mix for Asia. Fuel (2007). , 86-1365.

[36] Berchmans, H. J, & Hirata, S. Biodiesel production from crude Jatropha curcas L. seed oil with a high content of free fatty acids. Bioresour. Technol., (2008). , 99-1716.

[37] Demirbas, A. Progress and recent trends in biodiesel fuels. Energ Convers Manage (2009). , 50-14.

[38] Christi, Y. Biodiesel from microalgae. Biotechnol Adv. (2007). , 25-294.

[39] Fan, X, & Burton, R. Recent Development of Biodiesel Feedstocks and the Applications of Glycerol: A Review, The Open Fuels & Energy Science Journal (2009). , 2-100.

[40] Canakci, M, & Sanli, H. Biodiesel production from various feedstocks and their effects on the fuel properties. J. Ind. Microbiol. Biotechnol. (2008). , 35-431.

[41] Kulkarni, M. G, & Dalai, A. K. Waste cooking oil-An economical source for biodiesel: A review. Ind. Eng. Chem. Res. (2006). , 45-2901.

[42] He, H. Y, Wang, T, & Zhu, S. L. Continuous production of biodiesel fuel from vegetable oil using supercritical methanol process. Fuel (2007). , 86-442.

[43] Demirbas, A. Biodiesel from vegetable oils via transesterification in supercritical methanol. Energy Convers. Manage., (2002).

[44] Saka, S, & Kusdiana, D. Biodiesel fuel from rapeseed oil as prepared in supercritical methanol. Fuel (2001). , 80-225.

[45] Van Kasteren JMN and Nisworo APA process model to estimate the cost of industrial scale biodiesel production from waste cooking oil by supercritical transesterification. Resour. Conserv. Recy. (2007). , 50-442.

[46] Kusdiana, D, & Saka, S. Effects of water on biodiesel fuel production by supercritical methanol treatment. Bioresource Technol. (2004). , 91-289.

[47] Mata, T. M, Martins, A. A, & Caetano, A. S. Renew. Sustain. Energy Rev. (2010). , 14-217.

[48] Atabani, A. E, Silitonga, A. S, & Badruddin, I. A. Mahlia TMI, Masjuki HH and Mekhilef S. Renew. Sustain. Energy Rev. (2012). , 16-2070.

[49] Achten WMJVerchit L, Mathijs Franken YJ, Singh E, Aerts VP and Muys RB. Jatropha bio-diesel production and use. Biomass Bioenergy (2008). , 32(12), 1063-1084.

[50] Williams, M. A. Extraction of Lipids from Natural Sources, Chapter 5 in: Lipid Technologies and Applications, Edited by Gunstone FD and Padley FB, Marcel Dekker, New York, (1997). , 113.

[51] Mahanta, P, & Shrivastava, A. Technology development of bio-diesel as an energy alternative. Department of Mechanical Engineering Indian Institute of Technology. http://www.newagepublishers.com/samplechapter/001305.pdf;(2011). accessed 11 August 2012).

[52] Pramanik, K. Properties and use of Jatropha curcas oil and diesel fuel blends in compression ignition engine Renew. Energy. (2003). , 28-239.

[53] Narayan CM In: Proceedings on recent trends in automotive fuelsNagpur, India. (2002).

[54] Ramadhas, A. S. Jayaraj Sand Muraleedharan C. Use of Vegetable Oil as I.C. Engine Fuels: A Review. Int. J. Renew. Energy (2004). , 29-727.

[55] Ma, F, & Hanna, M. Biodiesel production: A review. Bioresourc. Technol. (1999). , 70-1.

[56] Ranganathan, S. V, Narasimhan, S. L, & Muthukumar, K. An overview of enzymatic production of biodiesel. Bioresourc. Technol. (2008). , 99-3975.

[57] Fukuda, H, Kondo, A, & Noda, H. Biodiesel Fuel Production by Transesterification of Oils. J. Biosci. Bioeng. (2001). , 92-405.

[58] Uzun, B. B, Putun, A. E, & Putun, E. Fast pyrolysis of soybean cake: Product yields and compositions. Bioresourc. Technol. (2006). , 97(4), 569-576.

[59] Sonntag NOVReactions of fats and fatty acids. Bailey's industrial oil and fat products, th edition, Edited by Swern D., John Wiley & Sons, New York, , 1, 99.

[60] Sharma, Y. C, Singh, B, & Upadhyay, S. N. Advancements in development and characterization of biodiesel: A review. Fuel (2008). , 87(12), 2355-2373.

[61] Sensoz, S, Angin, D, & Yorgun, S. Influence of particle size on the pyrolysis of rapeseed (Brassica napus L.): fuel properties of bio-oil. Biomass and Bioenerg. (2000). , 19-271.

[62] Sensoz, S, Demiral, I, & Gercüel, H. F. Olive bagasse (Olea europea L.) pyrolysis. Bioresourc. Technol. (2006). , 97(3), 429-436.

[63] Freedman, B, Pryde, E. H, & Mounts, T. L. Variables affecting the yields of fatty esters from transesterified vegetable oils. JAOCS (1984). , 61-1638.

[64] Demirbas, A. Biodiesel production from vegetable oils via catalytic and non-catalytic supercritical methanol transesterification methods. Prog Energy Combust Sci (2005). , 31-466.

[65] Al-widyan, M. I, & Al-shyoukh, A. O. Experimental evaluation of the transesterification of waste palm oil into biodiesel. Bioresour Technol (2002). , 85-253.

[66] Yong, W, Shiyi, O, Pengzhan, L, Feng, X, & Shuze, T. Comparison of two different processes to synthesize biodiesel by waste cooking oil. J Mol Catal A: Chem (2006). , 252-107.

[67] Lotero, E, Liu, Y, Lopez, D. E, Suwannakarn, K, Bruce, D. A, & Goodwin, J. G. Synthesis of biodiesel via acid catalysis. Ind Eng Chem Res (2005). , 44-5353.

[68] Ulf, S, Ricardo, S, & Matheus, V. R. Transesterification of vegetable oils: a review. J Braz Chem Soc (1998). , 9(1), 199-210.

[69] Raman, J. K, Sariah, A, Denis, P, Seng, C. E, & Pogaku, R. Production of biodiesel using immobilized lipase- a critical review. Crit Rev Biotechnol (2008). , 28-253.

[70] Nielsen, P. M, Brask, J, & Fjerbaek, L. Enzymatic biodiesel production: technical and economical considerations. Eur J Lipid Sci Technol (2008). , 110-692.

[71] Pinnarat, T, & Savage, P. E. Assessment of noncatalytic biodiesel synthesis using supercritical reaction conditions. Ind Eng Chem Res (2008). , 47-6801.

[72] Yin, J-Z, Xiao, M, & Song, J-B. Biodiesel from soybean oil in supercritical methanol with co-solvent. Energy Convers Manage (2008). , 49-908.

[73] Azcan, N, & Danisman, A. Alkali catalyzed transesterification of cottonseed oil by microwave irradiation. Fuel (2007). , 86-2639.

[74] Yoni, G, & Aharon, G. Continuous flow, circulating microwave system and its application in nanoparticle fabrication and biodiesel synthesis. J Phys Chem C (2008). , 112-8802.

[75] Thompson, L. H, & Doraiswamy, L. K. Sonochemistry: science and engineering. Ind Eng Chem Res (1999). , 38-1215.

[76] Colucci, J. A, Borrero, E. E, & Alape, F. Biodiesel from an alkaline transesterification reaction of soybean oil using ultrasonic mixing. JAOCS (2005). , 82(7), 525-530.

[77] Carmen, S, Vinatoru, M, Nishimura, R, & Maeda, Y. Fatty acids methyl esters from vegetable oil by means of ultrasonic energy. Ultrason Sonochem (2005). , 12-367.

[78] Sani, Y. M. Daud WMAW and Abdul Aziz AR. Solid acid-catalysed biodiesel produc‐ tion from algal oil- The Dual Advantage. Submitted.

[79] Van Eijcka, J, & Romijnk, H. Energy Policy, (2008). , 2008(36), 311-325.

[80] Meher, L. C, Sagar, D. V, & Naik, S. N. Technical aspects of biodiesel production by transesterification-A review. Renew. & Sust. Energy Rev. (2006). , 10-248.

[81] Demirbas, A, & Karsliöglu, S. Biodiesel production facilities from vegetable oils and animal fats. Energy Sources, Part A (2007). , 29-133.

[82] Qi, Z, Jie, C, Tiejun, W, & Ying, X. Review of biomass pyrolysis oil properties and upgrading research. Energy Conv. Manage., (2007). , 48-87.

[83] Maher, K. D, & Bressler, D. C. Pyrolysis of triglyceride materials for the production of renewable fuels and chemicals. Bioresour. Technol., (2007). , 98-2351.

[84] Chiaramonti, D, Bonini, D, & Fratini, E. Tondi, Gartner K, Bridgwater AV, Grimm HP, Soldaini I, Webster A and Baglioni P. Development of emulsions from biomass pyrolysis liquid and diesel and their use in engines-Part 1: emulsion production. Biomass Bioenerg, (2003). , 25, 85-99.

[85] Czernik, S, French, R, Feik, C, & Chornet, E. Hydrogen by catalytic steam reforming of liquid byproducts from biomass thermoconversion processes. Ind. Eng. Chem. Res. (2002). , 41-4209.

[86] Pindoria, R. V, Megaritis, A, Herod, A. A, & Kandiyoti, R. A two-stage fixed-bed reactor for direct hydrotreatment of volatiles from the hydropyrolysis of biomass: effect of catalyst temperature, pressure and catalyst ageing time on product characteristics. Fuel, (1998). , 77-1715.

[87] Pindoria, R. V, Lim, J. Y, Hawkes, J. E, Lazaro, M. J, Herod, A. A, & Kandiyoti, R. Structural characterization of biomass pyrolysis tars/oils from eucalyptus wood waste: effect of H-2 pressure and sample configuration. Fuel, (1997). , 1997(76), 1013-1023.

[88] Isayama, Y, & Saka, S. Biodiesel production by supercritical process with crude bio‐ methanol prepared by wood gasification. Bioresour. Technol., (2008). , 99-4775.

[89] Szybist, J. P, Kirby, S. R, & Boehman, A. L. NO_x emissions of alternative diesel fuels: A comparative analysis of biodiesel and FT diesel. Energy Fuels (2005). , 19-1484.

[90] Walker, K. Biodiesel from rapeseed. J. R. Agric. Soc. Engl. (1994). , 155-43.

[91] Hess, M. A, Haas, M. J, Foglia, T. A, & Marmer, W. N. Effect of antioxidant addition on NOx emissions from biodiesel. Energy Fuels (2005). , 19-1749.

[92] Carioca JOBBiofuels: Problems, challenges and perspectives. Biotechnol. J. (2010). , 5-260.

[93] Mcshane, P. C. Vegetable-oil based dielectric coolants, IEEE Industry Applications Magazine (2002). , 8-34.

[94] Agarwal, A. K. Biofuels (alcohols and biodiesel) applications as fuels for internal combustion engines Prog. Energy Combust. Sci. (2007). , 33-233.

[95] Abdullah, A. Z, Razali, N, Mootabadi, H, & Salamatinia, B. Critical technical areas for future improvement in biodiesel technologies. Environ. Res. Lett. (2007).

[96] Haas, M. J. Improving the economics of biodiesel production through the use of low value lipids as feedstocks: vegetable oil soapstock. Fuel Process. Technol., (2005). , 86, 1087-1096.

[97] Wang, Y, Ou, S, Liu, P, Xue, F, & Tang, S. Comparison of two different processes to synthesize biodiesel by waste cooking oil. J. Mol. Catal. A: Chem. (2006). , 252-107.

[98] Dorado, M. P, Cruz, F, Palomar, J. M, & Lopez, F. J. An approach to the economics of two vegetable oil-based biofuels in Spain. Renewable Energy (2006). , 31-1231.

[99] Zang, Y, Dube, M. A, Mclean, D. D, & Kates, M. Biodiesel production from waste cooking oil: 1. Process design and technological assessment. Bioresour. Technol., (2003). , 2003(90), 229-240.

Chapter 3

Prospects and Potential of Green Fuel from some Non Traditional Seed Oils Used as Biodiesel

Mushtaq Ahmad, Lee Keat Teong,
Muhammad Zafar, Shazia Sultana,
Haleema Sadia and Mir Ajab Khan

Additional information is available at the end of the chapter

1. Introduction

Today's diesel engines require a clean-burning, stable fuel that performs well under a variety of operating conditions. Biodiesel is the only alternative fuel that can be used directly in any existing, unmodified diesel engine. Because it has similar properties to petroleum diesel fuel, biodiesel can be blended in any ratio with petroleum diesel fuel. Many federal and state fleet vehicles in USA are already using biodiesel blends in their existing diesel engines (Harwood, 1981). The low emissions of biodiesel make it an ideal fuel for use in marine areas, national parks and forests, and heavily polluted cities. Biodiesel has many advantages as a transport fuel. For example, biodiesel can be produced from domestically grown oilseed plants. Producing biodiesel from domestic crops reduces the dependence on foreign petroleum, increases agricultural revenue, and creates jobs.

Presently world's energy needs are met through non-renewable resources such as petrochemicals, natural gas and coal. Since the demand and cost of petroleum based fuel is growing rapidly, and if present pattern of consumption continues, these resources will be depleted in near future. It is the need of time to explore alternative sources of fuel energy. An alternative fuel must be technically feasible, economically competitive, environmentally acceptable and easily available. Fatty acid methyl esters derived from renewable sources such as vegetable oils has gained importance as an alternative fuel for diesel engines. The edible oils such as soybean oil in USA, rapeseed oil in Europe and palm oil in countries with tropical climate such as Malaysia are being used for the production of biodiesel (Knothe, 2002).

1.1. Historical background

Biodiesel, an alternative diesel fuel, is made from renewable biological sources such as vegetable oils and animal fats. It is biodegradable and nontoxic, has low emission profiles and so far is environmentally beneficial (Krawczyk, 1996). Bio-diesel production is not something new, because the concept of using vegetable oil as fuel dates back to 1895. Rudolf Diesel developed the first diesel engine which was run with vegetable oil in 1900. The first engine was run using groundnut oil as fuel (Bijalwan et al., 2006). In 1911, Rudolf Diesel stated that the diesel engine can be fed with vegetable oil and would help considerably in the agricultural development of the countries which use it. In 1912, Rudolf Diesel said, the use of vegetable oils for engine fuels may seem insignificant today. But such oils may become in course of time as important as petroleum and the coal tar products of the present time (Babu and Devaradjany, 2003). After eight decades, the awareness about environment rose among the people to search for an alternative fuel that could burn with less pollution. Rudolf Diesel's prediction is becoming true today with more and more bio-diesel being used all over the world. With the advent of cheap petroleum, appropriate crude oil fractions were refined to serve as fuel and diesel fuels and diesel engines evolved together. In the 1930s and 1940s vegetable oils were used as diesel fuels from time to time, but usually only in emergency situations. Recently, because of increases in crude oil prices, limited resources of fossil oil and environmental concerns there has been a renewed focus on vegetable oils and animal fats to make biodiesel fuels. Continued and increasing use of petroleum will intensify local air pollution and magnify the global warming problems caused by CO_2 (Shay, 1993). In a particular case, such as the emission of pollutants in the closed environments of underground mines, biodiesel fuel has the potential to reduce the level of pollutants and the level of potential or probable carcinogens (Krawczyk, 1996).

Considerable research has been done on vegetable oils as diesel fuel. That research included palm oil, soybean oil, sunflower oil, coconut oil, rapeseed oil and tung oil. Animal fats, although mentioned frequently, have not been studied to the same extent as vegetable oils. Some methods applicable to vegetable oils are not applicable to animal fats because of natural property differences. Oil from algae, bacteria and fungi also has been investigated (Shay, 1993). Terpenes and latexes also were studied as diesel fuels. Microalgae have been examined as a source of methyl ester diesel fuel (Nagel and Lemke, 1990).

1.2. Sources of biodiesel

Alternative diesel fuels are made from natural, renewable sources such as vegetable oil and fats (Ratledge et al., 1985; Lee et al., 1995). There are more than 350 oil-bearing crops identified, among which only soybean, palm, sunflower, safflower,cottonseed, rapeseed and peanut oils are considered as potential alternative fuels for diesel engines (Pryor et al., 1982).

Vegetable oils are promising feedstocks for biodiesel production since they are renewable in nature, and can be produced on a large scale and environmentally friendly (Patil & Deng, 2009). Vegetable oils include edible and non-edible oils. More than 95% of biodiesel production feed stocks come from edible oils since they are mainly produced in many regions and the properties of biodiesel produced from these oils are much suitable to be used as diesel fuel substitute (Gui *et al.*, 2008). However, it may cause some problems such as the competition with the edible oil market, which increases both the cost of edible oils and biodiesel (Kansedo *et al.*, 2009).

In order to overcome these disadvantages, many researchers are interested in non-edible oils which are not suitable for human consumption because of the presence of some toxic components in the oil. Non edible oil crops can be grown in waste lands that are not suitable for food crops and the cost of cultivation is much lower because these crops can still sustain reasonably high yield without intensive care (Kumar *et al.*, 2007; Gui *et al.*, 2008)

Animal fats contain higher level of saturated fatty acids therefore they are solid at room temperature that may cause problems in the production process. Its cost is also higher than vegetable oils (Singh, 2009). The source of Biodiesel usually depends on the crops amenable to the regional climate. In the United States, soybean oil is the most commonly Biodiesel feedstock, whereas the rapeseed (canola) oil and palm oil are the most common source for Biodiesel, in Europe, and in tropical countries respectively (Knothe, 2002). A suitable source to produce Biodiesel should not competent with other applications that rise prices, for example pharmaceutical raw materials. But the demand for pharmaceutical raw material is lower than for fuel sources. As much as possible the Biodiesel source should fulfill two requirements: low production costs and large production scale. Refined oils have high production costs, but low production scale; on the other side, non-edible seeds, algae and sewerage have low production costs and are more available than refined or recycled oils. The oil percentage and the yield per hectare are important parameters to consider as Biodiesel source.

Algae can grow practically in every place where there is enough sunshine. Some algae can grow in saline water. The most significant difference of algal oil is in the yield and hence its biodiesel yield. According to some estimates, the yield (per acre) of oil from algae is over 200 times the yield from the best-performing plant/vegetable oils (Sheehan *et al.*, 1998b).

1.3. Biodiesel production

The seed oils usually contain free fatty acids, phospholipids, sterols, water, odorants and other impurities. Because of these the oil cannot be used as fuel directly. To overcome these problem the oil requires slight chemical modification mainly pyrolysis, microemulsion, dilution and transesterification. Pyrolysis is a method of conversion of one substance into another by mean of heat or by heat with the aid of the catalyst in the absence of air or oxygen (Sonntag, 1979). The process is simple, waste less, pollution free and effective compared with other cracking processes.

The vegetable oil is diluted with petroleum diesel to run the engine. Caterpillar Brazil, in 1980, used pre-combustion chamber engines with the mixture of 10% vegetable oil to maintain total power without any alteration or adjustment to the engine. At that point it was not practical to substitute 100% vegetable oil for diesel fuel, but a blend of 20% vegetable oil and 80% diesel fuel was successful. Some short-term experiments used up to a 50/50 ratio.

A micro emulsion define as a colloidal equilibrium dispersion of optically isotropic fluid microstructure with dimensions generally into 1–150 range formed spontaneously from two

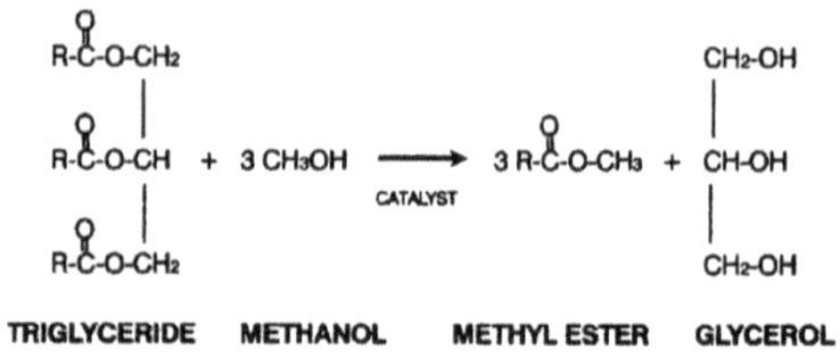

Figure 1. Transestrification Reaction

normally immiscible liquids and one and more ionic or more ionic amphiphiles (Schwab *et al.*, 1988). They can improve spray characteristics by explosive vaporization of the low boiling constituents in micelles (Pryde, 1984). The engine performances were the same for a microe-multion of 53% sunflower oil and the 25% blend of sunflower oil in diesel (Ziejewski *et al.*, 1983). A microemulsion prepared by blending soyabean oil, methanol, and 2-octanol and cetane improver in ratio of 52.7:13.3:33.3:1.0 also passed the 200 h EMA test (Goering, 1984).

Transesterification or alcoholysis is the displacement of alcohol from an ester by another in a process similar to hydrolysis, except than alcohol is used instead of water. This process has been widely used to reduce the high viscosity of triglycerides. A catalyst is usually used to improve the reaction rate and yield. Excess alcohol is used to shift the equilibrium toward the product because of reversible nature of reaction. For this purpose primary and secondary monohybrid aliphatic alcohols having 1-8 carbon atoms are used (Sprules and Price, 1950).

The main factors affecting transesterification are the alcohol to oil molar ratio, catalyst concentration, reaction temperature and reaction time. The methanol to oil ratio needs to be higher than stoichiometric ratio to drive the equilibrium to a maximum ester yields. The molar ratio is associated with the type of vegetable oil used. Ikwuagwu *et al.*, 2000 stated that molar ratio was 6: 1 for rubber seed oil. It was also undertaken sunflower oil was used (Vicente *et al.*, 2005). Catalysts are classified as alkali, acid and alkali-alcoholic. Transesterification of jojoba oil catalysed with sodium metoxide (Canoira *et al.*, 2006). Sodium hydroxide was also choosen to catalyse the transesterification of rubber seed oil beacause it is cheaper. Different homogeneous catalysts were used to transesterify sunflower oil to obtain 100% biodiesel yield by using sodium methoxide catalyst (Vicente *et al.*, 2005).

Transesterification consist of a number of consecutive, reversible reactions. It is usually reaction of vegetable or waste oil respectively with a low molecular weight alcohol, such as ethanol and methanol. During this process, the triglyceride molecule from vegetable oil is removed in the form of glycerin. The triglycerides are broken step wise into diglycerides, monoglyceride and finally converted into methyl esters and glycerol (Fig: 1).

There are various types of transesterification that includes based, acid and lipase catalyzed. The petroleum and other fossil fuels contain sulfur, ring molecules & aromatics while the biodiesel molecules are very simple hydrocarbon chains, containing no sulfur, ring molecules or aromatics. Biodiesel is thus essentially free of sulfur and aromatics. Biodiesel is made up of almost 10% oxygen, making it a naturally "oxygenated" fuel (Noureddini & Zhu).

Figure 2. Electric Oil Expeller

Figure 3. Mechanical Oil Expeller

Figure 4. Preparation of Methoxide catalyst

Figure 5. Mixing of catalyst

Figure 6. Separation of glycerin from Biodiesel

Figure 7. Biodiesel Filtration

Figure 8. Milk Thistle Flower

Figure 9. Milk Thistel Seeds

1.4. Biodiesel Scenario at Global Level

Use of bio-diesel is catching up all over the world especially in developed countries.

- In **Malaysia**, the tropical climate encourages production of bio-diesel from palm oil (Meher et al., 2006; Lam and Lee, 2011).

- The **United States** is contributing 25% of the world green house gases: i.e., oil and coal. We also need to reorganize its 70% of oil consumption is in transportation. The cost of bio-diesel is $3.00 a gallon. With the tax subsidy available in the law now, it could be sold for about $1.80. It is clearly known that the future depends on bio-fuels as replacement for fossil fuels. At present, USA uses 50 million gallons and **European countries** use 350 million gallons of bio-diesel annually. It is mixed with 20% of bio-diesel in fossil diesel.

- **France** is the country which uses 50% of bio-diesel mixed with diesel fuel.

- In **Zimbabwe**, 4 million jatropha has been planted in 2000 ha by the end of 1997.

- In **Nicaragua**, one million Jatropha curcas has been planted in 1000 ha. The harvest of pods reached 333000 tonnes in the 5th year with a seed of 5000 tonnes and the oil extracted was approximately 1600 tonnes per annum.

- In **Nepal**, 22.5 ha of area are planted with 40,000 rooted cuttings of Jatropha curcas. The rural women co-operative have been trained to extract oil, produce soap and use 30:70 mix (oil/kerosene) of oil and kerosene in stove without smoke (Bijalwan *et al.*, 2006; Paramathama *et al.*, 2007).

1.5. Non traditional seed oils

Various plant resources either edible or non-edible are used for biodiesel production. While in this study three oil seed plants i.e. wild safflower, safflower and milk thistle belonging to family Asteraceae were selected as non edible oil feed stocks for biodiesel production. These oil seeds are considered as nontraditional energy crops as these are un commonly cultivated and mainly found as weed. The raw material (crude oil) from these oil seeds can be used as a feed stock for biodiesel production.

Carthamus oxyacantha M. Bieb.

Carthamus oxyacantha M. B. (Figure 9) is a spiny-leaved annual herb up to 1.5 m tall commonly known as Wild Safflower. It is a hardy and xerophytic noxious weed of winter crops. Like other spiny plants in the genus Carthamus, this species is not eaten by livestock, enabling it to spread on grazing lands. It also competes with and reduces the yield of cereal crops. It is a valuable source of non edible and drying oil (28-29% oil content) from waste lands (Deshpande, 1952). However, it was almost eradicated through regular campaigns due to noxious weedy nature. Fruit an achene, obovate or elliptic, 3–5.5 mm long, 2–3.5 mm wide, 1.5–2 mm thick, truncate at apex, marginal notch at base, cross sectional outline broadly elliptic to slightly 4-sided. Glabrous, smooth and glossy, bone-white to ivory, less frequently beige, with densely distributed blotches and speckles in shades of brown. Scar subbasal, an outlined, diamond-shaped cavity containing a rough, vertical ridge. Pappus early deciduous, absent. Apex a round, rough, flat to uneven area, surrounded by irregularly edged black ring; style base deciduous. Embryo spatulate, cotyledons broad; endosperm absent. It is widely distributed in Afghanistan, Azerbaijan, India, Iran, Iraq, Kyrgyzstan, Pakistan, Tajikistan, Turkmenistan. *Carthamus oxycantha* seeds yields two types of oils: oleic oil and linoleic oil. Fatty acid oil composition of oleic oil is, palmitic acid 5-6%, stearic acid 1.5 -2%, oleic acid 74-80%, linoleic acid 13-18% and traces of linoleic acid and longer chain fatty acids. The fatty acid composition of linoleic oil, palmitic acid 5-8%, stearic acid 2-3%, oleic acid 8-30%, linoleic acid 67-89% and also traces of linoleic acid and longer chain fatty acids. *Carthamus oxycantha* fruit also contains proteins 20-25%, hull 60%, residual fat 2-15%. Flowers of *Carthamus oxycantha* contain two major pigments, the water soluble, yellow carthamidin and the formally important dye carthamin, flavonone which is orange red (Fernandez-Martinez *et al.*, 1993; Anjani, 2005). Flowers also contain 0.3-0.6% carthamin. Flavonoids, glycosides, sterols and serotonin derivatives have been identified from flowers and seeds (Figure 10) (Firestone, 1999). Two new glycosides, 2-O-methylglucopyranosyl-carthamoside and beta-D-fructofuranosyl carthamoside, along with the known compound 3', 4', 5, 7-tetrahydroxyflavanone have been isolated from *Carthamus oxyacantha* using recycling preparative HPLC. The structures of these compounds were established by mass spectrometric and extensive spectroscopic analysis (Hassan *et al.*, 2010). This oil seed plant is commonly found as noxious weed after harvesting of cash crop wheat. Throughout the world due to its weedy nature it is generally burnt after wheat harvesting, while in this study it is targeted as energy crop for biodiesel production.

Figure 10. Wild Safflower

Figure 11. Wild Safflower seeds

Carthamus tinctorious L. (Safflower)

Safflower is cultivated nontraditional seed oil crop which contains a higher percentage of essential unsaturated fatty acids and a lower percentage of saturated fatty acids than other vegetable seed oils. The oil, light colored and easily clarified, is used in liqueurs, candles, and as a drying oil in paints, linoleum, varnishes, and wax cloths. The flowers (Figure 11) have been the source of yellow and red dyes, largely replaced by synthetics, but still used in rouge. Annual thistle-like herb, branching above with a strong central stem to 1-2 m tall; leaves spiny, oblong or lanceolate, the upper ones clasping, minutely spinose-toothed; flowers in 1-6 heads per plant, 3-4.5 cm across, each head developing 20-60 seeds; corollas yellow, orange, white or red, surrounded by a cluster of leafy spiny bracts, which pass over gradually into the bracts of the involucre; achenes (fruits or seeds) (Figure 12) white, 7-9 mm long, shining. Many cultivars have been developed differing in flower color, degree of spininess, head size, oil content, resistance to disease and ease of harvest. Most common varieties have yellow or orange flowers, but red and white flowered varieties are known. Reported from the Central Asian and Near Eastern Center of Diversity, safflower thereof is reported to tolerate bacteria, disease, drought, frost, fungus, high pH, phage, salt, sand, rust, virus, wind, and wild. Wu and Jain (1977) discuss germplasm diversity in the World Collections of Safflower. ($2n$ = 24, 32). Believed to have originated in southern Asia and is known to have been cultivated in China, India, Pakistan, Persia and Egypt almost from prehistoric times. During Middle Ages it was cultivated in Italy, France, and Spain, and soon after discovery of America, the Spanish took it to Mexico and then to Venezuela and Colombia. It was introduced into United States in 1925 from the Mediterranean region and is now grown in all parts west of 100th meridian. Safflower grows in the temperate zone in areas where wheat and barley do well, and grows slowly during periods of cool short days in early part of season. Seedlings can withstand temperatures lower than many species; however, varieties differ greatly in their tolerance to frost; in general, frost damages budding and flowering thus reducing yields and quality. It thrives in heavy clays with good waterholding capacity, but will grow satisfactorily in deep sandy or clay loams with good drainage, and needs soil moisture from planting through flowering, Soils approaching neutral pH are best (Duke, 1978, 1979). Propagation is by seed, which are usually pretreated with insecticides and fungicides. Same machinery used for small grains may be used for planting, cultivation and harvesting. Seed should be planted in a soil prepared and completely free of seeds, when the soil temperature is about 4.4°C and the upper 10 cm of soil is moist. Seed germinates quickly at 15.5°C. Safflower matures in from 110-160 days from planting to harvest as a spring crop, as most of it is grown, and from 200 or more days as fall crop. It should be harvested when the plant is thoroughly dried. Since the seeds do not shatter easily, it may be harvested by direct combining. The crop is allowed to dry in the fields before threshing. Its average yields are 1,900 kg/ha, but yields above 4,500 kg/ha are not uncommon; in the Great Plains yields run about 850 kg/ha (C.S.I.R., 1948-1976). The world low production yield was 244 kg/ha in Israel, the international production yield was 789 kg/ha, and the world high production yield was 1,900 kg/ha in U.S.A. Yields higher than 4,000 kg/ha have been attained. Oil yields approach 50%, leaving a meal with ca 21% protein, 35% fiber, and 1-3% fat. Safflower is self-pollinated with some cross-pollination. Pollen and nectaries are abundant with insect working the flowers. Safflower is attacked by many fungi: *Alternaria carthami* (leaf spot and bud rot), *A. zinniae*, *Bremia lactucae*, *Cercospora carthami*, *Cercosporella carthami*, *Carthamus tinctorius* is only cultivated species of genus *Carthamus* commonly known as Safflower. It is cultivated since ancient times for not only the dye obtained from its flowers and medicinal uses but also for its seed oil and ornamental purposes.

Silybum marianum (L.) Gaertner

Milk thistle (*Silybum marianum* Gaertn.) (Figure 7) is a winter annual or a biennial. Its current distribution includes most temperate areas of the world. It is a broad-leaved species belonging to Asteraceae that reaches a height of 200–250 cm. Milk thistle is grown commercially as a medicinal plant in Europe, Egypt, China, and Argentina but it has been reported as a noxious weed in many other countries. Stems glabrous or slightly tomentose. Leaves: basal wing-petioled, blades 15–60 cm, margins coarsely lobed; cauline leaves clasping, progressively smaller and less divided, bases spiny, coiled, auriculate. Phyllary appendages spreading, ovate, 1–4 cm including long-tapered spine tips. Corollas 26–35 mm; tubes 13–25 mm, throats campanulate, 2–3 mm, lobes 5–9 mm. Cypselae brown and black spotted, 6–8 mm; pappus scales 15–20 mm. $2n = 34$. *Silybum marianum* is sometimes cultivated as an ornamental, a minor vegetable, or as a medicinal herb. Young shoots can be boiled and eaten like cabbage and young leaves can be added to salads. The seeds (Figure 8) can be used as a coffee substitute. Extracts of *S. marianum* are used as an herbal treatment for liver ailments. Milk thistle is toxic to livestock when consumed in large quantities, and it forms dense stands in pastures and rangelands. California reports up to 4 tons per acre in heavily infested areas. The leaves are very distinctive,

Figure 12. Safflower

Figure 13. Safflower seeds

with white marbling on the shiny green leaves. An annual or biennial, found in rough pasture, on grassy banks, in hedgerows and on waste ground. It is locally well-established and persistent, especially in coastal habitats in S. England, but is also a widespread casual. Lowland. Native of the Mediterranean region; naturalised or casual throughout much of

Europe and in N. America and Australia. The plant grows wild in Egypt on canal banks and in wet ground regions in the Nile Valley. The soil supporting this plant is fine-textured and moist. It occurs in two types, the most abundant has purple flowers while the least abundant has white flowers (Ahmad et al., 2008). Milk thistle is commonly found as a noxious weed in waste land and in along with cultivated filed of traditional crops. In this project this energy crop was first time reported as a feed stock for biodiesel production at global perspectives.

In this project these species were selected for biodiesel potential at global interest as renewable energy because of their oil which is non edible and species found as weeds on waste and marginal lands. The study conducted with aims to extract the seed oils from these resources for production of biodiesel through base catalyzed transesterification. Study may also confined to quality standards of biodiesel obtained from these species according to ASTM standards.

2. Methodology

The oil from these three resources was extracted by two methods;

1. Chemical method (Soxhlet Apparatus)

2. Mechanical method (Electric oil expeller) (Figure 1-2)

The oil seeds were oven-dried at 40°C over night and then ground with blender. 250ml of petroleum ether was poured into round bottom flask. Five gram of the sample was placed in the thimble and inserted in the centre of the extractor. The Soxhlet was heated at 60°C. When the solvent was boiling, the vapour rises through the vertical tube into the condenser at the top. The condensed liquid drips into the filter paper thimble in the centre, which contains the solid sample to be extracted. The extract seeps through the pores of the thimble and fills the siphon tube, where it flows back down into the round bottom flask. This process was allowed to continue for 3-4 hrs. Distinct layers of oil and petroleum ether appeared in round bottom flask. In this process of oil extraction, the solvent was recovered and reused. The resulting mixture containing oil was heated to evaporate solvent and weighed again to determine the amount of oil extracted. (AOAC, 1990). While in electric oil expelling method oil was extracted mechanically from seeds by using electric oil expeller (KEK P0015, 10127) and crude oil was collected in beakers for further processing. After an average of 5-6 turns, the oil is fully extracted from seeds.

2.1. Determination of free fatty acid number of seed oils

Free fatty acid content of oil seeds was determined by aqueous acid-base titration (Trajkovic et al., 1983). Two types of titration were performed i.e. blank titration and sample titration. In case of blank titration 0.14 g KOH was dissolved in 100 ml of distilled water to prepare 0.025M KOH solution and this solution was poured in burette. 10 ml of isopropyl alcohol and 2-3 drops of phenolphthalein were mixed in a conical flask and titrate it against 0.025 M KOH from burette until the color of solution became pink. Note the volume of KOH used. This was repeated three times to calculate mean volume of KOH used for blank titration. While in sample titration 9 ml isopropyl alcohol, 1 ml of wild safflower oil and 2-3 drops of phenolphathalein were taken in conical flask and titrate against 0.025M KOH from birette until end point i.e pink color appeared. Note the volume of KOH used and three readings were taken by repeating the same experiment to calculate the mean volume of KOH used to titrate the sample.

Acid number = (A-B) x C/D

A = Volume used in Sample/Actual titration, B = Volume used in Blank titration

C = Mass of Catalyst in g/l, D = Volume of oil used

2.2. Biodiesel synthesis

The method used for synthesis of fatty acid methyl esters (Biodiesel) from crude oil was alkali catalyzed transesterificaton (Ahmad *et al.*, 2010) (Figure 3,4,5 &6). There are numerous transesterification citations in the scientific and patent literature (Bradshaw and Meuly, 1944; Freedman *et al.*, 1984; Freedman *et al.*, 1986; Schwab *et al.*, 1987; Allen *et al.*, 1945; Trent, 1945; Tanaka *et al.*, 1981; Wimmer, 1992b; Ma *et al.*, 1998a; Ma *et al* 1998b; and Ma *et al.* 1999). Crude oil contains impurties which could affect the quality, yield and process of transesterification. The filtration of crude oil was done by using whattmann paper NO: 42 (See Plate 1). The filtered oil was heated up to 125 °C on hot plate (VELP Scientifica F20520166) in order to decompose triglycerides into monoglycerides and diglycerides. Transesterification of one liter oil (Plate 2) was carried out for the production of methyl esters by using different alkali catalysts (Ahmad *et al.*, 2011). Sodium hydroxide (NaOH) and potassium hydroxide (KOH) were used as catalyst. A specific amount of each alkali hydroxide (6.3 g for one liter oil) was added to methanol (200 ml) to make alkali methoxide which was used as a catalyst in reaction. The prepared methoxide was added to oil at 65°C and stirred for 35-40 min at 600 rpm.

After stirring the reaction mixture was kept overnight at room temperature to settle down distinct layers i.e. upper thin layer of soap, middle layer of FAME (fatty acid methyl ester) and the bottom dense layer of glycerin. These layers were then separated through separating glass funnel. Biodiesel washing is done with ordinary tap water in order to remove impurities and suspended particles. 3-4 washings were performed for complete clearance of biodiesel. Few drops of acetic acid were also added. The residual water was eliminated by treatment with anhydrous sodium sulphate (Na_2SO_4) followed by filtration.

3. Results and discussion

Catalyst	Wild Safflower	Safflower	Milk Thistle
NaOH	2.74	1.75	2.32
KOH	2.81	1.82	2.46

Table 1. Determination of FFA (%) contents through aqueous acid base titration

Catalyst	Catalyst concentration (g)	Wild Safflower (%)			Safflower (%)			Milk Thistle (%)		
		Biodiesel	Glycerin	Soap	Biodiesel	Glycerin	Soap	Biodiesel	Glycerin	Soap
NaOH	6.3	88	12	0	82	17	1	84	15	1
KOH	6.3	80	20	0	78	20	2	82	18	0

Table 2. yield of biodiesel and by-products by using various catalysts

Fuel Properties	Method	Wild Safflower	Safflower	Milk Thistle	HSD
Color	ASTM D-1500	2	2	2	2.0
Density @40°C Kg/L	ASTM D-1298	0.8980	0.8623	0.8990	0.8343
Kinematic Viscosity @ 40°C c St	ASTM D-445	6.45	6.13	6.23	4.223
Sulphur % wt	ASTM D-4294	0.1103	0.00041	0.0123	0.05
Total Acid No. mg KOH/gm	ASTM D-664	0.14	0.63	0.92	0.8
Flash Point °C (PMCC)	ASTM D-93	110	80	92	60-80
Pour Point °C	ASTM D-97	-12	-9	-6	-35 t0 -15
Distillation @ 90% recovery °C	ASTM D-86	358	352	354	360.4
Cloud Point °C	ASTM-2500	+15	+9	+7	-15 to 5
Calorific Value BTU/LB	ASTM-240	16977	16566	16472	20,400
Cetane Index	ASTM-976	50	52	51	46
Phosphorus % wt.	ASTM D-6728	-	-	-	-

Table 3. Fuel Properties of Biodiesel (B100) in comparison with HSD

The major share of all energy consumed worldwide comes from fossil sources (petroleum, coal and natural gas). However these sources are limited, and will be exhausted by the near future. Thus looking for alternative sources of new and renewable energy such as biomass is of vital importance. Alternative and renewable fuels have the potential to solve many of the current social problems and concerns, from air pollution to global warming to other environmental improvements and sustainability issues (MacLeana and Laveb, 2003).

During the decade of 1930s and 1940s, neat vegetable oils were used in diesel engines under an emergency situation (Ma and Hanna, 1999). Currently, most of the biodiesel is produced from the edible or vegetable oils using methanol and an alkaline catalyst such as sunflower

(Vicente *et al.*, 2004), canola (Singh *et al.*, 2006), palm (Darnoko and Cheryman, 2000; Cheng *et al.*, 2004), soybean oil (Encinar *et al.*, 2005) and waste vegetable oils (Felizardo *et al.*, 2006; Dorado *et al.*, 2002; Cetinkaya and Karaosmanolu, 2004). However, large amount of non-edible oils and fats are available such as safflower (Meka *et al.*, 2007), Pongame (Ahmad *et al.*, 2009), Sesame (Ahmad *et al.*, 2011), and tigernut oil (Ugheoke *et al.*, 2007) have been intensively investigated as potential low priced biodiesel sources. This study supports the production of biodiesel from non edible seed oils i.e. (wild safflower, safflower and milk thistle oil biodiesl as a viable sources of alternative to the diesel fuel.

3.1. FAMEs production

The percentage conversion of oil to biodiesel with NaOH and KOH at 65^{0}C is given in table 2. The results illustrated the greater oil to FAMEs conversion with NaOH as compared to KOH. The most common way to produce biodiesel is transesterification reaction in which triglycerides react with an alcohol to produce fatty acid mono-alkyl esters (Biodiesel) and glycerol. Methanol is the most common alcohol because of its low price compared to other alcohols. This reaction is referred as methanolysis. Generally transesterification is catalysed by a basic or an acid catalyst. However, the basic catalysts are the most commonly used in industry, because the process proves faster and the reaction conditions are moderated (Freedman *et al.*, 1984; Reid, 1911). In this project biodiesel was synthesized from wild safflower, safflower and milk thistle oil by base (NaOH and KOH) catalyzed transesterification with methanol. Most studies of the basic-catalysed transesterification of vegetable oils involve the calculations of the triglyceride conversion rate and the changes in product composition during reaction (Feuge and Gros, 1949; Freedman *et al.*, 1984, 1986; Schwab *et al.*, 1987; Peterson et al., 1991; Mittelbach and Trathnigg, 1990; Chang *et al.*, 1996; Mittelbach, 1996; Coteron *et al.*, 1997; Boocock *et al.*, 1998; Noureddini *et al.*, 1998; Vicente *et al.*, 1998; Darnoko and Cheryan, 2000).

3.2. Fuel properties

Biodiesel is characterized by their viscosity, density, cetane number, cloud and pour points, calorific value, distillation range, flash point, ash content, sulfur content, acid value, and phosphorus contents. These parameters are specified through the ASTM (American Standard Testing Methods) standards. This standard identifies the parameters the pure biodiesel (B100) must meet before being used as a pure fuel or being blended with petroleum-based diesel fuel. The properties of these oils methyl esters (B100) are given in table 3. These values are in the close range and comparable with high speed diesel (HSD).

The viscosity difference forms the basis of an analytical method, i.e. viscometry, applied to determine the conversion of vegetable oil to methyl ester. The viscosity difference between the componential triacylglycerols of vegetable oils and their corresponding methyl esters resulting from transesterification is approximately one digit (Knothe, 2001). Kinematic viscosity has been included in biodiesel standards (1.9-6.0 mm2/s in ASTM D6751 and 3.5-5.0 mm2/s in EN 14214) (Knothe, 2005). The viscosity of these oil biodiesel were near to ASTM standards. Biodiesels have a viscosity close to that of diesel fuels. As the oil temperature increases its viscosity decreases (Sarin & Sharma, 2007). The lower the viscosity of the biodiesel, the easier it is to pump and atomize and achieve finer droplets (Goodrum, 2007). The calorific value of edible and non-edible methyl ester was lower than that of diesel because of their oxygen content. The presence of oxygen in the biodiesel helps for a complete combustion of the fuels in the engine (Pramanik, 2003). Calorific value of these oil biodiesel were comparable to ASTM standard. The cetane number is one of the most commonly cited indicators of diesel fuel quality. It measures the readiness of the fuel to auto-ignite when injected into the engine. It is generally dependent on the composition of the fuel and can impact the engine's startability, noise level, and exhaust emissions. Cetane index of these three species were also in accordance with ASTM standards. The higher the cetane number, the more efficient the ignition is. Because of the higher oxygen content, biodiesel has a higher cetane number as compared to petroleum diesel. (Arjun *et al.*, 2008). Flash point is the important temperature specified for safety during transport, storage, and handling (Krisnangkura, 1992). Flash point of these oil methyl esters were found to be higher as compared to HSD. The flash point of bio-diesel is higher than the petro-diesel, which is safe for transport purpose. The ASTM standard for total acid number for pure biodiesel is 0.8 mg KOH/g. The TAN or acid value is the total amount of potassium hydroxide necessary to neutralize the free acids in biodiesel sample (Arjun *et al.*, 2008). Higher acid number could also cause degradation of rubber parts in older engines resulting in filter clogging The test result for the total acid number of these oil biodiesel were found to be ideal (Guo & Leung, 2003). Two important parameters for low-temperature applications of a fuel are cloud point (CP) and pour point (PP). The CP is the temperature at which wax first becomes visible when the fuel is cooled. The cloud point of methyl ester produced from these oils were found to be in accordance with ASTM standards. The PP is the temperature at which the amount of wax from solution is sufficient to gel the fuel; thus it is the lowest temperature at which the fuel can flow. The pour point of methyl ester produced from these oil were near to HSD. Biodiesel has a higher CP and PP compared to conventional diesel. The cloud points were affected by the presence of monoglycerides, however, the pour points were not affected. Moreover, the *cis* double bond present in the erucic acid of rapeseed oil hampered the lowering of the pour point of esters. The type of fatty acid branched chain available in the original oil has an impact on the pour point (Lee *et al.*, 1995). Biodiesel contains virtually trace amount of sulfur, so SO_2 emissions are reduced in direct proportion to the petrodiesel replacement (Demirbas, 2007). Sulpur contents in these three oil yielding plants were very low as compared to HSD.

4. Conclusion and recommendations

Based on above findings these three species of family Asteraceae have higher potential as a raw material source for biodiesel production at global interest and application. Following are some key recommendations which might be useful for production of raw material availability, production and consumption of biodiesel at global perspective;

1. In all developed countries, research and development has always played a vital role in profitable development of industry. In developed and some developing countries more and more R & D activities are being sponsored by the private sector and their Governments are assisting them and taking part in these activities by way of tax incentives and award schemes.

2. It is recommended that policies should be designed and incentives be offered by government to develop biodiesel companies and industries in the country.

3. Serious consideration should be given to establish a mega tree plantation for production of oil seeds in biodiesel application.

4. It is recommended that production of biodiesel to final use by consumer, quality should be given priority. Number of strategies should be given importance such as collection of seeds, extractions, processing, handling, storage and marketing. Therefore positive inspection system for all these sectors including agriculture, private sector and farming system.

5. In view of the present study as presented in this issue about the economic importance of national plants resources used for biodiesel production, research, development and cultivation efforts should be focused on these plants and identified other resources.

6. These three species are fast growing but cultivated on a small scale by rural farmers, could be produced on large scale for consumption and to be used as fuel. These species are more economical and need minimal quantity of water, fertilizer and pesticides. Such type of study on plant resources will make their data readily available for identifying promising species for future consideration for cultivation of biodiesel yielding crops.

7. It is proposed to further extend the project of bio-diesel. There is need to establish pilot projects to commercialize bio-diesel and set up its supply chain. The project may be extended step wise like conversion of vehicle fleets of designated departments on bio-diesel.

List of abbreviations

ASTM = American Society for Testing and Materials

FA = Fatty Acid

FAME = Fatty Acid Methyl Esters

FFA = Free Fatty Acid

EM = Engine Manufacturing Association

Author details

Mushtaq Ahmad[1*], Lee Keat Teong[2], Muhammad Zafar[1], Shazia Sultana[1], Haleema Sadia[1] and Mir Ajab Khan[1]

*Address all correspondence to: mushtaqflora@hotmail.com

1 Biofuel Lab., Department of Plant Sciences, Quaid-i-Azam University Islamabad, Pakistan

2 School of Chemical Engineering, University of Sains Malaysia, Malaysia

References

[1] Ahmad, M., A. Khan, M. Zafar and S. Sultana. 2010. Enviorrnment friendly renewable energy from Sesame biodiesel. *Energy Sources, Part A*. 32(2): 189-196.

[2] Ahmad, M., K. Ullah, M. A. Khan, S. Ali, M. Zafar and S. Sultana. 2011. Quantitative and ualitative analysis of sesame oil biodiesel. *Energy Sources, Part A*. 33: 1239-1249.

[3] Ahmad, M., M. A. Khan, A. Hasan, M. Zafar, and S. Sultana. 2008. Chemotaxonomic standardization of herbal drugs Milk thistle and Globe thistle. Asian J. of Chem. 6(20): 4443-4459.

[4] Ahmad, M., M. Zafar, M. A. Khan and S. Sultana. 2009. *Pongamia pinnata* as a biodiesel resource in Pakistan. *Energy Sources, Part A*. 31: 1436- 1442.

[5] Allen, H. D., G. Rock and W. A. Kline. 1945. Process for treating fats and fatty oils. US Patent, 2: 383-579.

[6] Anjani, (2005). Genetic variability and character association inwild safflower (*Carthamus oxycantha*).

[7] AOAC, 1990. Official methods of analysis, (13th edition). Association of OfficialAnalytical Chemists. Washington, DC.

[8] Arjun, B., L. K. Chhetri, C.Watts & M. R. Islam. 2008. Waste Cooking Oil as an Alternate Feedstock for Biodiesel Production. *Energies*.1: 3-18.

[9] Bijalwan, A., C. M. Sharma and V. K. Kediyal. 2006. Bio-diesel revolution. Science Reporter, January 2006. pp: 14–17.

[10] Boocock, D.G.B., S. K. Konar, V. Mao, C. Lee and S. Buligan. 1998. Fast formation of high-purity methyl esters from vegetable oils. *J. Am. Oil Chem. Soc.*, 75 (9): 1167–1172.

[11] Bradshaw, G. B. and W. C. Meuly. 1994. Preparation of detergents. US Patent, 2: 360-844.

[12] C.S.I.R. (Council of Scientific and Industrial Research). 1948-1976. The wealth of India. 11 vols. New Delhi.

[13]   Canoira, L., R. Alcantara, M. J. Garcia-Martinez and J. Carrasco. 2006. Biodiesel from Jojoba oil-wax: transesterification with methanol and properties as a fuel.*Biomass and Bioenergy*. 30: 76-81.

[14]   carthamosides from Carthamus oxycantha. *Natural Product Communication*. 5(3), 419-422.

[15]   Cetinkaya, M. and F. Karaosmanolu. 2004. Optimization of base-catalyzed transesterification reaction of used cooking oil. *Energ. Fuel*. 18 (6): 1888-1895.

[16]   Chang, D.Y. Z., J. H. Van Gerpen, I. Lee, L. A. Johnson, E. J. Hammond and S. J. Marley. 1996. Fuel properties and emissions of soybean oil esters as diesel fuel. *J. Am. Oil Chem. Soc.*, 73 (11): 1549–1555.

[17]   Cheng, S. F., Y. M. Choo, A. N. Ma and C.H. Chuah. 2004. Kinetics study on transesterification of Palm oil. *J. Oil Palm Res.*, 16 (2): 19-29.

[18]   Coteron, A., G. Vicente, M. Martinez and J. Aracil. 1997. Biodiesel production from vegetable oils. Influence of catalysts and operating conditions. In: Pandalai, S.G. (Ed.), Recent Res. Developments in Oil Chemistry, vol. 1. Transworld Research Network, India, pp. 109–114.

[19]   Darnoko, D. and M. Cheryan, 2000. Continuous production of palm methyl esters. *J. Am. Oil Chem. Soc.*, 77 (12): 1269–1272.

[20]   Deshpande, R. B. 1952. Wild safflower (*Carthamus oxyacantha*)- A possible oilseed crop for the desert and arid regions. *Indian J. Genet.*, 12: 10-14.

[21]   Dorado, M. P., E. Ballesteros, J. A. de Almeida, C. Schellert, H. P. Lö hrlein and R. Krause. 2002. Transesterification of karanja (*Pongamia pinnata*) oil by solid basic catalysts. *Am. Soc. Agr. Biol. Eng.*, 45 (3): 525-529.

[22]   Duke, J.A. 1978. The quest for tolerant germplasm. p. 1-61. In: ASA Special Symposium 32, Crop tolerance to suboptimal land conditions. Am. Soc. Agron. Madison, WI.

[23]   Encinar, J. M., F. Juan, J. F. Gonzalez, and A. Rodriguez-Reinares. 2005. Biodiesel from used frying oil: Variables affecting the yields and characteristics of the biodiesel. *Ind. Eng. Chem. Res.*, 44 (15): 5491-5499.

[24]   Felizardo, P., M. J. Correia, I. Raposo, J. F. Mendes, R. Berkemeier and J. M. Bordado. 2006. Production of biodiesel from waste frying oils. *Waste Manage.*, 26 (5): 487-494.

[25]   Fernandez-Martinez, M., Del-Rio, M. and de Haro, A. (1993). Survey of safflower (*Carthamus oxycantha* L.) germplasm for variants in fatty acid composition and other seed characters. *Euphytica*. 69,115-122.

[26]   Feuge, R.O. and A. T. Gros. 1949. Modification of vegetables oils. VII. Alkali-catalyzed interestification of peanut oil with ethanol. *J. Am. Oil Chem. Soc.* 26 (3): 97–102.

[27]   Firestone, D. (1999). Physical and chemical characteristics of oils, fats, and waxes. AOCS Press, Champaign, United States. pp 152-153.

[28] Freedman, B., E. H. Pryde and T. L. Mounts. 1984. Variables affecting the yields of fatty esters from transesterified vegetable oils. *J. Am. Oil Chem. Soc.*, 61(10): 1638–1643.

[29] Freedman, B., Kwolek, W. F., and E. H. Pryde. 1986. Quantitation in the analysis of transesterified soybean oil by capillary gas chromatography. *J. Am. Oil Chem. Soc.*, 63: 1370–1375.

[30] Goodrum, J.W. 2002. Volatility and boiling points of biodiesel from vegetable oils and tallow. *Biomass Bioenergy*. 22: 205–211.

[31] Gui, M. M., K. T. Lee and S. Bhatia. 2008. Feasibility of edible oil vs. non-edible oil vs. waste edible oil as biodiesel feedstock. *Energy*. 33: 1646– 1653.

[32] Guo, Y. and Y. C. Leung. 2003. Analysis on the biodiesel production using grease trap oils from restaurants, Macro Review. Japan, *Jpn. Macro Eng. Soc.*, 16: 421-426.

[33] Hassan, Z., Ahmed, V.U., Hussain Z.J., Zahoor A, Siddiqui, I.N., Rasool Z.M.. (2010). Two new

[34] Ikwuagwu, O. E., I. C. Ononogbu and O. U. Njoku. 2000. Production of biodiesel using rubber, Hevea Brasiliensis (kunth. Muell.) seed oil. *Ind. Crops and Product*. 12: 57-62.

[35] *Indian Journal of Agricultural Science*. 75(8), 516-518.

[36] Kansedo, J., K. T. Lee and S. Bhatia. 2009. Cerbera odollam (sea mango) oil as a promising non-edible feedstock for biodiesel production. *Fuel*. 88: 1148–1150.

[37] Knothe, G. 2000. Monitoring a progressing transesterification reaction by fiber optic NIR with correlation to ^{1}H NMR spectroscopy. *J. Am. Oil Chem. Soc.*, 77: 489–493.

[38] Knothe, G. 2001. Analytical methods used in the production and fuel quality assessment of biodiesel. *Am. Soc. Agr. Eng.*, 44 (2): 193-200.

[39] Knothe, G. 2002. Current perspectives on biodiesel. *INFORM.*, 13: 900–903.

[40] Knothe, G. 2005. Dependence of biodiesel fuel properties on the structure of fatty acid alkyl esters. *Fuel Process. Tech.*, 86: 1059-1070.

[41] Krawczyk, T. 1996. Biodiesel as alternative fuel makes in roads but hurdles remain. *INFORM.*, 7: 801-829.

[42] Kumar, A., A. Tiwari and H. Raheman. 2007. Biodiesel production from jatropha oil (Jatropha curcas) with high free fatty acids: an optimized process. *Biomass and Bioenergy*. 31: 569–575.

[43] Lam, M.K. and Lee, K. T., "Renewable and sustainable bioenergies production from palm oil mill effluent (POME): Win–win strategies toward better environmental protection", Biotechnology Advances, Elsevier, Vol. 29 (1) 2011, pp 124-141.

[44] Lee, I., L. A. Johnson and E. G. Hammond. 1995. Use of branched-chain esters to reduce the crystallization temperature of Biodiesel. *JAOCS.*, 72: 1155–60.

[45] Ma, F. and M. A. Hanna. 1999. Biodiesel production: a review. *Bioresource Technology*. 70: 1-15.

[46] Ma, F., L. D. Clements and M. A. Hanna. 1998a. The effects of catalyst free fatty acids and water on transesterificaton of beef tallow. *Trans. ASAE.*, 41: 1261-1264.

[47] Ma, F., L. D. Clements and M. A. Hanna. 1998b. Biodiesel fuel from animal fat. Ancillary studies on transesterificaton of beef tallow. *Ind. Eng. Chem Res.*, 37: 3768-3771.

[48] Ma, F., L. D. Clements and M. A. Hanna. 1999. The effects of mixing on transesterification of beef tallow. Bioresource *Technology*. 69: 289-293.

[49] MacLeana, H. and L. B. Laveb. 2003. Evaluating automobile fuel propulsion system technologies. *Prog .Energy Combust Sci.* 29: 1-69.

[50] Meher, L. C., S. Vidya, S. Dharmagadda and S. N. Naik. 2006. Optimization of alkali catalyzed transesterification of Pungamia pinnata oil for production of biodiesel. *Bioresour Technol.,* 97: 1392–1397.

[51] Meka, P. K., V. Tripathi and R. P. Singh. 2007. Synthesis of biodiesel fuel from safflower oil using various reaction parameters. *J. Oleo Sci.*, 56(1): 9-12.

[52] Mittelbach, M. and B. Trathnigg. 1990. Kinetics of alkaline catalyzed methanolysis of sunflower oil. *Fat Sci. Technol.*, 92 (4): 145–148.

[53] Nagel, N. and P. Lemke. 1990. Production of methyl fuel from miceoalgea. *Appl. Biochem. Biotechnol.* 24: 355-361. Calvin, M. 1985. Fuel oils from higher plants. *Ann. Proc. Phytochem. Soc. Eur.* 26: 147-160.

[54] Noureddini, H. and Zhu, D. 1997. Kinetics of transesterification of soybean oil. *J. Am. Oil Chem. Soc.* 74: 1457–1463.

[55] Paramathma, M., P. Venkatachalam and A. S. Rajan. 2007. Need for jatropha cultivation and bio-diesel production, We need to move to a biofuel future. *BioSpectr J.*, pp: 30–32.

[56] Patil, P. D. and S. Deng. 2009. Optimization of biodiesel production from edible and non edible vegetable oils. *Fuel.* 88: 1302–1306.

[57] Peterson, C.L., M. Feldman, R. Korus and D. L. Auld. 1991. Batch type transesterification process for winter rape oil. *Appl. Eng. Agric.*, 7 (6): 711–716.

[58] Pramanik, K. 2003. Properties and use of Jatropha curcas oil and diesel fuel blends incompression ignition engine. *Renewable Energy.* 28: 239– 248.

[59] Pryde, E. H. 1984. Vegetable oils as fuel alternatives—symposium overview. *JAOCS.* 61: 1609–1610.

[60] Pryor, R.W., M. A. Hanna, J. L. Schinstock and L. L. Bashford. 1982. Soybean oil fuel in a small diesel engine. *Transactions o f the ASAE.* 26: 333–338.

[61] Ratledge, S. and C. A. Boulton. 1985. Fats and oils. In: Comparative Biotechnology in industry,agriculture and medicine, New York: Pergamon Press, vol. 3. pp: 983–1003.

[62] Sarin, R. and M. Sharma. 2007. Jatropha Palm biodiesel blends: An optimum mix for Asia. *FUEL*. Vol. 86: 1365-1371.

[63] Schwab, A. W., G. J. Dykstra, E. Selke, S. C. Sorenson and E. H. Pryde. 1988. Diesel fuel from thermal decomposition of soybean oil. *JAOCS*. 65: 1781–1786.

[64] Shay, E. G. 1993. Diesel fuel from vegetable oils: status and opportunities. *Biomass and Bioenergy*. 4: 227-242.

[65] Sheehan, J., T. Dunahay, J. Benemann and P. Roessler. 1998b. A look back at the US Department of Energy's aquatic species program—biodiesel from Algae. National Renewable Energy Laboratory (NREL) Report: NREL/TP-580-24190, Golden, CO.

[66] Singh, A. B. H., J. Thompson and J. V. Gerpen. 2006. Process optimization of biodiesel production using different alkaline catalysts. *Appl. Eng. Agric.*, 22 (4): 597-600.

[67] Sonntag, N.O.V. 1979a. Structure and composition of fats and oils. Bailey's industrial oil and fat products, vol. 1, 4th edition, Swern, D. John Wiley and Sons, New York, p. 1-14.

[68] Tanaka, Y., A. Okabe and S. Ando. 1981. Method for the preparation of a lower alkyl ester of fatty acids. US Patent, 4: 305-590.

[69] Trent, W. R. 1945. Process of treating fatty glycerides. US Patent, 2: 383-632.

[70] Ugheoke, B. I., D. O. Patrick, H. M. Kefas and E.O. Onche. 2007. Determination of optimal catalyst concentration for maximum biodiesel yield from tigernut (*Cyperus Esculentus*) oil. *Leonardo J. Sci.*, 10: 131-136.

[71] Vicente, G., A. Coteron, M. Martı́nez and J. Aracil. 1998. Application of the factorial design of experiments and response surface methodology to optimize biodiesel production. *Ind. Crops Pro.*, 8 (1): 29–35.

[72] Wimmer, T. 1992b. Preparation of ester of fatty acids with short-chain alcohols. [In Austrian], pp: 349-571.

[73] Wu, K.K. and Jain, S.K. 1977. A note on germplasm diversity in the world collections of safflower. Econ. Bot. 31:72-75.

[74] Ziejewski, M. Z., K. R. Kaufman and G. L. Pratt. 1983. Vegetable oil as diesel fuel, USDA. Argic. Rev. Man., ARM-NC-28. pp: 106–111.

Chapter 4

Potential Production of Biofuel from Microalgae Biomass Produced in Wastewater

Rosana C. S. Schneider, Thiago R. Bjerk,
Pablo D. Gressler, Maiara P. Souza,
Valeriano A. Corbellini and Eduardo A. Lobo

Additional information is available at the end of the chapter

1. Introduction

Microalgae are the principal primary producers of oxygen in the world and exhibit enormous potential for biotechnological industries. Microalgae cultivation is an efficient option for wastewater bioremediation, and these microorganisms are particularly efficient at recovering high levels of nitrogen, inorganic phosphorus, and heavy metals from effluent. Furthermore, microalgae are responsible for the reduction of CO_2 from gaseous effluent and from the atmosphere. In general, the microalgae biomass can be used for the production of pigments, lipids, foods, and renewable energy [1].

Much of the biotechnological potential of microalgae is derived from the production of important compounds from their biomass. The biodiversity of the compounds derived from these microorganisms permits the development of new research and future technological advances that will produce as yet unknown benefits [2].

Microalgae grow in open systems (turf scrubber system, raceways, and tanks) and in closed systems (vertical (bubble column) or horizontal tubular photobioreactors, flat panels, biocoils, and bags). The closed systems favor the efficient control of the growth of these microorganisms because they allow for improved monitoring of the growth parameters [3-4].

Because microalgae contain a large amount of lipids, another important application of microalgae is biodiesel production [5]. In addition, after hydrolysis, the residual biomass can potentially be used for bioethanol production [6]. These options for microalgae uses are promising for reducing the environmental impact of a number of industries; however, there is a need for optimizing a number of parameters, such as increasing the lipid fraction and the availability of nutrients [7].

Notably, the microalgae biomass can produce biodiesel [5], bioethanol [6], biogas, biohydrogen [8-9] and bio-oils [10], as shown in Figure 1.

The productivity per unit area of microalgae is high compared to conventional processes for the production of raw materials for biofuels, and microalgae represent an important reserve of oil, carbohydrates, proteins, and other cellular substances that can be technologically exploited [2,11]. According to Brown *et al*. [12], 90-95% of the microalgae dry biomass is composed of proteins, carbohydrates, lipids, and minerals.

An advantage of culturing algae is that the application of pesticides is not required. Furthermore, after the extraction of the oil, by-products, such as proteins and the residual biomass, can be used as fertilizer [13]. Alternatively, the residual biomass can be fermented to produce bioethanol and biomethane [14]. Other applications include burning the biomass to produce energy [15].

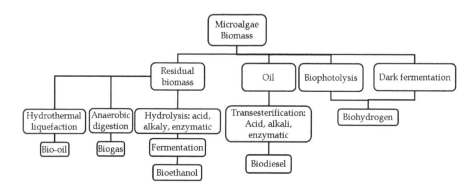

Figure 1. Diagram of the principal microalgae biomass transformation processes for biofuel production.

The cultivation of microalgae does not compete with other cropsfor space in agricultural areas, which immediately excludes them from the "biofuels versus food" controversy. Similar to other oil crops, microalgae exhibit a high oil productivity potential, which can reach up to 100,000 L he^{-1}. This productivity is excellent compared to more productive crops, such as palm, which yield 5,959 L he^{-1} and thus contribute to the alleviation of the environmental and economic problems associated with this industry[16].

Although the productivity of microalgae for biofuel production is lower than traditional methods, there is increasing interest and initiatives regarding the potential production of microalgae in conjunction with wastewater treatment, and a number of experts favor this option for microalgae production as the most plausible for commercial application in the short term [17].

2. Wastewater microalgae production

Photosynthetic microorganisms use pollutants as nutritional resources and grow in accordance with environmental conditions, such as light, temperature, pH, salinity, and the presence of inhibitors [18]. The eutrophication process (increases in nitrogen and inorganic phosphorus) of water can be used as a biological treatment when the microalgae grow in a controlled system. Furthermore, these microorganisms facilitate the removal of heavy metals and other organic contaminants from water [19-22].

In general, the use of microalgae can be combined with other treatment processes or as an additional step in the process to increase efficiency. Therefore, microalgae are an option for wastewater treatments that use processes such as oxidation [23], coagulation and flocculation [24], filtration [25], ozonation [26], chlorination [27], and reverse osmosis [28], among others. Treatments using these methods separately often prove efficient for the removal of pollutants; however, methods that are more practical, environmentally friendly, and produce less waste are desirable. In this case, the combination of traditional methods with microalgae bioremediation is promising [29]. The bioremediation process promoted by open systems, such as high rate algal ponds, combines microalgae production with wastewater treatment. In addition, the control of microalgae species, parasites, and natural biofloculation is important for cost reduction during the production of the microorganism [20, 30].

Many microalgae species grow under inhospitable conditions and present several possibilities for wastewater treatments. All microalgae production generates biomass, which must be used in a suitable manner [31-32].

Microalgae are typically cultivated in photobioreactors, such as open systems (turf scrubbers, open ponds, raceway ponds, and tanks) or closed system (tubular photobioreactors, flat panels, and coil systems). The closed systems allow for increased control of the environmental variables and are more effective at controlling the growth conditions. Therefore, the specific cultivation and input of CO_2 are more successful. However, open systems can be more efficient when using wastewater, and low energy costs are achieved for many microalgae species grown in effluents in open systems [33-35]. Because of the necessity for renewable energy and the constant search for efficient wastewater treatment systems at a low cost, the use of microalgae offers a system that combines wastewater bioremediation, CO_2 recovery, and biofuel production.

In turf scrubber systems, high rates of nutrient (phosphorus and nitrogen) removal are observed. This phenomenon was observed in the biomass retained in the prototype turf scrubber system used in three rivers in Chesapeake Bay, USA. The time of year was crucial for the bioremediation of excess nutrients in the river water, and the best results demonstrated the removal of 65% of the total nitrogen and up to 55% of the total phosphorus, both of which were fixed in the biomass [32].

Compared to other systems, such as tanks and photobioreactors (Fig. 2), the algae turf scrubber system is an alternative for the final treatment of wastewater. The turf scrubber system offers numerous advantageous characteristics, such as temperature control in regions with high solar incidence and the development of a microorganism community using microalgae, other bacteria, and fungi that promote nutrient removal. Under these conditions, it is possible to obtain biomass with the potential for producing biofuels. However, sufficient levels of oil in the biomass are an important consideration for the production of other biofuels, such as bioethanol, bio-oil, and biogas, among others, which would achieve the complete exploitation of the biomass.

Considering the possibility of using all the biomass, photobioreactors can be used to produce feedstock for biofuel, such as biodiesel and bioethanol, because the oil level of the biomass produced in closed systems is greater than in open systems. Table 1 shows the results obtained using a mixed system and a similar tubular photobioreactor with microalgae *Desmodesmus subspicatus* in the same effluent [36-37].

Figure 2. A) Mixed system prototype for microalgae production using a (1) scrubber, (2) tank, and (3) photobioreactor. B) Microalgae biomass in a mixed system separated by electroflotation [36].

Parameters	Mixed system		Photobioreactor	
	without CO_2	with CO_2	without CO_2	with CO_2
Cultivation Days	20	15	7	7
Maximum Cell Division (x10^6 cell mL^{-1})	25.48 ± 0.02	26.97 ± 0.21	8.49 ± 1.02	25.98± 1.57
Average Cell Division (K)	0.29 ± 0.48	0.16 ± 0.33	-0.12 ± 0.60	0.34 ± 0.40
Biomass (g L^{-1})	0.62 ± 0.11	0.72 ± 0.15	0.18 ± 5.65	1.41 ± 1.40
Lipids (%)	1.36 ± 0.29	6.07 ± 0.12	18.73 ± 0.25	12.00 ± 0.28

Table 1. Microalgae biomass growth and total lipids in a mixed system and a tubular photobioreactor [36-37].

The removal of nutrients from the effluent produced excellent results using the genus *Scenedesmus*, as shown in Table 2. Other studies have also produced promising results. According to Ai *et al.* [38], the cultivation of *Spirulina platensis* in photobioreactors was satisfactory because of the photosynthetic performance. The pH, temperature, and dissolved oxygen levels were controlled effectively; however, continuous operation was required to ensure the relia-bility of photosynthetic performance in the photobioreactor.

The cultivation of the diatom *Chaetoceros calcitrans* in photobioreactors exhibited high growth rates; the maximum specific growth rate (μ) achievable was 9.65×10^{-2} h^{-1} and 8.88×10^6 cells mL^{-1} in semicontinuous and batch systems, respectively. Even with a lower inci-dence of light, the results for the production of biomass were good [39].

The cultivation of microalgae *Chlorella* sp. in a semicontinuous photobioreactor produced a sat-isfactory level of biomass production (1.445 ± 0.015 g L^{-1} of dry cells). The growth, productivity and the amount of CO_2 removed obtained under conditions of increased control of the culture and a high concentration of inoculum using cells already adapted to the system increased the CO_2 assimilation[33]. The growth rate is also influenced by the concentration of microalgae un-til reaching an optimum concentration under the operational conditions used [40].

Therefore, microalgae can produce 3-10 times more energy per hectare than other land cul-tures and are associated with CO_2 mitigation and wastewater depollution [41]. Microalgae production is a promising alternative to land plants for reducing environmental impacts; however, the optimization of a number of the production parameters that are important for the viability of the process must be considered, such as the increase in lipid production [7].

Microalgae	System	Removal (%)	
		Nitrogen	Phosphorus
Melosira sp.; *Lygnbya* sp.; *Spirogyra* sp.; *Ulothrix* sp.; *Microspora* sp.; *Claophora* sp.; (seasonal succession) [32]	Turf scrubber	65	45-55
Chlorella sp.; *Euglena* sp.; *Spirogyra* sp.; *Scenedesmus* sp.; *Desmodesmus* sp.; *Pseudokirchneriella* sp.; *Phormidium* sp.; *Nitzschia* sp.[36]	Mix	99	65
Scenedesmus sp. [42]	Photobioreactor	98	98
Scenedesmus sp. [43]	Immobilized cell	70	94
Chlamydomonas sp. [44]	Photobioreactor	100	33
Scenedesmus obliquus [45]	Immobilized cell	100	-
Scenedesmus obliquus [46]	Photobioreactor	100	98

Table 2. Use of microalgae grown in different systems for the removal of nitrogen and phosphorus from wastewater.

The bioremediation of wastewater using microalgae is a promising option because it re-duces the application of the chemical compounds required in conventional mechanical methods, such as centrifugation, gravity settling, flotation, and tangential filtration [21].

The feasibility of using microalgae for bioremediation is directly related to the production of biofuels because of the high oil content. Without the high oil levels, using other bacteria for this purpose would be more advantageous because there are limitations to the removal of organic matter by microalgae. In the literature, emphasis is placed on the ability of microalgae to remove heavy metals from industrial effluents [47].

3. Biofuels

The term biofuel refers to solid, liquid, or gaseous fuels derived from renewable raw materials. The use of microalgal biomass for the production of energy involves the same procedures used for terrestrial biomass. Among the factors that influence the choice of the conversion process are the type and amount of raw material biomass, the type of energy desired, and the desired economic return from the product [30].

Microalgae have been investigated for the production of numerous biofuels including biodiesel, which is obtained by the extraction and transformation of the lipid material, bioethanol, which is produced from the sugars, starch, and carbohydrate residues in general, biogas, and bio-hydrogen, among others (Fig. 3) [8].

Between 1978 and 1996, the Office of Fuels Development at the U.S. Department of Energy developed extensive research programs to produce renewable fuels from algae. The main objective of the program, known as The Aquatic Species Program (ASP), was to produce biodiesel from algae with a high lipid content grown in tanks that utilize CO_2 waste from coal-based power plants. After nearly two decades, many advances have been made in manipulating the metabolism of algae and the engineering of microalgae production systems. The study included consideration of the production of fuels, such as methane gas, ethanol and biodiesel, and the direct burning of the algal biomass to produce steam or electricity [48].

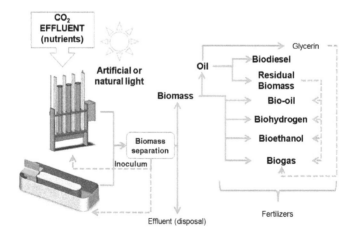

Figure 3. Utilization scheme for the microalgae biomass produced in wastewater.

3.1. Biodiesel

The choice of raw material is a critical factor contributing to the final cost of biodiesel andaccounts for 50-85% of the total cost of the fuel. Therefore, to minimize the cost of this biofuel, it is important to assess the raw material in terms of yield, quality, and the utilization of the by-products [49-50].

A positive aspect of the production of biodiesel from microalgae is the area of land needed for production. For example, to supply 50% of the fuel used by the transportation sector in the U.S. using palm oil, which is derived from a plant with a high oil yield per hectare, would require 24% of the total agricultural area available in the country. In contrast, if the oil from microalgae grown in photobioreactors was used, it would require only 1-3% of the total cultivation area [49].

The biochemical composition of the algal biomass can be manipulated through variations in the growth conditions, which can significantly alter the oil content and composition of the microorganism [51]. Biodiesel produced from microalgae has a fatty acid composition (14 to 22 carbon atoms) that is similar to the vegetable oils used for biodiesel production [51-52].

The biodiesel produced from microalgae contains unsaturated fatty acids [53], and when the biomass is obtained from wastewater and is composed of a mixture of microalgae genera, it can exhibit various fatty acids profiles. Bjerk [36] produced biodiesel using a mixed system containing the microalgae genera *Chlorella* sp., *Euglena* sp., *Spirogyra* sp., *Scenedesmus* sp., *Desmodesmus* sp., *Pseudokirchneriella* sp., *Phormidium* sp. (cyanobacteria), and *Nitzschia* sp., identified by microscopy in accordance with Bicudo and Menezes [54]. The CO_2 input, the stress exerted by the nutrient composition, and the existence of a screen to fix the filamentous algae contributed to differential growth and differences in the fatty acid profiles (Table 3). Consequently, the biodiesel produced was relatively stable in the presence of oxygen.

In this mixed system, a difference between the fatty acid profiles of the biomass obtained in the photobioreactor compared to the biomass obtained on the screen was observed. The biomass from the screen contained the filamentous algae genera, and the oil did not contain linoleic acid.

This observation is important for biodiesel production because the oil produced was less unsaturated. The iodine index reflects this trend; oils from species such as *Spirulina maxima and Nanochloropsis* sp. have iodine indices between 50 and 70 mg I_2 g^{-1} of oil, whereas in species such as *Dunaliella tertiolecta* and *Neochloris oleobundans*, the iodine index is greater than 100 mg I_2 g^{-1} of oil [56].

The composition and proportion of fatty acids in the microalgae oil depends on the species used, the nutritional composition of the medium, and other cultivation conditions [57].

Table 4 shows the microalgae commonly used for oil production. The literature lacks information regarding the iodine index or the composition of saturated and unsaturated fatty acids, which could help identify the appropriate microalgae species for biodiesel production. Information on numerous parameters is important, such as the oil unsaturation levels, the productivity of the microalgae in the respective effluents, the growth rate, and the total biomass composition. Using this information, a decision can be made regarding the economic and environmental feasibility of producing biodiesel and adequately allocating the waste.

Fatty acids*	without CO2 (%)	with CO2 (%)	with CO_2 (screen) (%)
Caprylic (C8:0)	0.05	0.08	-
Myristic (C14:0)	1.93	1.60	1.85
Pentadecanoic (C15:0)	0.50	0.44	0.52
Palmitoleic (C16:1)	1.28	2.02	4.20
Palmitic (C16:0)	29.58	24.68	32.50
Margaric (C17:0)	0.89	0.62	1.02
Linoleic (C 18:2)	15.12	9.51	-
Oleic (C 18:1n-9)	26.60	39.94	20.19
Estearic (C 18:0)	9.75	9.69	12.16
Araquidic (C 20:0)	0.70	1.43	1.72
Saturated and unsaturated not identified**	13.6	9.97	25.84

*The oil extraction method was adapted from the Bligh and Dyer (1959) method described by Gressler [37] using *Desmodesmussubspicatus* and the transesterification method described by Porte *et al.* [55] on a laboratorial scale.

Table 3. Relative proportion (%) of fatty acid methyl esters found in microalgae biomass cultivated in wastewater with and without CO_2 in a mixed system.

Among the microalgae shown in Table 4 that have an oil content that makes them competitive with land crops, twelve species (*Achnanthes sp., Chlorella sorokiniana, Chlorella sp., Chlorella vulgaris, Ellipsoidion sp., Neochloris oleoabundans, Nitzschia sp., Scenedemus quadricauda, Scenedemus sp., Schizochytrium sp., Skeletonoma costatum,* and *Skeletonoma sp.*) are from fresh water and can be investigated for the bioremediation of common urban and industrial effluents that do not have high salinity and contain pollutants that can be used as nutrients for the microorganisms. Because of their potential for oil production, a number of these microalgae species have been used for the production of biodiesel on a laboratory scale, although their potential industrial use associated with the bioremediation of industrial effluents is unknown. Studies using *Chlamydomonas* sp. [47] cultured in wastewater produced a rate of 18.4% oil and a fatty acid profile suitable for biodiesel production in addition to an excellent rate of nutrient removal (nitrogen and phosphorus).

Microalgae	Oil (%)	Microalgae	Oil (%)
Achnanthes sp.	44.5	*Nannochloris* sp.	20.0–35.0
Ankistrodesmus sp.	24.0–31.0	*Nannochloropsis oculata*	22.7–29.7
Botryococcus braunii	25–75	*Nannochloropsis* sp.	12.0-68.0
Chaetoceros calcitrans	39.8	*Neochloris oleoabundans*	35.0–54.0
Chaetoceros muelleri	33.6	*Nitzschia* sp.	45.0–47.0
Chlorella sorokiniana	19.3	*Phaeodactylum tricornutum*	18.7
Chlorella sp.	18.7–32	*Pavlova lutheri*	35.5 40.2
Chlorella vulgaris	19.2	*Pavlova salina*	30.9- 49.4
Chlorococcum sp.	19.3	*Phaeodactylum tricornutum*	18.0–57.0
Chlamydomonas sp.	18.4	*Synechocystis aquatilis*	18.5
Crypthecodinium cohnii	20.0	*Scenedemus quadricauda*	18.4
Cylindrotheca sp.	16–37	*Scenedemus* sp.	21.1
Dunaliella primolecta	23.0	*Schizochytrium* sp.	50.0–77.0
Ellipsoidion sp.	27.4	*Skeletonoma costatum*	21.0
Heterosigma sp.	39.9	*Skeletonoma* sp.	31.8
Isochrysis sp.	22.4-33	*Tetraselmis sueica*	15.0–23.0
Isochrysis galbana	7.0-40.0	*Thalassioria pseudonana*	20.6
Monallanthus salina	>20.0	*Thalassiosira* sp.	17.8

Adapted from [5,16,44,52,58-60], considering the values found under the respective production condition.

Table 4. Oil-producing microalgae with potential for biodiesel production.

3.2. Bioethanol

Bioethanol production from microalgae has received remarkable attention because of the high photosynthetic rates, the large biodiversity and variability of their biochemical composition, and the rapid biomass production exhibited by these microorganisms [1].

Furthermore, bioethanol derived from microalgae biomass is an option that demonstrates the greatest potential. John *et al*. [61] assessed microalgae biomass as a raw material for bioethanol production and argued that it is a sustainable alternative for the production of renewable biofuels. Examples of the genera of microalgae that fit the parameters for bioethanol production include the following: *Chlorella, Dunaliella, Chlamydomonas, Scenedesmus, Arthrospira,* and *Spirulina.* These microorganisms are suitable because they contain large amounts of starch and glycogen, which are essential factors for the production of bioethanol. The carbohydrate composition of these genera can be 70% of the biomass [62].

Traditionally, bioethanol is produced through the fermentation of sugar and starch, which are produced from different sources, such as sugarcane, maize, or a number of other grains [62].

After the oil extraction, the residual biomass contains carbohydrates that can be used for bioethanol production. This process represents a second-generation bioethanol and may be an alternative to the sugar cane ethanol produced in Brazil and corn or beet ethanol produced in other countries. The process requires pretreatment with a hydrolysis step before fermentation [63-65].

In bioethanol production, the processes vary depending on the type of biomass and involve the pretreatment of the biomass, saccharification, fermentation, and recovery of the product. The pretreatment of the biomass is a critical process because it is essential for the formation of the sugars used in the fermentation process (Table 5). Before the traditional fermentation process, acid hydrolysis is widely used for the conversion of carbohydrates from the cell wall into simple sugars. The acid pretreatment is efficient and involves low energy consumption [63].

Other techniques, such as enzymatic digestion [74] or gamma radiation [75], are interesting alternatives for increasing the chemical hydrolysis to render it more sustainable. Through analysis of the process in terms of energy, mass, and residue generation, it is possible to determine the best route. With enzymatic hydrolysis, the process can be renewable. Another technique for pretreatment of the biomass is hydrolysis mediated by fungi. Bjerk [36] investigated the *Aspergillus* genera for this purpose, and the bioethanol produced was monitored by gas chromatography using a headspace autosampler. The study demonstrated that seven strains (four isolates from *A. niger*, one from *A. terreus*, one from *A. fumigatus,* and one from *Aspergillus* sp.) were more efficient at hydrolyzing the residual biomass.

However, it is worth noting the importance of developing a well-designed and efficient system for the cultivation of these microorganisms, which can remove compounds that cause impurities in the final product. In addition, more studies should be undertaken to select strains that are resistant to adverse conditions, especially studies related to genetic engineering.

According to Yoon *et al*. [75], the use of gamma radiation is of potential interest for the hydrolysis of the microalgae biomass because compared to chemical or enzymatic digestion, gamma radiation raised the concentration of sugar reducers, and the saccharification yield was 0.235 g L^{-1} when gamma radiation was combined with acid hydrolysis. Acid hydrolysis alone produced a saccharification yield of only 0.017 g L^{-1}.

Microalgae	Pre treatment	Reaction condition		Fermenter	Bioethanol yield (%)	Ref.
		Temp. (°C)	Time (min)			
Chlamydomonas reinhardtii*	acid	110	30	Saccharomyces cerevisiae	29.2	[66]
Chlorococcum sp.	alkaline	120	30	Saccharomyces cerevisiae	26.1	[67]
	acid	140	30	Saccharomyces cerevisiae	10-35	[68]
Chlorococcum humicola	acid	160	15	Saccharomyces cerevisiae	52	[63]
Nizimuddinia zanardini**	acid	120	45	-	-	[69]
Kappaphycus alvarezii	acid	100	60	Saccharomyces cerevisiae	2.46	[70]
Scenedesmus obliquus***	acid	120	30	-	-	[71]
Spirogyra	alkaline	-	120	Saccharomyces cerevisiae	20	[72]
	enzymatic	-	-	Saccharomyces cerevisiae	4.42	[73]
	enzymatic	-	-	Zymomonas mobilis	9.7	

Glucose yield: * 58%; **70.2%; *** 14.7%

Table 5. Conditions of bioethanol production from microalgae.

3.3. Other biofuels

Several articles describe the thermochemical processing of algal biomass using gasification [63,76] liquefaction [77], pyrolysis [78], hydrogenation [79], and biochemical processing, such as fermentation [80-81]. However, engineering processes have not been investigated as a potential biotechnological method for the production of other biofuels from microalgae.

Currently, the energy derived from biomass is considered one of the best energy sources and can be converted into various forms depending on the need and the technology used, and biogas is chief among the forms of energy produced by biomass. [82].

Anaerobic digestion for biogas production is a promising energy route because it provides numerous environmental benefits. Biogas is produced through the anaerobic digestion of organic waste, drastically reducing the emission of greenhouse gases. As an added benefit, the by-products of fermentation, which are rich in nutrients, can be recycled for agricultural purposes. Adding anaerobic digestion to the use of biomass waste from which the oil has been removed produces an environmental gain and results in the complete exhaustion of the possible uses for the biomass. This strategy enables biomass waste to be an end-of-pipe technology for industrial processes that generate high amounts of organic matter containing phosphorus and nitrogen. A proposed system for this purpose is shown in Figure 4, which represents a simplification of the work performed by Chen *et al.* [83] and Ehimen *et al.* [84].

Therefore, using the residual microalgae biomass as a source of biogas is similar to other agricultural residue uses [85] in which the organic substrate is converted into biogas through anaerobic digestion, producing a gas mixture containing a higher percentage of carbon dioxide and methane [86].

The use of microalgae for biomethane production is significant because fermentation exhibits high stability and high conversion rates, which makes the process of bioenergy production more economically viable. For example, Feinberg (1984) (cited in Harun *et al.* [87]) considered exploiting *Tetraselmissuecica* for biomethane production in conjunction with the possibilities of producing other biofuels. The production of the following biofuels were proposed: biomethane alone (using total protein, carbohydrate, and lipids); biomethane and bioethanol (using carbohydrate for bioethanol production and protein and lipids for biomethane production); biomethane and biodiesel (using carbohydrate and protein for biomethane production and lipids for biodiesel production); and biomethane, biodiesel, and biomethanol (using carbohydrate for bioethanol production; lipids for biodiesel production, and proteins for biomethane production).

Harun *et al.* [47] also reported that the main factors influencing the process are the amount of the organic load, the temperature of the medium, the pH, and the retention time in the bioreactors, with long retention periods combined with high organic loads exhibiting greater effectiveness for biomethane production.

Converti *et al.* [82] demonstrated this effect, reporting the increased production of total biogas at 0.39 ± 0.02 m^3 kg^{-1} of dissolved organic carbon after 50 days of maturation and 0.30 ± 0.02 m^3 of biomethane.

When considering total biomass use, in addition to biogas, it is possible to produce biohydrogen and bio-oils using enzymatic and chemical processes.

The chemical processes that can be used for hydrogen production include gasification, partial oxidation of oil, and water electrolysis. In the literature, cyanobacteria are primarily used for the production of biohydrogen through a biological method, and the reaction is catalyzed by nitrogenases and hydrogenases [88]. Studies with *Anabaena* sp. also demonstrate that this biomass is promising for the production of biohydrogen and that adequate levels of air, water, minerals, and light are necessary because the process can be photosynthetic [9,89].

Bio-oil can be produced from any biomass, and for microalgae, a number of investigations have been performed using *Chlamydomonas, Chlorella, Scenedesmus* [90], *Chlorella vulgaris* [91-92], *Scenedesmus dimorphus, Spirulina platensis, Chlorogloeopsis fritschiiwer* [91], *Nannocloropsis oculata* [93], *Chlorella minutissima* [94], and *Dunaliella tertiolecta* [10].

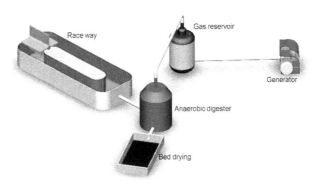

Figure 4. Anaerobic digestion of biomass waste in a unit of bioenergy production associated with an effluent treatment plant.

These initiatives highlight the potential use of hydrothermal liquefaction, which is a process that converts the biomass into bio-oil at a temperature range of 200-350°C and pressures of 15-20 MPa. According to Biller et al. [91], yields of 27-47% are possible, taking into account that microalgae can be produced using recycled nutrients, providing greater sustainability to the system.

A different bio-oil can be produced using pyrolysis in which the oil composition features compounds exhibiting boiling points lower than the hydrothermal liquefaction product [93]. In pyrolysis, the nitrogen content of the microalgae is converted into NOx during combustion. NOx is an undesirable emission that increases depending on the microalgae and their protein content; however, NOx emissions can be reduced by 42% using a hydrothermal pre-treatment process.

In terms of waste recovery, the use of *Dunaliella tertiolecta* cake under various catalyst dosage conditions, temperatures, and times were used in hydrothermal liquefaction, and the yield was 25.8% using 5% sodium carbonate as catalyst at 360°C [10].

Therefore, in addition to producing microalgae in urban or industrial effluents, it is possible that after the extraction of the oil for biodiesel production and the production of bioethanol from carbohydrates, biogas or bio-oil can be produced from the waste material.

4. Conclusions

This chapter reviews the initiatives for biofuel production from microalgae cultivated in wastewaters. The exploitation of the total microalgae biomass was considered, and the potential for biodiesel and bioethanol production was explored.

The various systems for microalgae production using wastewater and the consequences for biodiesel and bioethanol production were discussed in detail.

Microalgae have been used to produce biodiesel and bioethanol with excellent results; however, the use of microalgae must be expanded to include bioremediation combined with biofuel production. The commercial initiatives for this purpose will depend on the composition and volume of the effluent, on the selected microalgae species, and on the temperature and light conditions of the region. The initiatives will also depend on the particular biofuel of interest to the region or that required for local consumption. Therefore, each situation must be analyzed on an individual basis, and there is no single model; however, because of the wide biodiversity of microalgae and the extensive ongoing research capacity of many countries, it is likely that a conditions for viable microalgae production can be achieved anywhere.

Finally, it should be noted that microalgae that are adapted to the environment could produce biomass that, depending on the composition of cells, can be used as the raw material for the production of one or more biofuels.

The research and development of microalgae production in urban or industrial effluents involve principles of sustainable development, clean technology, and the ecology of the productive sectors, prioritizing preventive and remediation steps with the decreased use of energy and inputs. Therefore, there is an emphasis on the methods of treatment, the transformation processes, and the biotechnological products (biofuels), prioritizing the use of wastewater for biomass and bioenergy production. These developments will decrease the impact on activities of anthropogenic origin from the industrial, commercial and service sectors, among others.

Acknowledgements

The National Council of Technological and Scientific Development (Conselho Nacional de Desenvolvimento Científico e Tecnológico, CNPq), the National Council for the Improvement of Higher Education (Coordenação de Aperfeiçoamento de Pessoal de Nível Superior, CAPES) and the University of Santa Cruz do Sul Research Foundation (Fundo de Apoio à Pesquisa da Universidade de Santa Cruz do Sul, FAP/UNISC)

Author details

Rosana C. S. Schneider, Thiago R. Bjerk, Pablo D. Gressler, Maiara P. Souza, Valeriano A. Corbellini and Eduardo A. Lobo

Environmental Technology Post-Graduation Program, University of Santa Cruz do Sul, UNISC, Brazil

References

[1] Derner, R. B. et al. Microalgae, products and applications (Microalgas, produtos e aplicações). Ciência Rural 2006; 36(6) 1959-1967. http://dx.doi.org/10.1590/S0103-84782006000600050

[2] Richmond, A. Handbook of microalgal culture: biotechnology and applied phycology. Oxford: Blackwell Science, 2004.

[3] Borowitzka, M. A. Commercial production of microalgae: ponds, tanks, tubes and fermenters. Journal of Biotechnology1999; 70,(1-3) 313-321. http://dx.doi.org/ 10.1016/S0168-1656(99)00083-8

[4] Gressler, P., Schneider, R. C S., Corbellini, V. A., Bjerk T., Souza M. P., Zappe A., Lobo, E. A. Microalgas: aplicações em biorremediação e energia. In English: Microalgae: Aplications in bioremediation na energy. Caderno de Pesquisa, Série Biologia, 2012; 24, 1, 48-67. www.bioline.org.br/pdf?cp12004.

[5] Chisti, Y. Biodiesel from microalgae. Biotechnology Advances, v. 25, n. 3, p. 294-306, http://dx.doi.org/10.1016/j.biotechadv.2007.02.001, 2007.

[6] Harun, R. et al. Microalgal biomass as a cellulosic fermentation feedstock for bioethanol production. Renewable and Sustainable Energy Reviews 2010. http://dx.doi.org/10.1016/j.rser.2010.07.071

[7] Jiang, L. et al. Biomass and lipid production of marine microalgae using municipal wastewater and high concentration of CO2. Applied Energy 2011; 88(10) 3336-3341. http://dx.doi.org/ 10.1016/j.apenergy.2011.03.043

[8] Demirbas, M. F. Biofuels from algae for sustainable development. Applied Energy 2011;88(10) 3473-3480. http://dx.doi.org/10.1016/j.apenergy.2011.01.059

[9] Marques A. E. et al. Biohydrogen production by Anabaena sp. PCC 7120 wild-type and mutants under different conditions: Light, nickel, propane, carbon dioxide and nitrogen. Biomass and Bioenergy 2011;35(10) 4426-4434. http://dx.doi.org/10.1016/j.biombioe.2011.08.014

[10] Shuping, Z. et al. Production and characterization of bio-oil from hydrothermal liquefaction of microalgae Dunaliella tertiolecta cake. Energy 2010; 35(12) 5406-5411. http://dx.doi.org/10.1016/j.energy.2010.07.013,

[11] Lopes, E. J. Carbon dioxide sequestration in photobioreactors (Seqüestro de dióxido de carbono em fotobiorreatores). 2007. Thesis. Graduate Program in Chemical Engineering (Programa de Pós-Graduação: Engenharia Química-Mestrado e Doutorado) – School of Chemical Engineering at Campinas State University (Faculdade de Engenharia Química da Universidade Estadual de Campinas), Campinas, 2007.

[12] Brown, M. R. et al. Nutritional aspects of microalgae used in mariculture. Aquaculture 1997;151(1-4) 315-331. http://dx.doi.org/10.1016/S0044-8486(96)01501-3

[13] Spolaore, P. et al. Commercial applications of microalgae. Journal of Bioscience and Bioengineering 2006;101(2) 87-96. http://dx.doi.org/10.1263/jbb.101.87

[14] Hirano, A. et al. Temperature effect on continuous gasification of microalgal biomass: theoretical yield of methanol production and its energy balance. Catalysis Today 1998; 45 (1–4) 399–404. http://dx.doi.org/10.1016/S0920-5861(98)00275-2.

[15] Chen C. et al. Thermogravimetric analysis of microalgae combustion under different oxygen supply concentrations. Applied Energy 2011;88(9) 3189-3196.http://dx.doi.org/10.1016/j.apenergy.2011.03.003

[16] Demirbas, A, FatihDemirbas, M. Importance of algae oil as a source of biodiesel. Energy Conversion and Management 2011;52(1)163-170. http://dx.doi.org/10.1016/j.enconman.2010.06.055

[17] Harmelen, Van T, Oonk H. Microalgae biofixation processes: applications and potential contributions to greenhouse gas mitigation options.TNO:Eni Tecnologie;2006. http://www.fluxfarm.com/uploads/3/1/6/8/3168871/biofixation.pdf (accessed 18 July 2012).

[18] Farhadian, M. et al. In situ bioremediation of monoaromatic pollutants in groundwater: a review. Bioresource Technology 2008;99(13) 5296-308. http://dx.doi.org/10.1016/j.biortech.2007.10.025,

[19] Bashan, L. E. e Bashan, Y. Immobilized microalgae for removing pollutants: review of practical aspects. Bioresource Technology 2010;101(6) 1611-27. http://dx.doi.org/10.1016/j.biortech.2009.09.043,

[20] Park, J. B. K. et al. Wastewater treatment high rate algal ponds for biofuel production. Bioresource Technology 2011;102(1) 35-42. http://dx.doi.org/10.1016/j.biortech.2010.06.158

[21] Christenson, L, Sims, R. Production and harvesting of microalgae for wastewater treatment, biofuels, and bioproducts. Biotechnology Advances 2011;29(6) 686-702. http://dx.doi.org/10.1016/j.biotechadv.2011.05.015.

[22] Gattullo, C. E. et al. Removal of bisphenol A by the freshwater green alga Monoraphidiumbraunii and the role of natural organic matter. The Science of the Total Environment 2012; 416, 501-506. http://dx.doi.org/10.1016/j.biotechadv.2011.05.015

[23] Masroor, M. et al. An Overview of the Integration of Advanced Oxidation Technologies and Other Processes for Water and Wastewater Treatment. International Journal of Engineering 2009;3(2)120-146.

[24] Fuchs, W. et al. Influence of standard wastewater parameters and pre-flocculation on the fouling capacity during dead end membrane filtration of wastewater treatment effluents. Separation and Purification Technology2006;52(1)46-52. http://dx.doi.org/10.1016/j.seppur.2006.03.013.

[25] Chang, I.-S, Kim, S.-N. Wastewater treatment using membrane filtration—effect of biosolids concentration on cake resistance. Process Biochemistry 2005; 40(3-4) 1307-1314. http://dx.doi.org/10.1016/j.procbio.2004.06.019

[26] Almeida, E. et al. Industrial effluent treatment via oxidation in the presence of ozone. (Tratamento de efluentes industriais por processos oxidativos na presença de ozônio.) Quim. Nova 2004; 279(5) 818-824. http://dx.doi.org/10.1590/S0100-40422004000500023

[27] Daly, R. I. et al. Effect of chlorination on Microcystisaeruginosa cell integrity and subsequent microcystin release and degradation. Environmental Science & Technology 2007;41(12) 4447-4453.

[28] Caron, D. A et al. Harmful algae and their potential impacts on desalination operations off southern California. Water research 2010,44;(2) 385-416. http://dx.doi.org/10.1016/j.watres.2009.06.051, 2010

[29] Chinnasamy S, Bhatnagar A, Claxton R,. Biomass and bioenergy production potential of microalgae consortium in open and closed bioreactors using untreated carpet industry effluent as growth médium. Bioresource Technology 2010;101(17) 6751–6760. http://dx.doi.org/doi:10.1016/j.biortech.2010.03.094

[30] Brennan, L, Owende, P. Biofuels from microalgae—A review of technologies for production, processing, and extractions of biofuels and co-products. Renewable and Sustainable Energy Reviews 2010;(14)2 557-577. http://dx.doi.org/10.1016/j.rser.2009.10.009

[31] Pittman, J. K. et al. The potential of sustainable algal biofuel production using wastewater resources. Bioresource Technology 2010;102(1) 17-25. http://dx.doi.org/10.1016/j.biortech.2010.06.035

[32] Mulbry, W. et al. Toward scrubbing the bay: Nutrient removal using small algal turf scrubbers on Chesapeake Bay tributaries. Ecological Engineering 2010;36(4), 536-541. http://dx.doi.org/10.1016/j.ecoleng.2009.11.026

[33] Chiu, S.-Y. et al. Reduction of CO_2 by a high-density culture of Chlorella sp. in a semicontinuousphotobioreactor. Bioresource Technology 2008;99(9) 3389-3396. http://dx.doi.org/10.1016/j.biortech.2007.08.013

[34] Chiu, S.-Y. et al. The air-lift photobioreactors with flow patterning for high-density cultures of microalgae and carbon dioxide removal. Engineering in Life Sciences 2009;9(3) 254-260. http://dx.doi.org/10.1002/elsc.200800113

[35] Muñoz, R. et al. Biofilm photobioreactors for the treatment of industrial wastewaters. Journal of Hazardous Materials 2009;161(1) 29-34. http://dx.doi.org/10.1016/j.jhazmat.2008.03.018

[36] Bjerk, T. R. Microalgae cultive in fotobiorreator and joint reactor objectiving the bioremediation and production of biofuels (In Portuguese: Cultivo de microalgasemfo-

tobiorreator e reatormistovisando a biorremediação e produção de biocombustíveis). Universidade de Santa Cruz do Sul, Santa Cruz do Sul-Rio Grande do Sul. 2012.

[37] Gressler, P. D. Efficiency Evaluation of the Desmodesmus subspicatus (R. Chodat) E. Hegewald & A. Schmidt (Chlorophyceae) to grown in tubular photobioreactor with effluent from ETE-Unisc for bioremediation and energy generation (In Portuguese: Avaliação da eficiência de Desmodesmus subspicatus (R. Chodat) E. Hegewald & A. Schmidt (Chlorophyceae) cultivada em Fotobiorreator tubular com efluente da ETE-unisc, visando biorremediação e obtenção de energia). Universidade de Santa Cruz do Sul, Santa Cruz do Sul-Rio Grande do Sul. 2011.

[38] Ai, W. et al. Development of a ground-based space micro-algae photo-bioreactor. Advances in Space Research2008;41(5) 742–747. http://dx.doi.org/10.1016/j.asr. 2007.06.060.

[39] Krichnavaruk, S. et al. Optimal growth conditions and the cultivation of Chaetocero-scalcitrans in airlift photobioreactor. Chemical Engineering Journal 2005;105(3) 91-98. http://dx.doi.org/10.1016/j.cej.2004.10.002

[40] Vasumathi K.K. et al. Parameters influencing the design of photobioreactor for the growth of microalgae. Renewable and Sustainable Energy Reviews 2012;16(7) 5443-5450. http://dx.doi.org/10.1016/j.rser.2012.06.013

[41] Demirbas, A. Use of algae as biofuel sources. Energy Conversion and Management 2010; 51(12) 2738-2749. http://dx.doi.org/10.1016/j.enconman.2010.06.010.

[42] Xin, L. et al. Lipid accumulation and nutrient removal properties of a newly isolated freshwater microalga, Scenedesmus sp. LX1, growing in secondary effluent. New Biotechnology 2010;27(1) 59-63. http://dx.doi.org/10.1016/j.nbt.2009.11.006

[43] Fierro, S. et al. Nitrate and phosphate removal by chitosan immobilized Scenedesmus. Bioresource Technology 2008;99(5) 1274-1279. http://dx.doi.org/10.1016/j.biortech.2007.02.043

[44] Wu L,F. et al. The feasibility of biodiesel production by microalgae using industrial wastewater. Bioresource Technology 2012, 113 :14-18 .

[45] Ruiz-Marin, A. et al. Growth and nutrient removal in free and immobilized green algae in batch and semi-continuous cultures treating real wastewater. Bioresource Technology 2010;101(1) 58-64. http://dx.doi.org/10.1016/j.biortech.2009.02.076

[46] Martínez, M. E. et al. Nitrogen and phosphorus removal from urban wastewater by the microalga Scenedesmusobliquus. Bioresource Technology 2000;73(3) 263-272. http://dx.doi.org/10.1016/S0960-8524(99)00121-2

[47] Harun R. et al. Bioprocess engineering of microalgae to produce a variety of consumer products. Renewable and Sustainable Energy Reviews 2010; 14(3) 1037–1047. http://dx.doi.org/10.1016/j.rser.2009.11.004.

[48] Sheehan, J. et al. A Look Back at the US Department of Energy's Aquatic Species Program: biodiesel from algae. US Department of Energy's, Office of Fuels development:

NREL/TP-580-24190.1998. http://www.nrel.gov/docs/legosti/fy98/24190.pdf (accessed 01 august 2012).

[49] Soares, D. Assessmentof cell growthand lipids productivity ofmarine microalgaein differentcroppingsystems (In Portuguese: Avaliação do crescimentocelular e da produtividade de lipideos de microalgasmarinhasemdiferentes regimes de cultivo. 2010). 107 f. Thesis. Graduate Program in Biochemical Sciences. (Curso de Pós-Graduação em Ciências: Bioquímica – Mestrado e Doutorado) – Federal University of Parana (Universidade Federal do Paraná), Curitiba, 2010.

[50] Song, D. et al. Exploitation of oil-bearing microalgae for biodiesel. Chinese Journal of Biotechnology 2008; 24(3) 341-348. http://dx.doi.org/10.1016/S1872-2075(08)60016-3

[51] Qin J. Bio-hydrocarbons from algae—impacts of temperature, light and salinity on algae growth. Barton, Australia: Rural Industries Research and Development Corporation; 2005.

[52] MATA, T. M. et al. Microalgae for biodiesel production and other applications: A review. Renewable and Sustainable Energy Reviews2010;14(1) 217-232. http://dx.doi.org/10.1016/j.rser.2009.07.020

[53] Radmann, E. M., Costa, J. A. V. Lipid content and fatty acid composition of microalgae exposed to CO_2, SO_2 and NO (Conteúdo lipídico e composição de ácidos graxos de microalgas expostas aos gases CO_2, SO_2 e NO). Química Nova 2008;31(7) 1609–1612. http://dx.doi.org/10.1590/S0100-40422008000700002.

[54] Bicudo, C., E., M., Menezes, M. Genus of algae from inland waters of Brazil, identification and description key (Gênero de algas de águas continentais do Brasil, chave para identificação e descrição). 2. ed. São Carlos-SP: RIMA, 2006.

[55] Porte A. F, Schneider R. C. S, Kaercher J. A, Klamt R. A, Schmatz W. L, Silva W. L. T, Filho W. A. S. Sunflower biodiesel production and application in family farms in Brazil. Fuel 2010; 89(12) 3718–3724. http://dx.doi.org/10.1016/j.fuel.2010.07.025.

[56] Gouveia L, Oliveira A. C. Microalgae as a raw material for biofuels Production. J IndMicrobiolBiotechnol 2009;36 269–274. http://dx.doi.org/ 10.1007/s10295-008-0495-6.

[57] Guschina, I. A , Harwood, J. L. Lipids and lipid metabolism in eukaryotic algae. Progress in Lipid Research 2006;45(2) 160-186. http://dx.doi.org/ 10.1016/j.plipres.2006.01.001, 2006.

[58] Doan, T. T. Y. et al. Screening of marine microalgae for biodiesel feedstock. Biomass and Bioenergy 2011;35(7) 2534-2544. http://dx.doi.org/10.1016/j.biombioe.2011.02.021.

[59] Khan, S. A. et al. Prospects of biodiesel production from microalgae in India. Renewable and Sustainable Energy Reviews 2009;13(9) 2361–2372. http://dx.doi.org/10.1016/ j.rser.2009.04.005

[60] Kaiwan-arporn P, Hai P. D, Thu N. T, Annachhatre A. P. Cultivation of cyanobacte-
      ria for extraction of lipids. Biomass and Bioenergy 2012; 44 142-149. http://dx.doi.org/
      10.1016/j.biombioe.2012.04.017

[61] John R. P, Anisha G.S, Nampoothiri K. M, Pandey A. Micro and macroalgal biomass:
      A renewable source for bioethanol. Bioresource Technology2011; 102(1)
      186-193.http://dx.doi.org/10.1016/j.biortech.2010.06.139

[62] Harun, R, Danquah, M. K. Enzymatic hydrolysis of microalgae biomass for bioetha-
      nol production. Chemical Engineering Journal 2011b;168(3) 1079-1084. http://
      dx.doi.org/10.1016/j.cej.2011.01.088.

[63] Harun, R, Danquah, M. K. Influence of acid pre-treatment on microalgal biomass for
      bioethanol production. Process Biochemistry 2011a;46(1) 304-309. http://dx.doi.org/
      10.1016/j.procbio.2010.08.027

[64] Balat, M. Production of bioethanol from lignocellulosic materials via the biochemical
      pathway: A review. Energy Conversion and Management 2010;52(2) 858-875. http://
      dx.doi.org/10.1016/j.enconman.2010.08.013

[65] Goh, C. S, Lee, K. T. A visionary and conceptual macroalgae-based third-generation
      bioethanol (TGB) biorefinery in Sabah, Malaysia. as an underlay for renewable and
      sustainable development. Renewable and Sustainable Energy Reviews 2010;14(2)
      842-848. http://dx.doi.org/10.1016/j.rser.2009.10.001

[66] Nguyen MT, Choi SP, Lee J, Lee JH, S. S. Hydrothermal acid pretreatment of Chla-
      mydomonasreinhardtii biomass for ethanol production. J MicrobiolBiotechnol.
      2009;19(2) 161-166. http://dx.doi.org/10.4014/jmb.0810.578

[67] Harun, R., Jason, W. S. Y., Cherrington, T., &Danquah, M. K. Exploring alkaline pre-
      treatment of microalgal biomass for bioethanol production. Applied Energy
      2010;88(10) 3464-3467. http://dx.doi.org/10.1016/j.apenergy.2010.10.048

[68] Harun R, Liu B,Danquah M. K. Analysis of Process Configurations for Bioethanol
      Production from Microalgal Biomass. Progress in Biomass and Bioenergy Produc-
      tion:In Tech;2011. DOI: 10.5772/17468

[69] Yazdani, P., Karimi, K., &Taherzadeh, M. J. Improvement of Enzymatic Hydrolysis
      of A Marine Macro-Alga by Dilute Acid Hydrolysis Pretreatment. Bioenergy Tech-
      nology 2011, 186-191. . http://dx.doi.org/10.3384/ecp11057186

[70] Khambhaty, Y., Mody, K., Gandhi, M. R., Thampy, S., Maiti, P., Brahmbhatt, H., Es-
      waran, K., et al. (). Kappaphycusalvarezii as a source of bioethanol. Bioresource
      Technology2012;103(1) 180-185. . http://dx.doi.org/10.1016/j.biortech.2011.10.015

[71] Miranda, J. R., Passarinho, P. C., &Gouveia, L. Pre-treatment optimization of Scene-
      desmusobliquus microalga for bioethanol production. Bioresource Technology
      2012;104 342-8. http://dx.doi.org/doi:10.1016/j.biortech.2011.10.059

[72] Eshaq, F. S.; Ali, M. N.; Mohd, M. K. Spirogyra biomass a renewable source for biofuel (bioethanol) production. International Journal of Engineering Science and Technology 2010;2(12)7045-7054.

[73] Mushlihah, S., Sunarto, E., Irvansyah, M. Y., &Utami, R. S. Ethanol Production from Algae Spirogyra with Fermentation by Zymomonasmobilis and Saccharomyces cerevisiae. J. Basic. Appl. Sci. Res 2011;1(7), 589-593.

[74] Chen, R. et al. Use of an algal hydrolysate to improve enzymatic hydrolysis of lignocellulose. Bioresource Technology 2012;108(1) 149-154. http://dx.doi.org/10.1016/j.biortech.2011.12.143

[75] Yoon, M. et al. Improvement of saccharification process for bioethanol production from Undaria sp. by gamma irradiation. Radiation Physics and Chemistry 2012;81(8) 999-1002. http://dx.doi.org/10.1016/j.radphyschem.2011.11.035

[76] Elliott D.C., Sealock Jr. L.J., Chemical processing in high-pressure aqueous environments: Low temperature catalytic gasification. Chemical Engineering Research and Design 1996; 74(5): 563-566.

[77] Tsukahara K., Sawayama S. Liquid fuel production using microalgae. Jpn Petrol Inst 2005; 48(5) 251–259.

[78] Miao X. et al. Fast pyrolysis of microalgae to produce renewable fuels. J Anal Appl Pyrolysis 2004; 71(2) 855–863.

[79] Amin S. Review on biofuel oil and gas production processes from microalgae. Energy Conversion and Management 2009; 50(7) 1834-1840. http://dx.doi.org/10.1016/j.enconman.2009.03.001.

[80] Bently J.; Derby R. Ethanol. fuel cells: converging paths of opportunity. Renewable-FuelsAssociation 2008; http://www.ethanolrfa.org/objects/documents/129/rfa_fuel_cell_white_paper.pdf>. (Accessed 24 July, 2012.

[81] Xu, H. et al. High quality biodiesel production from a microalga Chlorella protothecoides by heterotrophic growth in fermenters. Journal of Biotechnology 2006;126(4) 499–507. http://dx.doi.org/10.1016/j.jbiotec.2006.05.002

[82] Converti, A. et al. Effect of temperature and nitrogen concentration on the growth and lipid content of Nannochloropsisoculata and Chlorella vulgaris for biodiesel production. Chemical Engineering and Processing: Process Intensification 2009;48(6) 1146-1151. http://dx.doi.org/10.1016/j.cep.2009.03.006

[83] Chen S., Chen B., Song D. Life-cycle energy production and emissions mitigation by comprehensive biogas–digestate utilization Bioresource Technology 2012; 114: 357-364. http://dx.doi.org/10.1016/j.biortech.2012.03.084.

[84] Ehimen E.A., Holm-Nielsen J.-B., Poulsen M., Boelsmand J.E. Influence of different pre-treatment routes on the anaerobic digestion of a filamentous algae. Renewable Energy 2013; 50: 476-480. http://dx.doi.org/10.1016/j.renene.2012.06.064.

[85] Bernard, O. Hurdles and challenges for modelling and control of microalgae for CO_2 mitigation and biofuel production. Journal of Process Control 2011; 21(10): 1378-1389. doi:10.1016/j.jprocont.2011.07.012

[86] Abdel-Raouf N., Al-Homaidan A.A., Ibraheem I.B.M. Microalgae and wastewater treatment. Saudi Journal of Biological Sciences 2012; 19(3): 257-275. http://dx.doi.org/10.1016/j.sjbs.2012.04.005

[87] Harun, R. et al. Technoeconomic analysis of an integrated microalgae photobioreactor, biodiesel and biogas production facility. Biomass and Bioenergy 2010; 35(1) 741-747. http://dx.doi.org/10.1016/j.biombioe.2010.10.007

[88] Tamagnini P., Leitão E., Oliveira P., Ferreira D., Pinto F., Harris D., Cyanobacterial hydrogenases. Diversity, regulation and applications. FEMS Microbiol Rev 2007; 31:692-720.

[89] Ferreira A.F., Marques A. C., Batista A.P., Marques P. A.S.S., Gouveia L., Silva C. M. Biological hydrogen production by Anabaena sp. e Yield,energy and CO_2 analysis including fermentative biomass recovery. International Journal of Hydrogen Energy 2012; 37: 179 -190. doi:10.1016/j.ijhydene.2011.09.056

[90] Bhatnagar A., Chinnasamy S., Singh M., Das K.C., Renewable biomass production by mixotrophic algae in the presence of various carbon sources and wastewaters. Applied Energy 2011; 88 (10): 3425-3431. http://dx.doi.org/10.1016/j.apenergy.2010.12.064

[91] Biller P., Ross A.B., Skill S.C., Lea-Langton A., Balasundaram B., Hall C., Riley R., Llewellyn C.A. Nutrient recycling of aqueous phase for microalgae cultivation from the hydrothermal liquefaction process Algal Research 2012; 1: 70-76. http://dx.doi.org/10.1016/j.algal.2012.02.002

[92] Tsukahara K., Kimura T., Minowa T., Sawayama S., Yagishita T., Inoue S., Hanaoka T., Usui Y., Ogi T., Microalgal cultivation in a solution recovered from the low-temperature catalytic gasification of the microalga. Journal of Bioscience and Bioengineering 2001; 91 (3) 311-313. http://dx.doi.org/10.1016/S1389-1723(01)80140-7

[93] Du Z., Mohr M., Ma X., Cheng Y., Lin X., Liu Y., Zhou W., Chen P., Ruan R. Hydrothermal pretreatment of microalgae for production of pyrolytic bio-oil with a low nitrogen content. Bioresource Technology 2012; 120: 13-18 http://dx.doi.org/10.1016/j.biortech.2012.06.007.

[94] Jena U., Vaidyanathan N., Chinnasamy S., Das K.C., Evaluation of microalgae cultivation using recovered aqueous co-product from thermochemical liquefaction of algal biomass, Bioresource Technology 2011; 102 (3): 3380–3387. doi:10.1016/j.biortech.2010.09.111

Chapter 5

Major Diseases of the Biofuel Plant, Physic Nut (*Jatropha curcas*)

Alexandre Reis Machado and Olinto Liparini Pereira

Additional information is available at the end of the chapter

1. Introduction

Worldwide, concern over the consequences of global warming has resulted in intensified searches for potential plants that couldsupply raw materials for producing renewable fuels. Therein, physic nut (*Jatropha curcas* L.) has gained attention as a perennial culture that produces seeds with high oil content and excellent properties. In addition to these attributes, many studies have describedphysic nut as a culture resistant to pests and disease. However, in recent years, the expansion of areas under cultivation has been accompanied by the appearance of various diseases. Thus, this chapter aims to provide information about the main diseases that occur in physic nut and their diagnosis and to encourage further research on disease control.

The existing literature contains various descriptions of the pathogens occurring in culture, most of which are caused by fungi, and of which we address the following: *Glomerella cingulata* (Ston.) Spauld. et Schrenk.;*Psathyrella subcorticalis* Speg.;*Schizophyllum alneum* L.;*Aecidium cnidscoli* P. Henn.; *Ramulariopsis cnidoscoli* Speg.;*Uromyces jatrophicola* P. Henn. (Viégas 1961);*Pestalotiopsis versicolor* Speg.(Phillips 1975);*Colletotrichum gloeosporioides* (Penz.)Sacc.; *Colletotrichum capsici* (Syd.) Butl.e Bisby.;*Passalora ajrekari* (Syd.) U. Braun (Freire & Parente 2006);*Phakopsora arthuriana* Buriticá & J.F. Hennen (Hennen et al. 2005);*Cochliobolus spicifer* Nelson (Mendes et al. 1998);*Cercospora jatrophicola* (Speg.) Chupp;*Cercospora jatrophigena*U. Braun; Pseudocercospora jatrophae-curcas (J.M. Yen) Deighton; Pseudocercospora jatrophae; Pseudocercospora jatropharum (Speg.) U. Braun (Crous & Braun 2003); and Elsinoë ja-trophae Bitanc. & Jenkins (Bitancourt & Jenkins 1951). Existing reports on pathogens include research on collar and root rot Nectria haematococca Berk. & Br. [Haematonectria haemato-cocca (Berk. & Broome) Samuels & Nirenberg], and its anamorph Fusarium solani (Martius) Appel & Wollenweber (Yue-kai et al. 2011), as well as Lasiodiplodia theobromae (Pat.) Griffon & Maubl (Latha et al. 2009; Pereira et al. 2009), Phytophthora palmivora var. palmivora (E.J. Butler) E.J. Butler (Erwin & Ribeiro 1996) and*Clitocybe tabescens* (Scop, ex Fr.) Bres (US-DA 1960).

2. Diseases

Although several descriptions of fungi exist, this chapter will discuss the most common and damaging diseases that affect physic nut, and draws on the following descriptions:

2.1. Anthracnose (figure 1)

Colletotrichum gloeosporioides (Penz.)Sacc.

Colletotrichum capsici (Syd.)Butl.and Bisby

This disease was first described in physic nut by the USDA (1960) in the USA, in Brazil by Viégas (1961), and later by Freire & Parente (2006) and Sá et al. (2011).Currently, the disease is present in all areas where physic nut is cultivated.

The most commonly observed symptoms are brown to black necrotic lesions that are irregularly shaped and appear on the edges and center of the leaf and which may contain a yellow halo. The lesions appear in the form of small, isolated points that coalesce and subsequently cause the complete destruction of the leaves. The fruit can also become infected, which leads to the appearance of dark brown lesions.

In addition to these symptoms, research in Mexico has indicated that the fungus *Colletotrichum capsici* caused stem canker and apical death of seedlings (Torres-Calzada et al. 2011).

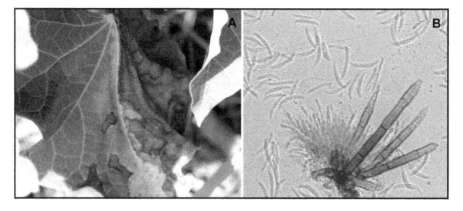

Figure 1. Anthracnose in *Jatropha curcas*. Symptoms on leaf (A). Curved conidia, dense conidiophores and septate setae of *Colletotrichum capsici* (B).

Colletotrichum is a fungus anamorph of the phylum Ascomycota and telemorph genus *Glomerella*. The species of this genus have the following characteristics: conidiomata that are acervular, subcuticular or epidermal, and may contain setae; conidiophores that are hyaline to brown; conidiogenous cells that are enteroblastic, phialidic and hyaline; conidia that are hyaline, aseptate (except prior to germination), straight or falcate, smooth and thin-walled; and appressoria that are brown, entirely or with crenate to irregular margins produced with germination of conidia (Sutton 1980).

Colletotrichum spp. is known to infect a large range of hosts and to cause various symptoms, the most common of which is anthracnose. This fungus can survive in seeds, crop residues, infected plants, and in soil as saprophytes. Although the disease occurs in various regions of the world, it is more severe in regions with a hot and humid climate (Agrios 2005).

So far, there are no recommendations for controlling this disease. Because of the damage it can cause to physic nut, this disease should be studied further.

2.2. Passalora leaf spot

Passalora ajrekari (Syd.) U. Braun

Passalora jatrophigena U. Braun & F.O. Freire

This disease was first described in Brazil by Braun & Freire (2004), and later by Freire & Parente (2006) in leaves of *Jatropha curcas* and *Jatropha podagrica*, and in others countries by Crous & Braun (2003).

The primary symptoms of this disease are rounded leaf lesions that are creamy to light brown in color, with a narrow dark brown halo, and later become limited by leaf veins and darken. Lesions measure 1-2 cm in diameter and rarely coalesce (Freire & Parente 2006).

The genus *Passalora* is a cercosporoid fungus, previously included in the genus *Cercospora* that has as its teleomorph the *Mycosphaerella*. Species share taxonomic characteristics such as branched, septate, smooth, hyaline to pigmentedhyphae; absent to well-developed stromata; solitary or fasciculate to synnematous conidiomataconidiophores, arising from stromata or hyphae, internal or superficial, pluriseptate, subhyaline to pigmented; conspicuous conidiogenous loci, with scars that are somewhat thickened and darkened; conidia that are solitary to catenate in simple or branched chains, amerosporous to scolecosporous, aseptate to pluriseptate, and pale to distinctly pigmented and hila that are somewhat thickened and darkened (Crous & Braun 2003).

Although it has been reported in several countries, to date this disease has not presented risk to physic nut cultivation.

2.3. Cercospora/Pseudocercosporaleaf spot (figure 2-3)

Cercospora jatrophicola (Speg.) Chupp,

Cercospora jatrophigena U. Braun

Pseudocercospora jatrophae-curcas (J.M. Yen) Deighton

Pseudocercospora jatrophae (*G.F. Atk.*) *A.K. Das & Chattopadh.*

Pseudocercospora jatropharum (Speg.) U. Braun

This disease manifests in the form of leaf spots that consist of well-delimited brown irregular necrotic spots (Dianese et al. 2010).

The genera mentioned above have the following taxonomic characteristics:

The genus *Cercospora* groups anamorphs of *Mycosphaerella* with hyphae that are colorless or near-colorous to pigmented, branched, septate, and smooth to faintly rough-walls. Stromata are lacking to well-developed, subhyaline to usually pigmented. Conidiophores are solitary to fasciculate, arising from internal hyphae or stromata, erect, subhyaline to pigmented. Conidiogenous loci (scars) are conspicuous, thickened and darkened. Conidia are solitary, scolecosporous, cylindrical-filiform, hyaline or subhyaline, mostly pluriseptate, and smooth and hila are thickened and darkened (Crous & Brown 2003).

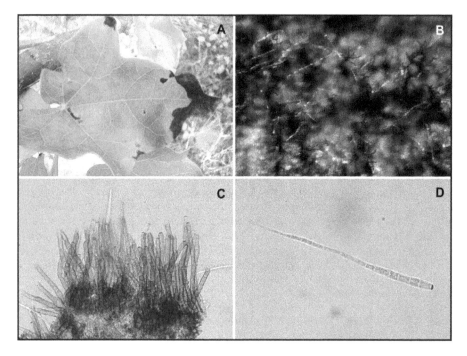

Figure 2. Cercospora leaf spot on *Jatropha curcas*. Necrotic symptoms on leaf (A); Fungal structures on leaf lesions (B); Pigmented conidiophores with conspicuous scars (C); Filiform conidia with conspicuous pigmented hilum (D).

The genus *Pseudocercospora* groups anamorphs of *Mycosphaerella* with basically pigmented conidiophores and inconspicuous, unthickened, not darkened conidiogenous loci; solitary, or catenulateconidia, aseptate to pluriseptate with conidial scars that are inconspicuous and not thickened (Crous & Brown 2003).

Crous & Braun (2003) cite the occurrence of five species of cercosporoid, indicated above, in the culture of physic nut. However, few studies have examined fungi cercosporoid in this crop. As a result, there is no information about favorable conditions, symptoms or disease control. To date, this disease has not presented risk to the cultivation of physic nut.

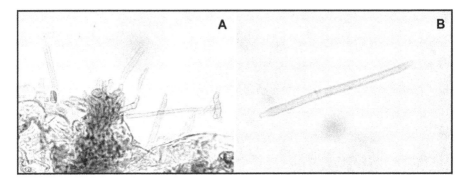

Figure 3. Pseudocercospora leaf spot on *Jatropha curcas*. Pigmented conidiophores with inconspicuous scars on coni-diogenous cells (C) and filiform pigmented conidia with inconspicuous hilum (D).

2.4. Powdery mildew (figure 4)

Pseudoidium jatrophae (Hosag., Siddappa, Vijay. & Udaiyan) U. Braun & R.T.A. Cook

The powdery mildew caused by the fungus *Pseudoidium jatrophae* (Braun & Cook 2012) was previously described as *Oidium heveae* Steim by Viégas (1961) in Braziland *Oidium jatrophae* Hosag., Siddappa, Vijay. & Udaiyan (Braun & Cook 2012)in India. This disease occurs commonly in physic nut plantations and it has been frequently observed in various regions of Brazil and the rest of world.

The most common symptoms of the disease are the production of abundant white or gray mycelia in leaves, petioles, stems, flowers and fruits (Dianese & Cargnin 2008). With the evolution of the disease, infected plants may show necrotic lesions, which cause leaf fall, underdevelopment, death of buds and young fruit deformation (Bedendo 2011).

The fungus that causes this disease is a typical biotrophic pathogen of the phylum Ascomycota, order Erysiphales. This pathogen may be characterized by white or grayish colonies, septate and branched mycelia; conidiophores that are erect or ascending, cylindrical, hyaline, septate and forming conidia singly; conidia that are usually large in proportion to the diameter of the conidiophores, simple, smooth, ellipsoid-ovoid doliiform, hyaline, single-celled (Braun & Cook 2012).

Figure 4. Powdery mildew on *Jatropha curcas*. Symptoms on petiole and stem (A);Symptoms on leaf (B) Leaf lesions on old infections (C); Symptoms on seedlings (D);Conidiophores (E); Conidia (F).

The disease generally favors warm temperatures, humidity of 75-80% and reduced light. Heavy rains are generally unfavorable to the pathogen (Furtado &Trindade 2005). In Brazil, the disease usually occurs in the dry season, apparently without causing extensive losses, because its occurrence coincides with the plants' period of natural defoliation (Saturnino et al. 2005).

Currently, there are no fungicides recommended for culture, but some studies cite that spraying sulfur fungicides works to control this fungus. Another measure is to control alternative hosts, especially plants of the family Euphorbiaceae (Furtado & Trindade 2005; Saturnino et al. 2005; Dias et al. 2007).

2.5. Rust (figure 5)

Phakopsora arthuriana Buriticá & Hennen

The first report of this disease in *Jatropha curcas*, described its cause as *Uredo jatrophicola* Arthur(Arthur 1915). In Brazil, this disease was first found in 1936 in São Paulo (Viégas 1945). Currently, it is widely distributed throughout Brazil (Dias et al. 2007) and several other countries.

The fungus that causes this disease was previously classified as *Phakopsora jatrophicola* (Arthur) Cummins; however, it was reclassified as *Phakopsora arthuriana* Buriticá & Hennen (Hennen et al. 2005).

The symptoms manifest in the leaves, initially in the form of small chlorotic points on the upper surface, which correspond to the underside of the leaf, and then small protruding pustules, which after breaking, release a powdery mass of uredospores of orange color, giving a ferruginous aspect. In severe infections, pustules coalesce to form necrotic spots, which are reddish brown and irregularly shaped and can destroy the leaf (Dias et al. 2007; Carneiro et al. 2009).

The *Phakopsora arthuriana* belongs to the phylum Basidiomycota, class Pucciniomycetes. It is characterized by uredinia hypophyllous, occasionally epiphyllous, in small groups opening by a pore, surrounded by numerous not septate paraphyses that project outside the host; urediniospores, ellipsoid, to obovoid, sessile, closely and finely echinulate, germ pores obscure; telia hypophyllous, suberpidermal in origin, closely around the uredinia; teliospores irregularly arranged, cuboid, ellipsoid to polygonal (Hennen et al. 2005).

Currently there are no fungicides recommended for this culture. However, according to Dias et al. (2007), protective copper fungicides can control this disease.

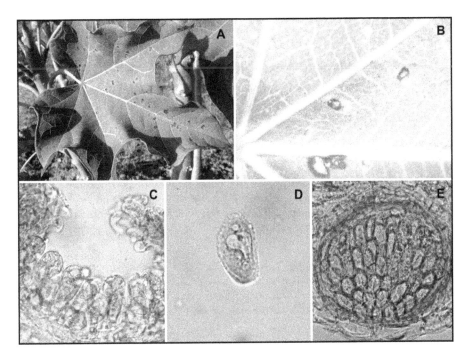

Figure 5. Rust disease on *Jatropha curcas*. Symptoms on adaxial leaf surfaces (A-B);Uredinia (C);Urediniospores (D); Telia with teliospores (E).

2.6. Stem canker and dieback (figure 6)

Lasiodiplodia theobromae(Pat.) Griffon & Maubl

The first report of this disease in Brazil was made byFreire & Parente (2006) and in Malaysia by Sulaiman & Thanarajoo (2012).

The disease manifests in the form of dieback that can progress until it takes over the trunk of the plant. Stem cankers have also been observed, causing necrotic lesions on branches and vascular discoloration. In Malaysia, disease incidence can be as high as 80% of a plantation (Freire &Parente 2006; Sulaiman & Thanarajoo 2012).

Figure 6. Dieback on *Jatropha curcas*. Symptoms observed in the field (A); Hyaline and pigmented conidia of *Lasiodiplodia theobromae* (B).

Characteristics of the *Lasiodiplodia* species commonly include the presence of paraphyses within the conidiomata pycnidial and conidia that are initially hyaline and aseptate. But in maturity, one median septum is formed, and the walls become dark brown with the formation of longitudinal striations due the deposition of melanin granules on the inner surface of the wall.

The identification of the *Lasiodiplodia* species based solely on morphological characteristics is not easy. Currently, it is known that what was initially identified as *Lasiodiplodia theobromae* is in fact a species complex (Alves et al. 2008). Thus, molecular studies are needed to correctlyidentify the pathogen, as was done by Thanarajoo & Sulaiman (2012).

Lasiodiplodia spp. is a fungus of the phylum Ascomycota, family Botryosphaeriaceae. Fungi in this family are known to survive as endophytes and demonstrate symptoms when plants are under some stress (Slippers & Wingfield 2007). Thus, many researchers see them as opportunistic pathogens.

Control of this disease can be achieved by pruning and destroying affected branches. Later plants should be brushed with copper fungicides or thiophanate methyl for injuries (Furtado & Trindade 2005). Additionally, balanced fertilization, soil analysis and sufficient levels of irrigation in regions with long periods of drought can aid in disease control.

2.7. Collar and root rot (figure 7)

Fusarium solani(Martius) Appel & Wollenweber

Lasiodiplodia theobromae(Pat.) Griffon & Maubl

Neoscytalidium dimidiatum(Penz.) Crous & Slippers

Macrophomina phaseolina(Tassi) Goid.

The first report of this disease in Brazil was made by Pereira et al. (2009), who identified it as being caused by *Lasiodiplodia theobromae*. In India, this same pathogen was reported by Latha et al. (2009), and *Macrophomina phaseolina*was reported by Patel et al. (2008). In China, Yue-Kai et al. (2011) identified the fungus *Fusarium solani*, and Machado et al. (in press) made the first description of *Neoscytalidium dimidiatum(Penz.). Crous & Slippers associated this pathogen with collar and root rot in physic nut in Brazil.*

All the pathogens mentioned above are typical soil fungi. They occur in a wide range of hosts, can be spread by seeds, and survive as parasites, saprophytes, endophytes or resistant structures, such as chlamydospores in Fusarium and Neoscytalidium or sclerotia in Macrophomina.

This disease has acquired great importance, because it can reduce productivity by causing the sudden death of plants and making cultivation areas unviable. The symptoms most commonly observed are wilting, leaf yellowing with subsequent fall, and cracks in the collar region. In the collar region, the appearance of black fungal structures in the bark of the plant has been consistently observed. Upon being removed from the soil, plant roots rot and the vascular system is affected by necrotic symptoms, ranging from light brown to black. Due to loss of support, the plants have often already fallen due to the wind.

The genus *Fusarium*has the following general characteristics: bright aerial mycelium, hyphae septate, conidiophores variable, single or grouped in sporodochia; conidia hyaline variable, principally of two kinds -multicellularmacroconidia, slightly curved or bent at the pointed ends and typically canoe-shaped; unicellular, ovoid or oblongmicroconidia, borne singly or in chains and also grouped in false heads, formed in mono or polyphialides; thick-walled chlamydospores are common in some species (Barnett & Hunter 1998).

The common characteristics of *Lasiodiplodia* species include the presence of paraphyses within the conidiomata pycnidial and initially hyaline and aseptateconidia. However, in maturity, one median septum is formed, and the walls become dark brown with the formation of longitudinal striations, due the deposition of melanin granules on the inner surface of the wall.

The genus *Neoscytalidium*is a group of fungi that produces synanamorph *Scytalidium*-like with septate and oblong to globosearthroconidia formed from aerial mycelia. Initially hyaline, with age, the arthroconidia become brown and with a thick wall. Commonly observed are pycnidia that are dark and globose immersed or superficially in a stroma that produces *Fusicoccum*-like conidia that are hyaline and ellipsoid to nearly fusiform. Dark septate conidia can also be observed.

Characteristics of the *Macrophomina* spp. generally include the formation of dark mycelia and abundant production of sclerotia in PDA. Eventually, the formation of conidiomata pycnidial can be observed, with the release of hyaline conidia with apical mucoid appendages.

In areas prone to prolonged dry seasons, a higher incidence of collar and root rot has been observed. Therefore, it is believed that the water stress is the main factor that predisposes plants to this disease.

The above-mentioned pathogens are difficult to control, due to the fact that they survive in soil. Therefore, to reduce disease incidence, it is first necessary to provide water and fertilizer balanced for proper plant development. When transplanting seedlings to the field, all forms of injury should be avoided. Another control measure would be to use healthy propagative material as well as seed treatments.

Figure 7. Collar and root rot on *Jatropha curcas*. Wilting symptoms observed in the field (**A**);Detail of thecollar rot (**B**);Detail ofroot rot (**C**);Macroconidia of *Fusariumsolani*(**D**); Pigmented and hyaline conidia of *Lasiodiplodia theobromae*(**E**);Arthroconidia of *Neoscytalidium dimidiatum*(**F**);*Fusicoccum*-like conidia of *Neoscytalidium dimidiatum*(**G**); Conidia of *Macrophomina phaseolina*(**H**). Sclerotia of *Macrophomina phaseolina* produced on sterilized Pine twigs in culture (**I**).

2.8. Yellow mosaic

In addition to the several fungal diseases mentioned, there is also yellow mosaic, a disease caused by a strain of the virus *Indian Cassava Mosaic Virus* (Gao et al. 2010). This disease, detected in physic nut plantations in India, causes mosaic, reduced leaf size, leaf distortion, blistering and stunting of diseased plants. The disease is transmitted by the vector *Bemisia tabaci* in a non-persistent manner, but not through mechanical inoculation or seeds (Narayanaet al. 2006).

3. Seed associated fungi

Seeds propagate the majority of cultures worldwide. These cultures are vulnerable to infection by several pathogens that can survive in seeds. These pathogens may cause reduction of seed germination, as well as deformation, discoloration, reductions in size and weight, and deterioration during storage. They can further contribute to rotting roots, damping-off, necrosis in leaves, and the spread of diseases across long distances. Consequently, these diseases cause losses valued at billions of dollars (Neergaard 1977; Agarwal & Sinclair 1997). To date, few studies have addressed the seed pathology of physic nut, and there is no information available about the losses that seed pathogens cause in this culture. But, follows below the major pathogens and saprophytic fungi associated with seeds.

Macrophomina phaseolina (Figure 8)

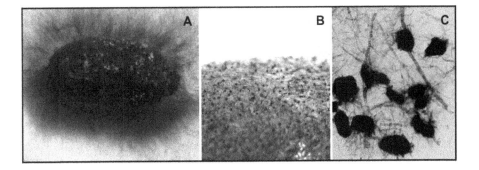

Figure 8. *Macrophomina phaseolina*on *Jatropha curcas* seed. Seed covered by mycelium (**A**); Detail of sclerotia on seed (**B**);Black sclerotia (**C**).

Colletotrichum capsici (Figure 9)

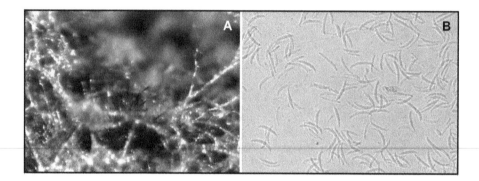

Figure 9. *Colletotrichum capsici* on *Jatropha curcas* seed. Conidiomata with black setae on seed surface (**A**). Curved asseptate conidia (**B**).

Fusarium sp. (Figure 10)

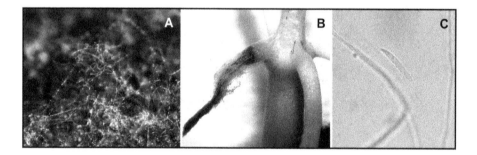

Figure 10. *Fusarium* sp. on *Jatropha curcas* seed. Seed covered by hyaline mycelium (**A**); Radicle with necrotic lesion (**B**); Macroconidia (**C**).

Lasiodiplodia theobromae (Figure 11)

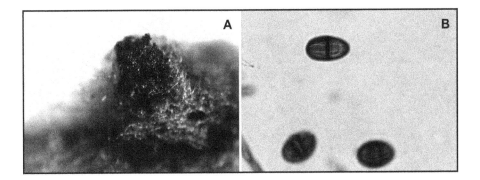

Figure 11. *Lasiodiplodia theobromae* on *Jatropha curcas* seed. Conidiomata producing a black cirrus of conidia on seed surfasse (**A**); Detail of mature conidia (**B**).

Curvularia sp. (Figure 12)

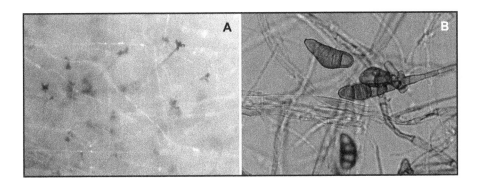

Figure 12. *Curvularia* sp. on *Jatropha curcas* seed. Mycelium and conidiophores producing conidia on seed surface (**A**); Dark septate conidia (**B**).

Other fungi commonly associated with Jatropha curcas seeds (Figure 13)

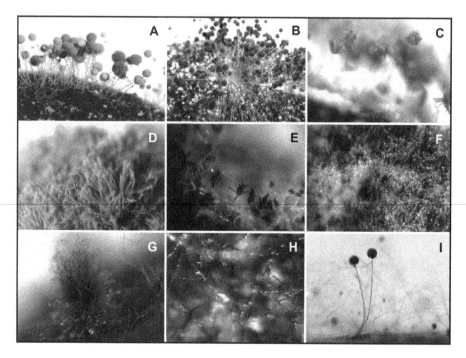

Figure 13. Genera of fungi often observed on Jatropha curcas seeds: Aspergillus (**A-C**); Penicillium (**D**);Stachybotrys (**E**); Acremonium (**F**); Chaetomium (**G**);Alternaria (**H**); Rhizopus (**I**).

Although there are no recommendations for fungicide useon physic nut, treatments can be administered by soaking seeds for 20 minutes in a solution of 1 liter of formaldehyde 40% diluted in 240 liters of water (Massola and Bedendo, 2005). This treatment is indicated for the seeds of *Ricinus communis* L., but it also works well for physic nut.

4. Conclusion

Despite the fact that most literature considered physic nut as resistant to pests and diseases, this review emphasizes the diversity of pathogens associated with this plant and the damage that they may cause. Most of these diseases may become a serious problem for Brazilian farmers, due to its severity and the lack of registered chemical products for these pathogens. Studies should be carried out in order to know the environmental conditions that favor to these diseases on *J. curcas*, as well as the development of control strategies and resistant varieties.

Acknowledgements

The authors wish to thank **FAPEMIG**, **CNPq** and **CAPES** for financial support of the work.

Author details

Alexandre Reis Machado and Olinto Liparini Pereira

*Address all correspondence to: alexandrerm.agro@yahoo.com.br

*Address all correspondence to: oliparini@ufv.br

Departamento de Fitopatologia, Universidade Federal de Viçosa, Viçosa, MG, Brazil

References

[1] Agarwal VK, Sinclair JB(1997). Principles of Seed Pathology. nd edn, Boca Raton, Florida: CRC Press, Inc., 539 pp.

[2] Agrios GN(2005). Plant pathology. Burlington, MA: Elsevier Academic, 922p.

[3] Alves, A., Crous, P. W., Correia, A., & Phillips, A. J. L. (2008). Morphological and mo‐ lecular data reveal cryptic species in *Lasiodiplodia theobromae*. Fungal Diversity , 28, 1-13.

[4] Arthur JC(1915). Uredinales of Porto Rico based on collections by F. L. Steven. s. My‐ cologia , 7, 168-196.

[5] Barnett HL, Hunter BB(1998). Illustrated genera of imperfect fungi. ed. St. Paul, MN: American Phytopatological Society Press, 218p.

[6] Bedendo IP(2011). Oídios.IN: Amorim L, Rezende JAM, Bergamin Filho A. Manual de Fitopatologia: princípios e conceitos. São Paulo: Agronômica Ceres, , 1

[7] Bitancourt AA, Jenkins AE ((1951).) Estudos sôbre as Miriangiales. II. Vinte novas espécies de Elsinoaceas neotropicais. Arquivos do Instituto Biológico ., 20, 1-28.

[8] Braun, U., & Cook, R. T. A. (2012). Taxonomic Manual of the Erysiphales (Powdery Mildews). CBS Biodiversity Series , 11(11), 1-707.

[9] Braun, U., & Freire, F. C. O. (2004). Some cercosporoid hyphomycetes from Brazil-III. Cryptogamie Mycologie, , 25(3), 221-246.

[10] Carneiro, S. M. T. P. G., Ramos, A. L. M., Romano, E., Marianowski, T., & de Oliveira, J. P. (2009). Ocorrência de *Phakopsora jatrophicola* em pinhão manso no estado do Par‐ aná. Summa Phytopathologica

[11]   Crous, P. W., & Braun, U. (2003). *Mycosphaerella* and its anamorphs: 1. Names pub-lished in Cercospora and Passalora. CBS Biodiversity Series , 1, 1-569.

[12]   Dianese, A. C., & Cargnin, A. (2008). Ocorrência de *Oidium* sp. em pinhão manso (*Jatropha curcas* L.) em Planaltina, DF.Documentos 231. Planaltina, DF: Embrapa Cer-rados, 15p.

[13]   Dianese AC, Dianese JC, dos Santos Junior JDG(2010). New records for the brazilian cerrado of leaf pathogens on *Jatropha curcas*. Boletim de Pesquisa e Desenvolvimento 293. Planaltina, DF: Embrapa Cerrados, 13p.

[14]   Dias, L. A. S., Leme, L. P., Laviola, B. G., Pallini, A., Pereira, O. L., Dias, D. C. F. S., Carvalho, M., Manfio, Santos., Sousa, L. C. A., Oliveira, T. S., & Pretti, L. A. (2007). Cultivo de pinhão manso (*Jatropha curcas* L.) para produção de óleo combustível. Vi-çosa, MG: Editora UFV.

[15]   Erwin DC, And Ribeiro OK(1996). *Phytophthora* Diseases Worldwide. APS Press, St. Paul, Minnesota, 562 pages.

[16]   Freire FCO, Parente, GB(2006). As doenças das Jatrofas (*Jatropha curcas* L. e *J. podagri-ca* Hook.) no estado do Ceará. C. omunicado Técnico 120. Embrapa,, 4.

[17]   Furtado EL, Trindade DR(2005). Doenças da seringueira (*Hevea* spp.). IN: Kimati H, Amorim L, Rezende JAM, Bergamin Filho A, Camargo LEA. Manual de Fitopatolo-gia: Doenças das Plantas Cultivadas. São Paulo: editora Agronômica Ceres, , 2

[18]   Gao, S., Qu, J., Chua, N., & Ye, J. (2010). A new strain of *Indian cassava mosaic virus* causes a mosaic disease in the biodiesel crop *Jatropha curcas*. Archives of Virolgy , 155, 607-612.

[19]   Hennen JF, Figueiredo MB, De Carvalho Jr AA, Hennen PG(2005). Catalogue of plant rust fungi (Uredinales) of Brazil. 490p.

[20]   Latha, P., Prakasam, V., Kamalakannan, A., Gopalakrishnan, C., Raguchander, T., Paramathma, M., & Samiyappan, R. (2009). First report of *Lasiodiplodia theobromae* (Pat.) Griffon & Maubl causing root rot and collar rot disease of physic nut (*Jatropha curcas* L.) in Ind. ia. Australasian Plant Disease Notes, 4, 19-20.

[21]   Machado AR, Pinho DB, Dutra DC, Pereira OL (In press) Collar and root rot caused by *Neoscytalidium dimidiatum* in the biofuel plant *Jatropha curcas*. *Plant Disease*.

[22]   Massola NS, Bedendo IP(2005). Doenças da Mamoneira (*Ricinus communis*). IN: Ki-mati H, Amorim L, Rezende JAM, Bergamin Filho A, Camargo LEA. Manual de Fito-patologia: Doenças das Plantas Cultivadas. São Paulo: editora Agronômica Ceres, 2 v.

[23]   Mendes, M. A. S., Silva, V. L., Dianese, J. C., Ferreira, M. A. S. V., Santos, C. E. N., Gomes, Neto. E., Urben, A. F., & Castro, C. (1998). Fungos em plantas no Brasil. Bra-sília.Embrapa-SPI/Embrapa Cenargen. 555p.

[24] Narayana DSA, Shankarappa KS, Govindappa MR, Prameela HA, Rao MRG, Rangaswamy KT(2006). Natural occurrence of Jatropha mosaic virus disease in India.Current Science , 91(5), 584-586.

[25] Neergaard, P. (1977). Seed pathology. London: The MacMillan Press.

[26] Patel DS, Patel SI, Patel RL(2008). A new report on root rot of *Jaropha curcas* caused by *Macrophomina phaseolina* from Gujarat, India. *J Mycol Pl Pathol* , 38(2), 356-357.

[27] Pereira OL, Dutra DC, Dias LAS. (2009). *Lasiodiplodia theobromae* is the causal agent of a damaging root and collar rot disease on the biofuel plant *Jatropha curcas* in Brazil. *Australasian Plant Disease Notes*, 4, 120-123.

[28] Phillips, S. (1975). A new record of *Pestalotiopsis versicolor* on the leaves of *Jatropha curcas*. Indian Phytopathology 28: 546.

[29] Sá DAC, Santos GRS, Furtado GQ, Erasmo EAL, Nascimento IR(2011). Transporte, patogenicidade e transmissibilidade de fungos associados às sementes de pinhão manso. Revista Brasileira de Sementes, , 33(4), 663-670.

[30] Saturnino, H. M., Pacheco, D. D., Kakida, J., Tominaga, N., & Gonçalves, N. P. (2005). Cultura do pinhão-manso (*Jatropha curcas* L.).Informe Agropecuário , 26, 44-78.

[31] Slippers, B., & Wingfield, . (2007). Botryosphaeriaceae as endophytes and latent pathogens of woody plants: diversity, ecology and impact. *Fungal Biology Reviews*, 21, 90-106.

[32] Sulaiman, R., & Thanarajoo, S. S. (2012). First report of *Lasiodiplodia theobromae*causing stem canker of *Jatropha curcas* in Malaysia. Plant Disease 96(5):767.

[33] Sutton BC(1980). The Coelomycetes, Fungi Imperfecti with acervuli, pycnidia and stromata.Commonwealth Mycological Institute,Kew, U.K.

[34] Torres-Calzada, C., Tapia-Tussell, R., Nexticapan-Garcez, A., Matin-Mex, R., Quijano-Ramayo, A., Cortés-Velázquez, A., Higuera-Ciapara, I., & Perez-Brito, D. (2011). First report of *Colletotrichum capsici* causing anthracnose in *Jatropha curcas* in Yucatan, Mexico. New Disease Reports 23(6).

[35] USDA(1960). Index of Plant Diseases in the United States. U.S.D.A. Agric. Handb. , 165, 1-531.

[36] Viégas AP(1945). Alguns fungos do Brasil IV: Uredinales. Campinas: Bragantia, n.1, , 5, 7-8.

[37] Viégas AP(1961). Índice de fungos da América do Sul. Campinas. Instituto Agronômico, 919p.

[38] Yue-Kai, W., Guo-Teng, O., & Jin-Yong, Y. (2011). First report of *Nectria haematococca* causing root rot disease of physic nut (*Jatropha curcas*) in China. *Australasian Plant Disease Notes* , 6(1), 39-42.

Biodiesel Production

Biodiesel: Production, Characterization, Metallic Corrosion and Analytical Methods for Contaminants

Rodrigo A. A. Munoz, David M. Fernandes,
Douglas Q. Santos, Tatielli G. G. Barbosa and
Raquel M. F. Sousa

Additional information is available at the end of the chapter

1. Introduction

In face of recent changes in the edaphoclimatic conditions (climate and soil) occurring worldwide, it has been necessary reflections on the need to exploit natural resources in a sustainable manner. Sustainability is a systemic concept, relating to the continuity of economic, social, cultural and environmental aspects of human society. It proposes to be a means of configuring the civilization and human activity so that the society, its members and its economies can fulfill its needs and express its greatest potential at present, while preserving biodiversity and natural ecosystems, planning and acting to achieve pro-efficiency in maintaining undefined these ideals [1].

According to the International Energy Outlook -2011 (IEO 2011), the energy consumption in the world was 505 quadrillion of British thermal units (Btu) in 2008, while in 2006 this consumption was 472 quadrillion Btu. According to 2011 bulletin, the outlook for energy consumption in 2020 will be 619 quadrillion Btu, and 770 quadrillion Btu in 2035 for the countries of the Organization for Economic Cooperation and Development (OECD) [2]. This gradual increase in worldwide energy consumption increases the search for renewable energy, once the existing conventional sources are exhaustible ones, such as for example the oil. In this sense, biofuels have appeared, such as ethanol (bioethanol) and biodiesel, which emerged very strongly due to large government incentives.

Within the energy issues, Brazil, for its favorable natural conditions, presents a huge potential for the production of biofuels, especially ethanol and biodiesel, a fact which makes it a strategic country in relation to the sustainability of such market. There is an effort to consolidate energy from renewable sources, as well as the use of byproducts from these industries, with the creation of programs and incentives by the federal government, such as Proálcool [3].

This text reports a brief historic background of biodiesel in Brazil, methods for biodiesel production indicating the main raw materials, and the regulated physical-chemical properties for the quality control of biodiesel discussing the consequences of cases of non-conformity and their regulated methods by European, American and Brazilian norms. A special topic is dedicated to metallic corrosion which is closely related to storage stability of biodiesel, and finally a comprehensive review of analytical methods developed for monitoring contaminants (glycerol and trace metal) in biodiesel is presented.

1.1. Historic background of biodiesel in Brazil

The first use of vegetable oil in diesel engine was tested at the request of the French government with the intention of stimulating energy self-sufficiency in its colonies in Africa, minimizing the costs relating to imports of coal and liquid fuels. The oil selected for the tests was from peanut, whose culture was abundant in tropical countries. The diesel engine produced by the French company Otto, powered by peanut oil, was presented at the Paris Exhibition in 1900. Other experiments conducted by Rudolph Diesel were held in St. Petersburg with locomotives powered by castor oil and animal oils. In both cases the results were very good and the engines showed good performance.[3] In chapter "Liquid Fuel" from Diesel's book "Die Entstehung des Dieselmotors" (The Emergence of Diesel Engines)[4], it mentions:

For completeness, it is important that, back in 1900, vegetable oils were already being used successfully in diesel engines. During the Paris Exhibition of 1900, the French company Otto demonstrated the operation of a small diesel engine with peanut oil. This experiment was so successful that only some of the people present realized the circumstances in which it was conducted. The engine, which had been built to consume oil, was operated with vegetable oil without any modification. It was also observed that the consumption of vegetable oil resulted in a use of heat literally identical to the oil.

Vegetable oils were also used as emergency fuel, among other applications, during the Second World War. For example, Brazil has banned the export of cottonseed oil because this product could be used to replace imports of diesel oil. Reductions in imports of liquid fuels were also reported in Argentina, which required greater commercial exploitation of vegetable oils. China produced "diesel", lubricating oils, "gasoline" and "kerosene", the last two by cracking processes, from tung oil and other raw material oilseeds. However, the requirements of war forced installation of cracking units of unusual technology base. Quickly, research activities with new oil sources were expanded, but with the subsequent decline in the price of crude oil barrel (post World War II), reaching more affordable price, these researches were abandoned, as happened in India [5].

The use of vegetable oil as an alternative renewable fuel to compete with diesel oil was proposed in the early 1980. The more advanced study with the sunflower oil happened in South Africa because of diesel oil embargo, and the first International Conference on Plants and Vegetable Oils was held in Fargo, North Dakota, in August 1982 [6].

In Brazil, since the 1930s, efforts have been made to incorporate renewable fuels in the energy matrix, which are mainly accomplished by government authorities, universities and research institutes. In addition to the PRO-ALCOHOL (created in 1975, this program was designed to ensure the supply of ethanol from sugarcane in the process of replacing gasoline, as government initiative to tackle the successive increases in oil prices), it was established the Production Plan of Vegetable Oils for Energy Purposes (PRO-OIL), which from the 1980s was named as the National Program of Vegetable Oils for Energy Purposes. This program was designed to generate a significant surplus of vegetable oils, able to make their production costs competitive with mineral oil. It was envisaged by legislation a mixture of 30% of vegetable oil in diesel oil, with prospects for full replacement in the long term. At this time, it was proposed as a technological alternative the transesterification or alcoholysis of vegetable oils, highlighting the studies conducted at Federal University of Ceará, using different sources of vegetable oils such as soybean, babassu, peanut, cottonseed and sunflower, among others. Unfortunately, this program was abandoned by the government in 1986, when the oil price fell again in the international market along with the high cost of production and the crushing of oilseeds, which were decisive factors for the slowdown of the program. However, even after the end of PRO-OIL, there was a considerable progress in researches on the production and use of biodiesel in Brazil, which were conducted in different universities and research centers, particularly the registration of the first Brazilian patent deposited by Chemical Engineer Expedito José de Sá Parente [7]. In accordance with Parente he did not developed a new method: *The transesterification process has been known for many years. What I have patented was the production of esters for use as fuel in diesel cycle engines, which is entirely different from what Rudolf Diesel did. Modern engines could not run for a long time using a vegetable oil under the conditions tested by Diesel.* Table 1 shows the evolution of fuels in Brazil since the 1970s.

In Brazil, especially from the year 2005, when there was the beginning of the National Biodiesel Program, these surveys were intensified due to growing importance in using this material, and three years later, in January became mandatory the addition of 2% (v/v) of biodiesel to commercial diesel, whose mixture was called B2 (B for blend). In March 2008, the National Energy Research Council (CNPE) determined the mandatory addition of 3% (v/v) from July. Since January 1, 2010, the diesel oil sold in Brazil contains 5% biodiesel. This rule was established by Resolution number 6/2009 of the National Energy Policy Council (CNPE), published in the Official Gazette (DOU) in October 26, 2009, which increased from 4% to 5% the mandatory blend percentage of biodiesel to diesel oil. Until mid-2011, two more increases were conducted over, and at that time the addition was 5% (v/v) of biodiesel to diesel, and such addition was put forward to year 2011, previously proposed for the year 2013, due mainly to the increase in diesel consumption and the consequent increase of the fleet of cars using this fuel.

Year	Event
1973	First Oil Shock
1974	Creation of the Pro-Alcohol
1977	Addition of 4.5% Ethanol to Gasoline
1979	Addition of 15% Ethanol to Gasoline
1980	Second Oil Shock
1983	Alcohol cars account for over 90% of total sales
1985	Percentage of Alcohol added to gasoline reaches 22%
1989	Oil prices fall and gasoline equates with alcohol
1992	Rio 92: Signing of the milestone about climate change
90's	Alcohol comes to represent 20-25% of Gasoline
2005	Founded PNPB
2007	Third Oil Shock
January 2008	Beginning of mandatory B2
March 2008	CNPE determines the mandatory use of B3 from July 2008
April 2008	Alcohol Consumption equates to Gasoline
July 2009	Validity of B4
January 2010	Validity of B5

Table 1. Evolution of fuels in Brazil since the 1970s.

This same law, published on January 13, 2005, introduced the biodiesel in the Brazilian energy matrix and expanded the administrative competence of the National Agency of Petroleum and Natural Gas (ANP), which became, since then, the National Agency of Petroleum, Natural Gas and Biofuels. Since the publication of the abovementioned law, ANP took over the assignment to regulate and supervise the activities related to production, quality control, distribution, sale and marketing of biodiesel and diesel-biodiesel blend [8]. The continued rise in the percentage of biodiesel added to diesel demonstrates the success of the National Program for Production and Use of Biodiesel and the experience accumulated by Brazil in production and large-scale use of biofuels, especially the biodiesel.

1.2. Aspects of the biodiesel market in Brazil

According to the Monthly Bulletin of Renewable Fuels of the Ministry of Mines and Energy (MME), based on deliveries of auctions promoted by ANP, it shows that the estimated production in May 2012 was 173,000 m^3. The total of the year up to this month, the cumulative production was 958,000 m^3. The installed production capacity, in May 2012, stood at 6.092 million m^3/year (507,000 m^3/month) of which 88% is related to companies holding the social seal [9]. The Social Fuel Seal is a component of identification created from the Decree No.

5297 of December 6, 2004, awarded by MDA to biodiesel producer who meets the criteria described in Instruction No. 01 of February 19, 2009. The Seal gives to the possessor the character of promoter for social inclusion of family farmers classified in the National Program of Family Agriculture (PRONAF).

2. Raw materials for biodiesel production

In general, biodiesel can be produced from any source having oil, either of animal or vegetable source; however, to ensure the quality of the final product, some factors must be observed, as, perennial crops, oil content and preferably with no potential for the food industry, as occurs primarily for soybean in Brazil. Furthermore, it should be noted the productivity per unit area, the agronomic balance and other aspects related to the life cycle of the plant (seasonal). These raw materials, along with the production processes, depend on the region concerned. The economic, environmental and social diversities have given distinct regional motivations for the production and consumption. Due to the favorable conditions of climate and soil, Brazil presents numerous possibilities for use as raw material for the biodiesel industry. It can be highlighted the use of soybean, castor, palm, cottonseed, sunflower, macauba, rapeseed, jatropha, animal fat (tallow) and residual oils, among which, the latter presents itself as an excellent alternative, because it is a material whose reuse at industrial level was insignificant until its application in the biodiesel industry. Oil contents of some of the oil seeds used in Brazil are shown in Figure 1.

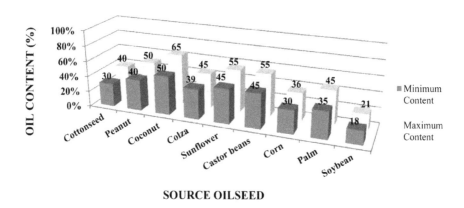

SOURCE OILSEED

Figure 1. Oil content of some of the seeds used in Brazil.

In Brazil the most relevant is the use of soybean for biodiesel production, followed by beef tallow and cottonseed oil as described by Monthly Bulletin of Biodiesel (published by ANP - reference month May 2012) as shown in Figure 2.

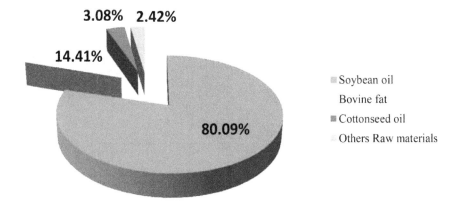

Figure 2. Oil sources used for biodiesel production in Brazil.

In Brazil, other species of oilseeds are cultivated and they are distributed by region as shown in Table 2.

Region	Raw materials that are available
Southeast	Soybean, castor, cottonseed, sunflower, jatropha and macauba
South	Soybean, corn, rapeseed, sunflower, jatropha and cottonseed
Centralwest	Soybean, castor, cottonseed, sunflower, jatropha, macauba, palm and animal fat
North	Palm, babassu and soybean
Northeast	Babassu, soybean, castor, palm, jatropha, cottonseed and coconut

Table 2. Distribution of oilseeds in Brazilian regions. Adapted from ref. [7].

Despite the many advantages in the use of biodiesel, there are two serious problems from the viewpoint of production of vegetable oils which can retard or impede the use of derivatives of vegetable oils as fuel. The first problem is the amount of oil produced. Some of the key raw materials used in Brazil and the corresponding harvest months and oil yields (t/ha) are listed in Table 3. The second problem is the quality of the extracted oil. Oils with high concentrations of polyunsaturated fatty acids are undesirable for biodiesel production, for decreasing their stability to oxidation [12]. Generally, the more unsaturated acids, such as linoleic and linolenic acids, respectively, with two and three unsaturations, are more susceptible to oxidation. In addition, these oils induce a higher carbon deposit than oils with high content of monounsaturated or saturated fatty acids as occurs to the palm oil.

Name	Scientific name	Harvest months/ years	Yield (ton/ha)
Peanut	*Arachis hypogea L.*	3	0.6-0.8
Cottonseed	*Gossipum Hirsutum L.*	3	0.1-0.2
Babassu	*Orbiguya phalenta Mart.*	12	0.1-0.3
Rapeseed	*Brassica napus L.*	3	0.5-0.9
Coconut	*Cocos mucifera L.*	12	1.3-1.9
Palm	*Elaesis guineenses L.*	12	3-6
Sunflower	*Helianthus annus L.*	3	0.5-1.9
Castor beans	*Ricinus communis L.*	3	0.5-0.9
Jatropha	*Jatropha Curcas*	12	0.5-2.5*
Soybean	*Glycine Max (L.) Merril*	3	0.2-0.4

Table 3. Name, Scientific name, harvests months and yields of raw materials for biodiesel production in Brazil. Adapted from refs. [10] and [11]*.

2.1. Cottonseed

The cotton is currently produced by more than 60 countries. China, India, U.S.A., Pakistan and Brazil emerge as major producers. Together, they account for 80% of world production of around 23.3 million tons. Only these five countries have more than 22 million hectares of planted areas, out of 35 million on average that are planted in different points of the planet [13]. The cottonseed oil shows 13-44% oleic acid (18:1) and 33-59% linoleic acid (18:2) [14], and gossypol, the main byproduct of the cotton industry, is a polyphenolic compound, yellow color, toxic, non-interfering in the biodiesel production process [15].

2.2. Palm

Over 70% of the world's oils come from 4 plant species: soybean, palm, sunflower and rapeseed, and currently, the palm oil (known in Brazil as "dendê") is the most traded oil in the world. It is used in the food industry in margarine, ice cream, biscuits, pies, and others. It is also used in the industry of hygiene and cleanliness in the composition of soaps, detergents and cosmetics. In the chemical industry, it is part of the lubricant composition and can also be used as biofuel [16].

In Brazil, the rise of the use of this oilseed is so great that the Federal Government launched in 2010, the Program for Sustainable Production of Oil Palm, which aims to regulate the expansion of oil production and offer tools to ensure a production in sustainable environmental and social bases [16].

This oilseed has the largest oil content among the raw materials for the biodiesel production reaching the yield of 6000 kg/hectare/year. This yield is about 15 times larger than the main raw material in Brazil, the soybean.

The palm oil presents approximately 35-47% palmitic acid (16:0) and 36-47% oleic acid (18:1) [14], and the main problem in using it as fuel is its high viscosity ($\approx$ 20 Cst), about 10 times greater than oil diesel ($\approx$ 2 Cts) (ASTM D445).

2.3. Rapeseed

The rapeseed (*Brassica napus*) is a winter oilseed of Brassica genus, from Cruciferae family and it has 39-45% oil with excellent quality for fatty acid composition. It is the main oil used for biodiesel production in Europe, and is usually known as "colza". It started in Canada in 1974 and with plant breeding developed a seed without erucic acid, a substance which limited its use for human consumption [14]. It represents, in percentage, 15.32% of world production of vegetable oils, considering the 2011/2012 crop, behind only the production of palm oil (33.14%) and soybean (27.28%) [17]. The rapeseed oil contains mostly oleic acid (18:1), 62% followed by 32% linoleic acid (18:2) [18].

2.4. Jatropha

The jatropha (*Jatropha curcas*) is a drought-resistant plant, multipurpose, having a large amount of oil compared to other inedible plants [19]. The highest incidence and the center of diversity of this genus are the Central America and the Caribbean Islands. The genus has over 210 species growing in different regions of the globe [20-21]. This oleaginous has a different composition when it comes to saturated fatty acids [22-23], due to the presence of 16.4% palmitic acid (16:0) against 2.3% soybean and 7% corn [24]. The jatropha contains mostly 40.3% linoleic acid (18:1), 37% linolenic acid (18:2) which indicates a good quality of biodiesel to be produced from this source because the lower the amount of raw material unsaturation, the better is the stability to oxidation [25].

2.5. Soybean

In 2010/2011 harvest, Brazil produced about 75 million tons of soybeans, 65% of this production was exported, equivalent to 48.7 million tons (29.1 million tons of grain, 13.7 million tons of bran and 1.6 million tons of oil). China is the main destination of Brazilian exports of grain (66%) and oil (60%) and the European Union is responsible for the majority of imported bran (70%) [26]. In 2011/12 harvest, the soybean had a stake in percentage terms in the world production of vegetable oils of 27.28% and is the second after only to oil palm, with 33.14% [17].

In Brazil, in May 2012, the soybean had a participation of 80.09% in the production of biodiesel. Taking into account that the biodiesel production in Brazil in that same month was 188,367 x 10^3 m^3 (ANP) then about 150,863 x 10^3 m^3 of the biodiesel produced was derived from soybean oil. This fact is due to the growth of agricultural productivity of soybean and also the technical-industrial feasibility in this oilseed cultivation.

The soybean has about 19.0-30.0% of oleic acid (18:1) and 44.0-62.0% of linoleic acid (18:2) [14].

One aspect widely discussed in relation to this oilseed is the competition between the production for the biofuel market and food industry. In 2011/12 harvest, Brazil exported approximately 44.9% of the soybeans produced [10], due mainly to the increased buying power of emerging countries such as China and India and a small part (about 8%) was for national biodiesel production. China in this season was the largest buyer of Brazilian soybeans (66%) and processed refined oil (60%).

Regarding to soybean oil, Brazil in 2010/11 harvest produced about $7,434 \times 10^3$ tons, and $5,495 \times 10^3$ tons (74.85%) were used for domestic market and $1,758 \times 10^3$ tons (23.95%) for export [26]. In Brazil, this situation has fluctuated over the months, due to the offseason of soybeans and also because many other oilseeds arose that can be intended for biodiesel production, such as jatropha, cottonseed, moringa, among other.

2.6. Waste cooking oil

This raw material represents currently a potential use for biodiesel production worldwide since it is a residue from cooking process of industries, restaurants, bars, among other outlets. The high cost of biodiesel in the national and international market is mainly due to the soaring cost of industrialized vegetable oils and virgin oils [27,28].

The biodiesel made from these raw materials is most often cheaper than from sources such as soybean, rapeseed and sunflower, but the quality of the oil used must be monitored to obtain a high quality biodiesel. The free fatty acid (FFA) content and water content (moisture content) are the main parameters to be analyzed. For a reaction catalyzed by alkali, a FFA content of less than 3% is required, since the use of oil with a high FFA content produces the hydrolysis of triglycerides at high temperatures during the frying process [29]. The residual oil has as striking characteristic a high acidity index and it is previously necessary a transesterification reaction, an acid catalysis using a strong acid (generally H_2SO_4) to promote the esterification of fatty acids and thereby reduction of FFA to less than 1% [27,28]. The residual cooking oil has mostly in its composition, oleic acid (18:1) and linoleic acid (18:2) [30].

The use of recycled cooking oil for Biodiesel production has been relevant in recent times, and the flight KL705, originated from Amsterdam with destination to Rio de Janeiro, was carried out in part with this raw material, fueling a Boeing 777-200 [31].

2.7. Moringa oleifera

The *Moringa oleifera* belongs to the family Moringaceae, consisting of only one genus (Moringa) and fourteen known species, native to northern India. This species is found in natural conditions in India, Africa, Asia, Arabia, South America and the Pacific and Caribbean Islands [32]. Moringa is a multipurpose plant from the leaves to the seeds, showing different properties. The Moringa leaves are the source of a diet rich in protein for both humans and animals. Moreover, oil extraction from its seed enables the utilization of this raw material to produce biodiesel having a remarkable oxidation resistance, with shelf time between 4 and 5 years [33]. The seeds of *Moringa Oleifera* contain between 33 and 41% m/m oil [33]. This value is considered a good natural emollient for cosmetic, based on tactile properties, a lower natural occurrence of color and odor besides a high concentration of oleic acid (>73%) [34]. Recently studies showing the potential of Moringa oil extracted from seeds of India and Pakistan for the biodiesel production were published [35,36]. The Moringa has in its composition about 7% of palmitic acid (16:1) and 78% of oleic acid (18:1) [32]. Variations in oil content within countries and species are attributed to possible changes in environmental and geological conditions of the regions [37].

3. Methods for biodiesel production

Vegetable oils and animal fats contain, in addition to triacylglycerols, free fatty acids, phospholipids, sterols, water and other impurities. These compounds give special properties to these raw materials which prevent their use directly as a fuel, as, for example, the blockage point and the high viscosity [38]. These problems can be overcome with adjustments to the compression engines; however, such possibility makes the process very expensive and often impractical. The adaptation of the fuel becomes more interesting and it can be used directly to the existing fleet, without adjusting existing technologies by those that cause chemical modifications (cracking, esterification and transesterification). In the last decade some reviews on the different methods for biodiesel production were reported in the literature [39-47].

The transesterification (also called alcoholysis) is generally a term used to describe the important class of organic reactions in which a triacylglycerol reacts with alcohol in catalytic environment and becomes ester and glycerol [39]. The overall process is a sequence of three consecutive and reversible reactions, in which the mono-and diacylglycerols are formed as intermediates of reaction. In the transesterification reaction three moles of alcohol are needed for each mol of triacylglycerol. In practice, it is used an excess of alcohol in order to increase the ester productivity (shifting the reaction towards the product side) and to allow the separation of the obtained glycerol (the impure fraction containing glycerol is called glycerin).

Glycerin makes the oil more dense and viscous, therefore, during the transesterification process, the glycerin is removed resulting in a product of lower viscosity. In the transesterification, the reaction of biodiesel production creates alkyl esters and glycerol, and the glycerol layer is denser than the ester one and it deposits at the bottom of the reactor [39]. The process is based on the stoichiometric reaction of the alkyl glycerol with alcohol, which in most cases has short organic chain in the presence of a catalyst [39-42]. Alcohols such as methanol, ethanol, propanol or butanol can be used in the transesterification reaction and the (produced) monoesters are known as methyl, ethyl, propyl and butyl esters, respectively. The raw materials used as triacylglycerol source for biodiesel production were discussed previously in the text (Section 2).

The technology for the biodiesel production prevailing in the world is methyl transesterification, where, vegetable oils or animal fat are mixed with methanol that, along with a catalyst, produce the biodiesel. This option occurs mainly because of high ethanol cost and operational facilities. The advantages of biodiesel production using methanol include: ester phase separation (biodiesel) of glycerol that occurs instantaneously; the alcohol recovery that is completed and can be returned to the process; the obtaining of glycerol which is feedstock to the chemical industry; the synthesis is more attractive under the industrial viewpoint, because it is faster and cheaper than the others.

In Brazil, the enterprises that are in operation adopt the methylic route; however, due to ethanol production in Brazil is established beyond favorable environmental factors, there are enterprises that adopt the ethylic route (at industrial level). The ethyl route is recognized as more ecologically friendly, because ethanol is a renewable source. Production of ethyl ester is slightly more complex and requires more steps and the use of specific centrifugal pump sand optimized for a good separation of glycerol from esters. The more alcohol is added to the oil, the faster is the conversion to ester, however, an excess can stabilize the emulsion and complicate the separation of glycerol.

The catalysis used for biodiesel production can be chemical or biological, homogeneous or heterogeneous, acidic or basic. The process for producing biodiesel by basic catalysis is faster than the process of acid catalysis, the biodiesel produced presents less corrosivity, [43,44] and the most used catalysis are the potassium hydroxides, sodium hydroxide, potassium methoxide and methoxide sodium. In Brazil, KOH is more expensive than NaOH, however, there is less soap formation using KOH [45]. The base catalysis is the preferred procedure when there are oils with low water content and low acidity index. On the other hand, if oils and fats have a high content of free fatty acids, it is recommended a pre-treatment or the acid catalysis, followed by an alkaline transesterification [46].

In general, the transesterification of oils or fats can be affected by several factors such as [48]: presence of free fatty acids, moisture [50], type of alcohol used, the molar ratio of alcohol/oil, concentration and type of catalyst, time, temperature of reaction [49] and the intensity of shaking [41].

The content of free fatty acids and moisture are important parameters to determine the feasibility of the basic process of transesterification of vegetable oils and animal fats, requiring low levels of free fatty acids and moisture in the raw material for higher conversion efficiency. Studies have shown [43] that the transesterification of beef tallow catalyzed by NaOH in the presence of free fatty acids (FFA) and water has its yield compromised. Therefore, the raw material that has a high content of free fatty acids can be purified by saponification or using acid catalysis for previous esterification of these acids [51].

The production of ethyl esters via basic catalysis becomes more difficult compared with the production of methyl esters, due to the undesired formation of stable emulsion during the reaction. Methanol and ethanol are immiscible in triglyceride at room temperature, and the reaction media are commonly kept under mechanical stirring to increase mass transfer. In methanolysis, these emulsions easily form two layers: a bottom one rich in glycerol and a top one rich in esters; in ethanolysis the phases are more stable complicating the separation and purification of esters [39,50].

The molar ratio of alcohol/oil is one of the most important variables affecting the yield of esters in the transesterification reaction. The transesterification is a reaction in equilibrium, which requires an excess of alcohol to drive the reaction in the formation of esters. For a maximum conversion of esters, the molar ratio should be greater than or equal to 6:1 [39], but a very high molar ratio of alcohol/oil interferes with the separation of glycerol since its solubility increase takes place. When glycerin remains in solution, it favors the reaction equilibrium for the reactants and then decreasing the yield of esters [52]. Some researchers [42,53] have studied the molar ratios between 3:1 and 15:1 with ethanol, and they have observed arise in the yield of esters with molar ratio increase up to a value 12:1. The best results were between 9:1 and 12:1 [48].

The catalyst concentration is another factor extremely important for the alkaline transesterification [53] since the excessive addition of catalyst can promote an acidity reduction, but, on the other hand, it leads to the formation of soap, hampering the separation of glycerol from esters and consequently decreasing the yield of the reaction.

The catalysts most used during transesterification are the alkoxides [54,55], hydroxides [56] and carbonates of sodium or potassium. The alkoxides of alkali cations such as potassium methoxide (CH_3ONa) are the most reactive catalysts, since they exhibit high yields (> 98%) in a short reaction time (30 min), even at low molar concentrations (0.5 mol L^{-1}). The greater efficiency of the catalyst CH_3ONa relative to NaOH is described by Freedman and colleagues [48,54] which had a similar conversion of oil at concentrations of 1% NaOH and 0.5% CH_3ONa, while in the work developed by Ma and colleagues [45], NaOH showed better performance in ester than CH_3ONain the transesterification of beef tallow. Vincent and colleagues [57] reported good yields obtained with methoxide catalyst, but higher conversion rate was obtained with NaOH and lower with CH_3OK at 65° C, in methanol/oil ratio of 6:1 and catalyst concentration of 1%. Hydroxides of alkaline cations (KOH and NaOH) are more accessible in price than the respective alkoxides, but are less reactive.

Stirring of the reaction medium is an important factor in the transesterification process. Stirring should be intense to transfer amounts of mass of triglycerides from oil phase to the interface with methanol, as the reaction mixture is heterogeneous, consisting of two phases. In this case, the greater the stirring the higher is the mass transesterification [58]. Ma and colleagues [45] added NaOH and MeOH to the tallow beef melted in a reactor and found that after a certain reaction time without stirring anything occurred, suggesting the need of stirring for the reaction be initiated.

Some oils and fats, which may be used as raw materials for the production of biodiesel, have high levels of free fatty acids. The presence of free fatty acids impairs the synthesis of biodiesel via homogeneous basic catalysis [59]. In this sense, heterogeneous acid catalysts, which simultaneously promote reactions of alcoholysis of triglycerides and esterification of free fatty acids, present themselves as promising substitutes of homogeneous basic catalysts [60]. Moreover, such catalysts have the advantages inherent to heterogeneous catalysis, such as significantly reducing the number of purification steps of products as well as the possibility of reuse and enable the production of biofuel by a continuous process with fixed bed reactors [60].

For the production of biodiesel there are other technological routes besides the transesterification, as for example, the esterification catalyzed by an acid, preferably the sulphonic or sulfuric acids. The obtained yield is very high (99%), but the reaction is slow, requiring high temperatures (above 100°C) over 3 hours to reach the mentioned output [45,46,61,62]. Furthermore, it is necessary to use a large excess of alcohol to ensure a high yield reaction. The acid catalysis is suitable for oils with high content of free fatty acids and moisture. In this case the process is the esterification of free acids and not the transesterification of triacylglycerol.

The enzymatic catalysis (biological) allows the simple recovery of glycerol, the transesterification of triglycerols with high content of fatty acids, the total esterification of free fatty acids, and the use of mild conditions in the process, with yields of at least 90%, making it a commercially viable alternative. In this type of catalyst there are no side reactions that result in by products, which reduce the expenditure of further purification. Some enzymes require cofactors, metal ions or organic compounds (coenzymes). These co-factors will influence the activity of biological catalyst [47]. The advantages of the process are: lack of aqueous alkaline waste, lower production of other contaminants, greater selectivity and good yields. The main drawbacks of this methodology are the high cost of pure enzymes, the high cost of extraction and purification process of macromolecules and their instability in solution, which represent an obstacle to the recovery of the biocatalyst after its use [47]. On the other hand, the immobilization of enzymes allows their reuse, reducing the process cost. In the case of biocatalysis in nonaqueous media, immobilization also results in improvement in enzyme activity. Thus, many processes of transesterification using immobilized lipases have been developed [51,63,64].

Another possible and effective reaction is the supercritical transesterification using methanol, which allows a conversion of 60-90% in only 1 minute and more than 95% in 4 minutes. The better reaction conditions are: temperature of 350°C, pressure of 30 MPa and the volumetric ratio between methanol and oil of 42:1 for 240 seconds. The supercritical treatment of lipids with suitable solvent such as methanol depends on the relationship between temperature, pressure and thermo physical properties such as dielectric constant, viscosity, specific mass (density) and polarity [65,66]. The process is attractive since it overcomes problems such as oil/fat waste that is rich in free fatty acids and also the problem of the presence of water that often favors the formation of soap. However, side reactions involving the unsaturated esters occur when the reaction temperature exceeds 300°C, resulting in loss of material. There is also a critical residence time value in high temperature above which the efficiency decreases [51,67].

The H-BIO process was developed to introduce the processing of renewable raw materials in the scheme of petroleum refining and allow the use of existing facilities. The vegetable or animal oil is blended with fractions of petroleum diesel to be converted to units of Hydrotreating (HDT), which are employed in refineries, especially for reducing the sulfur content and improving the diesel oil quality, adjusting the fuel characteristics to the legal specifications. It was already conducted tests with up to 30% vegetable oil in the HDT load, mixed with diesel fractions, generating a product that has the same characteristics as petroleum diesel, but the use of such a high proportion of vegetable oil, in existing industrial HDT units meet operational constraints and limitations of some equipment that were not rated for such task in its original design. A patent of this method was registered (INPI PI0900789-0 A2) and a summary of this method is available at the webpage of Petrobras (Brazilian Oil Company) [68].

4. Physical-chemical properties of biodiesel

Because of the importance of biodiesel and regulations for its use in the country, the concern of the ANP (National Agency of Petroleum and Natural Gas and Biofuels) is to ensure a quality fuel in any situation, through the establishment of quality standards for biodiesel. The federal law 11.097/2005 establishes in the country the introduction of minimum percentages of mixture of biodiesel to diesel, and also the monitoring of such insertion. The Brazilian specification of biodiesel is similar to European and American ones, with some flexibility to meet the characteristics of domestic raw materials. This specification is issued by ANP decree. This agency has developed standards in recent years with respect to biodiesel, among them there are the resolutions 15, 41 and 42. Resolution No. 42 of ANP [69] provides a specification for biodiesel (B100) according to the provisions contained in the Technical Regulation No. 4/2004, part that composes such Resolution. B100 can be added to diesel oil in proportions defined by volume, sold by various economic agents authorized throughout the national territory. Any vegetable oil can be used as fuel for diesel engines, but some oils have better performance in terms of their thermodynamic properties [70].

Visual observation of biodiesel formed which should be presented clear and free of impurities, either in suspension, precipitated material or any other. This parameter is adopted only by ANP and it is a simple but important test which is performed in a tube without graduation because this parameter is a qualitative indication of quality.

Tables 4 and 5 present the specifications for biodiesel in accordance with Brazilian, European and American standards and the corresponding analytical methods, respectively.

The specific mass (density) is connected with the molecular structure, i.e. the longer the carbon chain of the ester alkyl, the greater is the density which is, however, reduced by the presence of unsaturations. It is an important parameter for the vehicle injection system. Biodiesel has a specific mass greater than the diesel. This parameter is variable according to the raw material, alcohol excess, among others; very high values may indicate contamination with soap and/or vegetable oil, and alcohol excess causes a decrease in density. The European standard presents as specification limit the values of 860-900 kg m^{-3}, according to the manual method EN ISO 3675 using glass hydrometers and the automatic method EN ISO 12 185 using digital densimeters, and the later with better repeatability. ASTM D6751 norm does not consider the specific mass as a measure of quality of biodiesels. ANP provides the specification range with values between 850 to 900 kg m^{-3}, which adopts the European standard methods, in addition to ASTM D1298 (manual) and ASTM D4052 (automatic), corresponding to NBR 7148 and NBR 14065, respectively. The Brazilian specification differs from European only for this parameter. However, as the reference temperature in Brazil is 20°C and assuming that the lower the temperature of the test, the higher is the density, it can be concluded that the European specification is more restrictive.

The kinematic viscosity (measurement of internal resistance of liquid flow) is an important parameter for the vehicle injection system and fuel pumping system. It depends on the efficiency of the process (reduction of viscosity of raw material). The viscosity reaches high levels with polymerization processes and/or thermal or oxidative degradation. The EN 14214 standard provides an acceptable range from 3.5 to 5.0 mm²/s (EN ISO 3104 method), while the ASTM D6751 standard allows a broader range from 1.9 to 6.0 mm²/s (ASTM D445 method). The Brazilian standard adopts, besides the methods already mentioned, also the ABNT NBR 10441 method, with the allowed viscosity range from 3.0 to 6.0 mm²/s. The bottleneck for this parameter is the upper limit of the specification which is the value of 6.0 mm²/s for ANP and ASTM, while the standard EN 14214 has a limit 5.0 mm²/s, which may restrict the use of some raw materials as for example the biodiesel from castor beans.

The flash point corresponds to the lowest temperature at which the product generates steam enough to ignite when a flame is applied under controlled conditions. This analysis measures the power of self-igniting the fuel and is essential for safety in stocking, handling, transport and storage of fuel. Biodiesel has a flash point much higher than diesel, and low flash point is commonly linked to the presence of alcohol residue in the process. ANP states that the flash point presents at least 100°C and adopts the EN ISO 3679 method, the same of European standard, ASTM D93, the same used by ASTM D6751, and also recommends ABTN NBR 14598.

Water and sediment in biodiesel are usually higher than in diesel, as biodiesel is hygroscopic. The water may generate an unwanted reaction, producing free fatty acids, growth of microorganisms, corrosion and malfunctions of the engine [71]. ASTM D6751 standard adopted this parameter as a quality control for biodiesel using ASTM D2709 method, while ANP and European standard recommends coulometric method (Karl Fischer) EN ISO 12937, and the Brazilian standard also recommends ASTM D6304 method. Comparing the methods, it seems that the coulometric method has greater sensitivity, higher repeatability and lower response time compared to the volumetric method (ASTM D2709).

The total contamination is mainly originated from the raw material, from soaps formed during the process and from unsaponifiables such as wax, hydrocarbons, carotenoids, vitamins and cholesterol (animal origin and used oils). The unsaponifiables present higher boiling point and create waste and soaps in engines and which result in sulphated ashes and finally in abrasion. ASTM standard has not adopted this parameter that is indicated by the European and Brazilian standards, which recommend the same method of analysis described by EN 12662, with maximum specification limit of mg kg⁻¹.

Property	Unit	Limits		
		ANP 07/2008	EN 14214	ASTM D6751
Aspect	---	Limpid and without impurities	---	---
Density	kg/m³	850-900 (20°C)	860-900 (15°C)	---
Kinematic viscosity (40°C)	mm²/s	3.0-6.0	3.5-5.0	1.9-6.0
Water and sediment, max.	%vol.	---	---	0.050
Flash point, min.	°C	100	101	130
Distillation, 90% recovered vol., max.	°C	---	---	360
Carbon residue, max.	% mass	0.050 (100% of the sample)	0.3 (10% Distillation residual)	0.05 (100% of the sample)
Sulfated ash, max.	% mass	0.020	0.02	0.020
Sulfur content, max.	mg/kg	10	10	15
Copper strip corrosion, 3h at 50°C, max.	Rating	1	1	3
Cetane number	---	Note	51 (min.)	47 (min.)
Cold soak filterability	°C	By region	By region	---
Pour point	°C	---	By region	---
Cloud point	°C	---	---	Note
Sodium and potassium, max.	mg/kg	5	5	5
Calcium and magnesium, max.	mg/kg	5	5	---
Phosphorus content, max.	mg/kg	10	4.0	10
Total contamination, max.	mg/kg	24	24	---
Ester content, min.	% mass	96.5%	96.5%	---
Acid value, max.	mg KOH/g	0.50	0.5	0.5
Free glycerol, max.	% mass	0.02	0.02	0.020
Total glycerol, max.	% mass	0.25	0.25	0.240
Monoglycerides	% mass	0.80 (max.)	0.80 (max.)	---
Diglycerides	% mass	0.20 (max.)	0.20 (max.)	---
Triglycerides	% mass	0.20 (max.)	0.20 (max.)	---
Methanol or Ethanol, max.	% mass	0.20	0.20	0.20
Iodine value	g I₂/100 g	Note	120 (max.)	---
Oxidation stability at 110 °C, min.	H	6	6	3
Water content, max.	mg/kg	380	500	---
Linolenic acid	% mass	---	12 (max.)	---
Polyunsaturated methyl esters (with more than four double bonds)	% mass	---	1 (max.)	---

Table 4. Comparison of limits for the quality control of biodiesel in accordance with Brazilian, European and American standards.

The ester content is the main property as it indicates the degree of purity of the biodiesel produced and the efficiency of the production process used. It depends on the amount of unsaponifiables in the raw material and process variables (time, temperature, agitation speed, molar ratio, catalyst concentration, type of catalyst, water, free fatty acids, and alcohol). The low content of esters indicates a low yield of the transesterification reaction, i.e., most of the triacylglycerols has not reacted, which can cause difficulties in combustion and carbonization of engine cylinders. This parameter is required by the ANP and EN 14214 standards, whose minimum percentage is 96.5 % wt. using the EN ISO 14103 method. Brazilian standard also recommends ABNT NBR 15342 method for biodiesel from animal origin or for blends in which there is the presence of biodiesel from castor beans.

The carbon residue indicates the tendency of a fuel to form carbon deposits in engines. These residues are deposited in the nozzles and other parts of the engine, reducing their useful life. They correspond to the amount of triacylglycerols, soaps, leftover of catalyst and unsaponifiables, which are present in the final biodiesel. American and Brazilian standards adopt the same ASTM D4530 method, also having the same specification limit, no more than 0.050 % wt. Now the European standard indicates EN ISO 10370 method in which the maximum allowed is 0.3 % wt.

Sulphated ashes cause saturation of filters and wear on various parts of the engine and may be present in the form of abrasive solids, soluble metal soaps and catalyst residues. The maximum content of sulphated ashes in biodiesel is 0.020 % wt. set by the EN 14214 standard (EN ISO 3987 method), ANP (EN ISO 3987, ABNT NBR 6294 and ASTM D874 methods) and by ASTM D6751 (ASTM D874 method).

The sulfur content derived from the raw material generates toxic emissions affecting the performance of vehicle emission control system. Brazilian standard adopts methods of analysis described by ASTM D5453 (molecular fluorescence), EN ISO 20 846 (also by molecular fluorescence) and EN ISO 20 884 (dispersive X-ray fluorescence), with the maximum acceptable limit of 50 mg kg^{-1}. As to ASTM D6751, the maximum limit is 15 mg kg^{-1} (ASTM D5453 method). The most restrictive is the European standard which recommends the methods given by EN ISO 20846 and EN ISO 20884, with the maximum value of 10 mg kg^{-1}.

Group I (Na + K) and group II (Ca + Mg) metal ions cause the formation of deposits of insoluble soaps, as well as catalyze polymerization reactions. They derive from catalysts employed in the production of biodiesel, as KOH, NaOH and / or CH_3ONa or CH_3OK. Calcium and magnesium may also be present as impurities in the NaOH or KOH used.

Phosphorus can damage catalytic converters used in emission control systems of the engine. They comes mainly from the raw material and, eventually, from residues of the phosphoric acid used in the neutralization. Phosphorus is determined in biodiesel via optical emission spectroscopy with inductively coupled plasma. Both the European standard (maximum 4.0 mg kg^{-1}) as the Brazilian (maximum 10.0 mg kg^{-1}) recommends the analytical method EN ISO 14107. The Brazilian standard also recommends the method ABNT NBR 15553 and ASTM D4951. The latter is also indicated in the American standard (maximum 10.0 mg kg^{-1}).

Property	Methods			EN 14214	ASTM D6751
	ABNT NBR	ASTM D	EN/ISO		
	ANP 07/2008				
Aspect	---	---	---	---	---
Density	7148 14065	1298 4052	EN ISO 3675 EN ISO 12185	EN ISO 3675 EN ISO 12185	---
Kinematic viscosity (40°C)	10441	445	EN ISO 3104	EN ISO 3104	D445
Water and sediment	---	---	---	---	D2709
Flash point	14598	93	EN ISO 3679	EN ISO 2719 EN ISO 3679	D93
Distillation, 90% recovered vol.	---	---	---	---	D1160
Carbon residue	---	4530	---	EN ISO 10370	D4530
Sulfated ash	6294	874	EN ISO 3987	EN ISO 3987	D874
Sulfur content	15867	5453	EN ISO 20846 EN ISO 20884	EN ISO 20846 EN ISO 20884	D5453
Copper strip corrosion, 3h at 50°C	14359	130	EN ISO 2160	EN ISO 2160	D130
Cetane number	---	613 6890	EN ISO 5165	EN ISO 5165	D613
Cold soak filterability	14747	6371	EN 116	---	D7501
Pour point	---	---	---	---	---
Cloud point	---	---	---	---	D2500
Sodium and potassium	15553 15554 15555 15556	---	EN 14108 EN 14109 EN 14538	EN 14108 EN 14109 EN 14538	EN 14538
Calcium and magnesium	15553 15556	---	EN 14538	EN 14538	---
Phosphorus content.	15553	4951	EN 14107	EN 14107	D4951
Total contamination	---	---	EN 12662	EN 12662	---
Ester content	15342	---	EN 14103	EN 14103	---
Acid value	14448	664	EN 14104	EN 14104	D664
Free glycerol	15341	6584	EN 14105 EN 14106	EN 14105 EN 14106	D6584
Total glycerol	15344	6584	EN 14105	EN 14105	D6584
Monoglycerides	15342 15344 15908	6584	EN 14105	EN 14105	---
Diglycerides	15342 15344 15908	6584	EN 14105	EN 14105	---
Triglycerides	15342 15344 15908	6584	EN 14105	EN 14105	---
Methanol or Ethanol	15343	---	EN 14110	EN 14110	EN 14110
Iodine value	---	---	EN 14111	EN 14111	---
Oxidation stability at 110 °C	---	---	EN 14112	EN 15751 EN 14112	EN 15751
Water content	---	6304	EN ISO 12937	EN ISO 12937	---
Linolenic acid	---	---	---	EN 14103	---

Table 5. Established methods for the quality control of biodiesel in accordance with Brazilian, European and American standards.

Corrosivity to copper (copper strip corrosion) due to sulfur compounds, as well as due to free fatty acids, can lead to problems of corrosion in storage tanks and some engine parts. As acids are included in this parameter, it keeps a relationship with the acid index. The maximum acceptable values are degree 1, for standard EN 14214 and ANP, and degree 3 for ASTM D6751 (the degree number corresponds to a visual comparison with the copper strip established by ASTM D4951).

The acid number can increase or accelerate corrosion of the engine. It also measures the presence of free fatty acids and other acids and is related to the quality of the process. In injection systems that work at higher temperatures, faster biodiesel degradation may occur increasing the level of acidity and causing problems in filters. All standards have adopted as the specification limit for that parameter the value of 0.5 mg de KOH/g. The ANP standard recommended the methods ABNT NBR 14448, ASTM D664 and EN ISO 14104, and these last two methods are also adopted by European standard and American standard, respectively.

The cetane number measures the ignition quality of fuel. A low cetane number indicates a poorer ignition, which can form deposits and wear on pistons as well as provide greater fuel consumption. It depends upon the feedstock besides the oxygenate content in biodiesel. The number of cetane is measured with the aid of a special motor and the cetane index is calculated. The cetane index is a useful tool to estimate the number of cetanes according to ASTM standard [51].

The content of total glycerol and free glycerol as well as monoglyceride, diglyceride and triglyceride, reflects the quality of biodiesel. A high content of these can cause problems ranging from the formation of crystals, crusts inside the fuel storage tank, contributing to the formation or deposit of waste on pistons, injectors, valves, rings (of segments), filters, up to clogging the nozzles, decreasing the engine life. These are intermediate products of the process that ended up not reacting. The free glycerin (a byproduct) depends on the efficiency of separation of esters/glycerin.

The presence of residual alcohol in the biodiesel may cause corrosion on items of aluminum and zinc, and also can influence the flash point, reduce the cetane number and decrease the lubricity of the engine [51]. The alcohol content is determined by the chromatographic method EN ISO 14110 indicated by the standards EN 14214, ASTM D6751 and ANP. The Brazilian standard also recommends the method ABNT NBR 15343.

The iodine value is related to viscosity and cetane number, and indicates the quantitative degree of unsaturation of esters that form the biodiesel. The method suggested by the Brazilian and European standards is the same EN ISO 14111, though the latter is the only one to provide a limit to the parameter of 120 g I_2/100 g.

The oxidation stability determines the degradation of biodiesel, and mixtures thereof. It is related to the time required to degrade the biodiesel under controlled heating and in the presence of oxygen. The method adopted by the ANP standard and also EN 14214 is given in EN 14112, where the specification limit for both is 6 h at least (induction time by Rancimat method). For ASTM D6751 standard this limit is 3 hours and the recommended method is EN 15751.

In addition to the mentioned properties, other parameters may also be important and relate directly to the raw material used in the production of biodiesel: the soap content, the boiling range of esters, the peroxide index (that express the oil oxidation degree), the filterability (that express the difficulty to filter the oil before injecting into the engine) and the gum content, which expresses the amount of gums formed by polymerization of unsaturated oil components during the combustion [51].

In Brazil, the National Program for Production and Use of Biodiesel in one of its aspects provides the Program for Engine Test and Experiment, which is coordinated by the Ministry of Science and Technology (MCT). This program establishes test performances on vehicles and stationary engines and the gradual use increase of the blend biodiesel/diesel and evaluating its consequent technical feasibility. Currently in Brazil, B5 blend is used and it is extremely important to carry out tests to ensure commercial guarantees to vehicles moved with this blend and the quality of the blend commercialized. Currently in Brazil, B5 blend is used and test performing is extremely important to ensure commercial guarantees to vehicles moved with this blend and the quality of the blend commercialized.

At low temperatures, biodiesel tends to partially solidify or increase its viscosity (fluidity loss) inhibiting or even stopping the flow of fuel with the consequent clogging of filters, damaging the starting system of the engine. In tropical countries like Brazil, where most states have high temperatures throughout the year, the effects of these engines operating at low temperatures are minimized, and such problems are mainly present in countries in North America and Europe. Some properties analyzed in such conditions are: cloud point (ASTM D2500), which is the temperature at which, in a process of fuel cooling, it is observed the formation of first crystals, cold filter plugging point (EN 116), which is the temperature where the fuel ceases to be filtered when a cooling occurs and pour point (EN 3016) which is the temperature that the fuel loses its fluidity at established test conditions.

Thus, a major problem is the biodiesels from raw materials that have a high content of saturated compounds in its chain, as occurs for the tallow and the palm oil [72]. Biodiesels from raw materials of animal source when compared to those of vegetable origin have higher pour point, cold filter plugging point and cloud point.

Despite having standards for performing tests, there are no fixed and established values for these parameters. In Europe, they are measured for each country in terms of its climate. In U.S.A., they are dependent on the seasons and climate and in Brazil they depend on the states and months of the year, as shown in Table 6.

Despite the problems that can arise with the use of this fuel in cold seasons and countries with low average temperature, the biodiesel has many advantages over petroleum diesel as, it is a renewable fuel with lower emissions of particulate matters, polycyclic aromatic hydrocarbons, and sulfur compounds; storage and handling of this material is safer, since it has a flash point much higher than petroleum diesel, and also a higher flammability point which ensures a greater safety in loading/unloading and handling of this material by the drivers and operators [72].

Brazilian Federation Units	Maximum limit, °C											
	JAN	FEB	MAR	APR	MAY	JUN	JUL	AUG	SEP	OCT	NOV	DEC
SP - MG – MS	14	14	14	12	8	8	8	8	8	12	14	14
GO - DF - MT - ES – RJ	14	14	14	14	10	10	10	10	10	14	14	14
PR - SC - RS	14	14	14	10	5	5	5	5	5	10	14	14

Table 6. Maximum limits of cold filter plugging point for the Brazilian Federation states according to the months of the year in accordance with the ANP resolution number 14 (May 2012). For the other federation states (not included in the table), the value remains in 19°C.

5. Metallic corrosion in biodiesel

Although biodiesel has many properties that assist in good yield as a fuel as a relatively high flash point and good lubricating properties compared to diesel, some of these proper‐ ties facilitate its self-oxidation, and oxidation of the metallic materials which they are in con‐ tact. The metallic corrosion becomes extremely important since many of the engine parts are composed of variety metals such as aluminum, copper, stainless steel and alloys. The per‐ cent of aluminum in engine parts includes piston (100%), cylinder heads (70%), and engine blocks (19%). Pumps and injectors are composed of copper and its alloys. Parts composed of stainless steel include fuel filter, valve bodies, nozzle and pump ring [73-75]. The metallic corrosion may occur due to the following factors:

a) biodiesel is an ester so makes hydrogen bonds with water; then it becomes much more hygroscopic compared to diesel which is composed by hydrocarbons. Water acts on the cor‐ rosion of metallic materials, or it causes the hydrolysis of biodiesel, resulting in fatty acids and glycerol which increase metallic corrosion, or it promotes microbial growth and thereby microbial corrosion [73,75-77].

b) the presence of impurities like water, methanol, free glycerol, free fatty acid, catalyst resi‐ dues (Na and K) due to incomplete conversion or inadequate purification can also result in metallic corrosion [73,75].

c) due to its good lubricity, biodiesel dissolves more metallic parts than diesel, and these trace metals in solution enhance biodiesel degradation and promote metallic corrosion [73,75].

d) metals into biodiesel like brass, copper and aluminum act as catalysts for biodiesel oxida‐ tion. Therefore, the acid number of biodiesel increases proportionally with the corrosion rate for different metals [78-81].

Generally, the corrosion tests are performed by immersion tests with metallic coupons in bi‐ odiesel [79,80,82,83]. After immersion, the weight of the coupons was measured and the cor‐ rosion was analyzed by measurement of corrosion rate, according Equation 1:

$$Corrosion\ rate = \frac{W\ x\ 534}{D\ x\ T\ x\ A} \tag{1}$$

Where:

Corrosion rate = mpy (stands for mils (0.001 inch) per year);

W = weight loss (mg);

D = density (g cm^{-3});

A = exposed surface area (square inch);

T = exposure time (h).

Kaul et al. [80] studied the corrosivity of biodiesel from different oil sources by static immersion tests using metallic (aluminum alloy) piston for 300 days at 15 to 40 ° C. In this study the authors observed that the corrosion rate varied with the chemical composition of each oleaginous. The corrosion tests were performed with *Jatropha curcas*, *Karanja*, *Madhuca indica* and *Salvadora* biodiesels. Corrosion rates values were 0.0117, 0.0058, 0.0058, and 0.1236 mpy, respectively. The *Salvadora oleoides* biodiesel presented the highest rate due to its higher content of total sulfur (1200 ppm), while other biodiesels presented lower concentrations of total sulfur (around 1 ppm), except *Madhuca indica* biodiesel (164.8 ppm). *Jatropha curcas* biodiesel was slightly more corrosive than *Karanja* and *Madhuca indica* biodiesels because of its elevated concentration of C18:2 (19–41%) fatty acid, which is more prone to oxidation due to presence of two double bonds.

The metal corrosivity also depends on the nature of the metal exposed to the biofuel. Haseeb et al. [82] performed static immersion tests with coupons of copper and leaded bronze (87% Cu, 6% Sn, 6% Pb) in palm biodiesel at room temperature (25-30 °C) for 840 h; the corrosion rates for copper and bronze were 0.042 and 0.018 mpy, respectively. Additionally, at 60 °C in 2640 h, the corrosion rates of both metals were relatively higher, 0.053 mpy for copper and 0.023 mpy for bronze. The corrosion resistance of bronze was believed to be related to the presence of alloying elements such as tin (Sn) in the alloy. These results clearly show that copper is more prone to corrosion by biodiesel.

Similar behavior was observed by Fazal et al. [83], which performed static immersion tests and evaluated the corrosion rates of copper, brass (Cu: 58.5 %; Zn:41.5), aluminum, cast iron (C:3 %; Si:1.84 %; Mn: 0.82 %; P:0.098%; S:0.089; Fe: balance) at room temperature (25-27 °C) and for 120 days in palm biodiesel. The authors verified that copper presented higher corrosion rates (0.39278 mpy) followed by brass (0.209898 mpy), which contained zinc its composition that probably reduced its corrosion. Aluminum presented higher corrosion rates (0.173055 mpy) than cast iron (0.112232 mpy). This is accordance with Geller et al. [84], which reported that copper alloys are more prone to corrosion in biodiesel than ferrous alloys.

Recently, Hu et al. [85] compared the corrosion rates of several metals in rapeseed biodiesel and proposed corrosion mechanisms through static immersion tests at 43 °C for 60 days. The obtained corrosion rates of copper, carbon steel, aluminum, and stainless steel were 0.02334,

0.01819, 0.00324, and 0.00087 mmy, respectively (which correspond to 0.9336, 0.7276, 0.1296, and 0.0348 mpy, respectively). This study indicates that copper and carbon steel presented higher corrosion rates than aluminum and stainless steel.

Using scanning electron microscope with energy dispersive X-ray analysis and X-ray photoelectron spectroscopy to analyze the effects of biodiesel on the corrosion of different metallic materials, Hu et al. [85] reported that the corrosion process of metal surfaces in biodiesel was mainly attributed to the chemical corrosion and the products after corrosion were primarily fatty acid salts or metal oxides, depending on the studied metal. Elements of copper and iron are catalysts for the decomposition of biodiesel because they enabled various chemical reactions to easily take place. According to Hu et al. [85], metals were oxidized by oxygen and active oxygen atom dissolved in biodiesel, resulting in the formation of metal oxides (CuO, Cu_2O, Fe_2O_3, etc.). Copper and carbon steel were easily oxidized but aluminum and stainless steel were protected by films of metal oxide and then their corrosion rates were lower. The protective metal oxide layer prevented the metal surface from contact with the oxygen and atom oxygen as well as from contact with the oil sample.

Fazal et al. [86] has also found that the oxygen concentration increases with increasing temperatures, which may explain the high corrosion rates at higher temperatures. However, water and fatty acids are also responsible for metallic corrosion and need to be considered [80].

Similarly to Hu et al. [85] which carried out static immersion tests at 80 °C for 600 h in rapeseed biodiesel, Norouzi et al. [78] verified corrosion rates of 0.9 mpy for copper and 0.35 mpy for aluminum and confirmed that increase in temperature enhanced corrosion rates.

Another limiting factor that should be considered is the stirring. Fazal et al. [79] obtained corrosion rates of 0.586, 0.202, and 0.015 mpy for copper, aluminum, stainless steel, respectively, in immersion tests at 80 °C for 50 days under 250 rpm stirring in palm biodiesel. The corrosion rates presented superior values than those obtained in static immersion tests. Table 7 presents corrosion rates of different metallic materials in various biodiesels.

Analyzing the data in Table 7, palm biodiesel is less corrosive than rapeseed biodiesel. This different behavior can be correlated by the chemical composition of each biodiesel; rapeseed biodiesel presents 68.821 % oleic acid (C18:1) and 19.5927 % linoleic acid (C18:2) [78], whilst palm biodiesel presents 41.8 % oleic acid (C18:1) and 9.10 % linoleic acid (C18:2) [87].

The content of metal released to biodiesel during corrosion can be quantified. In the static immersion tests in rapeseed biodiesel performed by Hu et al. [85], 41.088 mg L^{-1} Cu (copper coupon), 3.544 mg L^{-1} Fe (carbon steel coupon) and 2.756 mg L^{-1} Fe and 9.02 mg L^{-1} Cr (stainless steel coupon) were obtained. On the other hand, Haseeb et al. [80] quantified lower amounts of metals in palm biodiesel (copper: 5 ppm Cu, bronze: 5 ppm Cu, 4 ppm Pb and 10 ppm Zn) because this biodiesel is less corrosive. According to McCormick et al. [88] the high concentration of metals in biodiesel generates higher oxidation of the biofuel. Biodiesel containing more than 6 ppm of metal exhibits very short OSI (oil stability index) induction time [88]. The negative effect of the presence of metal contaminants on the biodiesel oxidation stability was also reported by other authors who performed experiments adding organometallic standards or powdered metals into different biodiesels [25,89-94].

Operation	Biodiesel	METALS (mpy)							Ref.
		Aluminum	Copper	Bronze	Carbon steel	Stainless steel	Brass	Cast iron	
300 days 15-40 °C static	*Jatropha curcas*	0.0117	-	-	-	-	-	-	[80]
	Karanja,	0.0058	-	-	-	-	-	-	[80]
	Madhuca	0.0058	-	-	-	-	-	-	[80]
	Salvadora	0.1236	-	-	-	-	-	-	[80]
60 days 43 °C static	Rapeesed	0.1296	0.9336	-	0.7276	0.0348	-	-	[85]
25 days 80 °C static	Rapeesed	~0.35	~0.9	-	-	-	-	-	[78]
50 days 80 °C 250 rpm	Palm	0.202	0.586	-	-	0.015	-	-	[78]
120 days 25-27 °C static	Palm	0.173055	0.39278	-	-	-	0.209898	0.112232	[83]
35 days 23-30 °C static	Palm	-	0.042	0.018	-	-	-	-	[82]
		-	0.053	0.023	-	-	-	-	[82]

Table 7. Corrosion rate of some metals in different biodiesels.

Fazal et al. [79] and Hu et al. [85] have reported that biodiesel is more corrosive than diesel oil (based on the corrosion rates in different metallic materials). Figure 3 displays data from both reports [79,85] on the corrosion rate of copper, aluminum, stainless steel, and carbon steel in biodiesel and diesel. See that the first work only evaluated the corrosion of copper, aluminum, and stainless steel [79], while the second work also evaluated all the previous materials and in addition carbon steel [85]. The higher corrosion rates were also confirmed by the metal release verified in diesel and biodiesel, especially copper and iron [85]. Figure 3 also indicates that biodiesel presents more corrosive behavior than diesel oil and copper is not compatible with biodiesel.

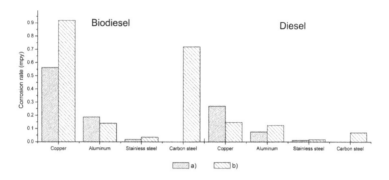

Figure 3. Corrosion rate of metals in biodiesel and diesel oil: a) adapted from Ref. [79]; b) adapted from ref. [85].

An alternative way to reduce the corrosive behavior of biodiesel is to use it as blends with diesel oil. According to Norouzi et al. [78], the greater the amount of diesel in blends, the lower corrosion rate as well as the lower total acid number (TAN). Figure 4 shows the variation of (a) corrosion rate of aluminum and copper coupons and of (b) TAN number of biodiesels exposed to these metallic materials through static immersion tests.

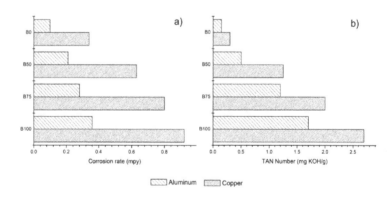

Figure 4. a) Corrosion rate of the Al and Cu and b) TAN of the rapeseed biodiesel-diesel blends in contact with both metals (adapted from ref. [78]).

There is a clear relationship between corrosion and TAN of biodiesel; the higher the corrosion rates the higher the TAN numbers. These results show that copper was more corrosive than aluminum and that as long the biodiesel-diesel blend was richer in biodiesel, higher corrosion rates were verified and consequently higher TAN numbers.

Some studies reported that the corrosiveness of biodiesel can be reduced by using corrosion inhibitors or antioxidants. Corrosion inhibitors act by the formation of adsorbed monolayer films at the metal-solution interface. The common corrosion inhibitors in oil are imidazolines, primary amines, diamines, amino-amines, oxyalkylated amines, naphthaneic acid, phosphate esters, dodecyl benzene sulfonic acids, etc [95]. Amine inhibitors can reduce the dissolution of metal oxide layers into biodiesel by forming stable oxide layers on the metal surface [95]. Fazal et al. [95] reported the inhibition effect on the corrosion of grey cast iron (C: 3%, Si: 1.84%, Mn: 0.82%, P: 0.098%, S: 0.089%, Fe: balance) immersed (static) in biodiesel containing 100 ppm of ethylenediamine (EDA), n-butylamine (nBA), tert-butylamine (TBA) at room temperature for 1200 h. The corrosion rates in each case are shown in Figure 5. The results indicate that EDA is the most powerful corrosion inhibitor under the experimental conditions. However, the analysis of fuel properties revealed that the biodiesel containing EDA was more degraded.

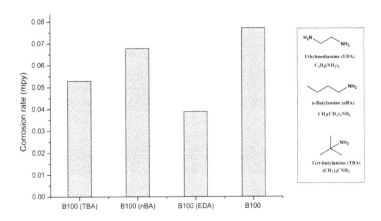

Figure 5. Corrosion rate of cast iron (duplicate and respective standard deviation) in palm biodiesel with and without corrosion inhibitors (EDA, nBA and TBA) and their respective chemical structures (adapted from ref. [95]).

Antioxidants can be classified into two groups: chain breakers and hydroperoxide decomposers. The chain breakers have two most common types of antioxidants: phenolic and amine-types. These antioxidants present a highly labile hydrogen that is able to abstract a peroxy radical and thus stops oxidation reactions in the ester chain [96]. The chain breakers antioxidants can be re-classified into natural antioxidants (tocopherols present in vegetables oils) and synthetic antioxidants such as butyl-hydroxytoluene (BHT), tert-butylhydroquinone (TBHQ), butyl-hydroxyanisol (BHA), pyrogallol (PY), and propyl gallate (PG) [96].

Liang et al. [97] have reported that synthethic antioxidants presented superior antioxidant activity in palm biodiesel than natural antioxidants. Other authors investigated BHT, BHA, TBHQ, and natural antioxidants as potential antioxidants in soybean oil biodiesel and TBHQ presented a superior antioxidant activity [98,99]. Jain and Sharma compared the efficiency of eight synthetic antioxidants in different biodiesels and they concluded that only three antioxidants significantly increased the stability of biodiesel in the order of TBHQ > PY

> PG [96]. The addition of antioxidants can also overcome the low oxidation stability of bio-diesels promoted by metal contamination (experiments performed by adding organometal-lic standards) in order to re-establish the required 6-hour induction time [25,90-94]. This statement is especially essential considering the presence of metals in biodiesel due to corro-sion of containers and engine components.

Almeida et al. [100] evaluated the effect of the synthethic antioxidant TBHQ on the corrosive character of biodiesel against copper coupon through static immersion tests. Due to the strong catalytic effect of copper towards biodiesel oxidation, the oxidation stability of bio-diesel decreased tremendously after 24 h of exposure even in the presence of TBHQ. The copper content in biodiesel continuously increased with the exposure time and the metal concentration was much higher in the non-stabilized biodiesel. Then, the presence of TBHQ decreased the corrosion rate of the copper coupon and the authors claimed that the antioxi-dant may have acted as a corrosion inhibitor through the formation af a protective layer on the metallic surface. Performing mass-spectrometry (MS) and MS-MS analysis of deteriorat-ed biodiesel containing TBHQ (after the metal corrosion for 24 h and 168 h), the authors have also identified the formation of new molecules of high molecular weight formed by the association between oxidized antioxidants molecules and free radicals of long-chain mole-cules (fatty acid derivatives). These results give light to possible side-reactions between phe-nolic antioxidants and esters molecules of biodiesel under oxidative conditions.

6. Analytical methods for contaminants in biodiesel

6.1. Free and total glycerol

The determination of the content of total glycerol and free glycerol in biodiesel is required by the different regulatory agencies and upper limits are established as listed in Table 4. Glycerol is the major by-product in the biodiesel production and its removal is necessary. A high content of total glycerol (sum glycerol, mono-, di- and triacylglycerols) and free glycer-ol can cause problems ranging from the formation of deposits in injectors, sediments inside the fuel storage tank, reducing the engine life. The recommended method for the determina-tion of total and free glycerol in biodiesel is gas chromatography (GC) (Table 5). Both ASTM and EN methods require a derivatization step (silylation reaction) using N-methyl-N-trime-thylsilyltrifluoroacetamide (MSTFA) with a few differences in the standard solution concen-trations and column temperature conditions. Table 8 and 9 present a comprehensive list of analytical methods developed for the determination of free glycerol and total glycerol, re-spectively. Generally, all analytical methods listed in Tables 8 and 9 required a sample prep-aration step. GC methods often required a derivatization step (modification of a functional group of the analyte such as the silylation reaction) which improves the chromatographic performance. Derivatization reagents (when required) are cited and in a separate column other sample preparation steps (analyte extraction procedures) are detailed in accordance with information contained in the literature.

Method	Derivatization reagents	Analyte extraction procedures	Detection limit	Ref.
Enzymatic assay	No	2 g sample + 6.0 mL of 0.1 mol L^{-1} HCl	Not cited	[101]
GC-FID/ MS	BSTFA	No	10^{-4}-10^{-5} % wt	[102]
HPSEC	No	No	Not cited	[103]
GC-FID	MSTFA/ pyridine	No	No cited	[104]
GC-FID/ MS	BSTFA	No	10^{-4} % wt	[105]
HPLC-PAD	No	4 g sample + 45 g H_2O + 50 mL hexane stirred for 30 min at 40°C, 2 h standing	1 µg g-1	[106]
GPC-RI	No	PTFE membrane filtration	Not cited	[107]
Greenhill enzymatic assay	No	Not cited	5 ppm	[108]
UV-Vis	$NaIO_4$, acetylacetone	1 g sample + 4 mL hexane + 2 mL distilled water + 2 mL ethanol; 5 min vortex-stirring, 15 min centrifugation at 2000 rpm	Not cited	[109]
HPSEC	No	No	Not cited	[110]
HPLC-RI	No	4-20 g sample + 4.5 mL distilled water; 30 min stirring, 2 h standing	4×10^{-4} % wt	[111]
GC-FID	MSTFA/ pyridine	No	Not cited	[112-114]
CE-UV-Vis-DAD	$NaIO_4$	200 mg sample + 800 mg water + 200 µL chloroform; 10 min vortex-stirring, 15 min centrifugation at 2000 rpm	4.3 mg L^{-1}	[115]
SFC-MS-UV-ELSD	No	Methanol dissolution	Not cited	[116]
IC-PAD	No	5 g sample + 45 g distilled water; 5 min shaking, 5 min standing	7×10^{-5} % wt	[117]
UV-Vis-DAD	$NaIO_4$, acetylacetone	Aqueous extraction, 30 min heating	0.011 % wt	[118]
Cyclic voltammetry	No	2 g sample + 6.0 mL water; 5 min vortex, 10 min centrifugation; C18 filtration	2.3 mg L^{-1}	[119]
FIA-UV-Vis	KIO_4, acetylacetone	1 g sample + 4 mL distilled water; 30 min shaking, 5 min centrifugation at 3000 rpm	4×10^{-4} % wt	[120]

Method	Derivatization reagents	Analyte extraction procedures	Detection limit	Ref.
Enzymatic assay with amperometry	No	400 µL sample + 800 µL distilled water + 800 µL ethanol + 1600 µL heptane; 2 min vortex-stirring and centrifugation	1×10^{-5} % wt	[121]
Enzymatic assay with amperometry	No	400 µL sample + 800 µL distilled water + 800 µL ethanol + 1600 µL heptane; 2 min vortex-stirring and centrifugation	0.013 % wt	[122]
Titration	$NaIO_4$, $C_2H_4(OH)_2$	Aqueous Extraction	9×10^{-4} % wt	[123]
GC-MS	BSTFA, TMCS, MSTFA	No	0.04 µg mL^{-1}	[124]
HPAE-PAD	No	Aqueous extraction	0.5 µg kg-1	[125]
UV-Vis	$NaIO_4$, acetylacetone	Aqueous extraction	0.5 mg L^{-1}	[126]
UV-Vis	No	Solid phase extraction	0.004 % wt	[127]
FIA-Amperometry	No	250 mg sample + 5.0 mL water; 5 min vortex, 10 min centrifugation	5 mg kg^{-1}	[128]
Enzymatic assay with colorimetric detection	No	400 µL sample + 800 µL ethanol + 800 µL distilled water + 1600 µL heptane; 1 min vortex-mixing, 2 min centrifugation	7.1×10^{-6} % wt	[129]
FIA-PAD	No	1.0 g sample + 4.0 mL water; 5 min vortex-stirring and centrifugation	44.2 µg L^{-1}	[130]
Cyclic voltammetry	No	Not cited	33 µmol L^{-1}	[131]
GC-FID-MS	MSTFA	No	0.053 % wt	[132]
GC-FID	MSTFA	No	0.02-0.09 % wt	[133]

CE: Capillary Electrophoresis; DAD: Diode Array Detector; ELSD: Evaporative Light Scattering Detector; FIA: Flow Injection Analysis; FID: Flame Ionization Detector; GC: Gas Chromatography; GPC: Gel Permeation Chromatography; HPAE: High Performance Anion Exchange Chromatography; HPLC: High Performance Liquid Chromatography; HPSEC: High Performance Size Exclusion Chromatography; IC: Ion Chromatography; IR: Infrared Spectrophotometry; PAD: Pulsed Amperometric Detection; RI: Differential Refractive Index Detector; SEC: Size Exclusion Chromatography; SFC: Supercritical Fluid Chromatography; UV-Vis: Ultraviolet Spectrophotometry; BSTFA: N, O-bis (trimethylsilyl)-trifluoroacetamide; MSTFA: N-methyl-N-(trimethylsilyl)-trifluoroacetamide; PTFE: Polytetrafluoroethylene; TMCS: Trimethylchlorosilane.

Table 8. Methods for free glycerol determination in biodiesel.

Method	Derivatization reagents	Sample preparation	Detection limit	Ref.
Enzymatic assay	No	Saponification; Solid phase extraction with C8	Not cited	[101]
Greenhill enzymatic assay	No	Saponification	75 ppm	[108]
GC-FID	MSTFA	No	Not cited	[114]
IC-PAD	No	Saponification; 5 g sample + 45 g distilled water; 5 min shaking, 5 min standing	7×10^{-5} % wt	[117]
UV-Vis-DAD	$NaIO_4$, acetylacetone	Saponification; Aqueous extraction, 30 min heating	0.064 % wt	[118]
Enzymatic assay with amperometry	No	Transesterification; 0.1 g sample + 3.9 mL distilled water + 5.0 mL heptane; 2 min vortex-stirring and centrifugation	1×10^{-5} % wt	[121]
Titration	CH_3NaO, $NaIO_4$, $C_2H_4(OH)_2$	Aqueous Extraction	0.0046 % wt	[123]
HPLC-UV-Vis	9,9-dimetoxifluorene	No	0.05 % wt	[134]
HPSEC-PAD	No	Saponification	0.5 µg kg^{-1}	[125]
UV-Vis	$NaIO_4$, acetylacetone	Saponification; Aqueous extraction	1.4 mg L^{-1}	[126]
Enzymatic assay with colorimetric detection		Transesterification; 0.5 g sample + 2.0 mL distilled water + 2.0 mL heptane; 1 min vortex-mixing, 2 min centrifugation	7.1×10^{-6} % wt	[129]
GC-FID-MS	MSTFA	No	2.458 % w/w	[132]
GC-FID	MSTFA	No	0.15-0.69 % wt.	[133]

Table 9. Methods for total glycerol determination in biodiesel.

Several GC methods coupled with flame ionization (FID) or mass spectrometric (MS) detectors are listed in Tables 8 and 9 including works published since the early Nineties. The most recent contribution reported a new GC method with reduced sample preparation and analysis time (25 min elution) that can be applied for a wide range of oilseed-derived biodiesels including biodiesel from tallow, babassu, and palm kernel, which contain shorter chain fatty acids and then cannot be accurately analyzed by the EN 14105 method [133]. The EN14105 (and also the ASTM 6584 method) is based on the GC-FID method developed by

Plank and Lorbeer [104], who reported the simultaneous determination of glycerol, mono-, di-, and triacylglycerides in C18 methylic esters produced from vegetable oils by the silylation of the free hydroxyl groups using MSTFA. Analytical methods employing different modes of liquid chromatography coupled with varied detectors were also reported but in less extension. The first HPLC method employed pulsed-amperometric detection and thus it was not necessary the derivatization step [106]. Additionally, this HPLC method allows the determination of residual alcohol (methanol or ethanol) [106]. A high performance size exclusion chromatography method was developed for the simultaneous determination of the total amounts of mono-, di-, and triacylglycerides, fatty acid methyl esters, free glycerol and methanol [110]. The method is simple, robust, relatively fast, and required minimal sample treatment (no derivatization steps); however, the SEC columns are quite expensive. Preliminary results using a supercritical fluid chromatography coupled with three different detectors (MS, UV, and ELSD) were reported [116]. Separation of fatty acid methyl esters, free fatty acids, and glycerol was obtained in less than 5 min [116].

The oxidation reaction of glycerol with periodate is the basis of the first spectrophotometric method developed for glycerol determination in biodiesel [109]. The oxidation of glycerol results in the formation of formaldehyde, which is reacted with acetylacetone leading to the formation of 3,5-diacetyl-1,4-dihydrolutidine that can be measured at 410 nm [109]. Other spectrophotometric methods using a similar approach were reported in the literature [118,120,126,127]. The great advantage of the spectrophometric methods is their low-cost, rapidness, accuracy and moderate sensitivity. A capillary electrophoresis method was developed based on the UV detection of iodate generated by the oxidation of glycerol by periodate [115]. Similarly, glycerol in biodiesel was determined by alkaline titration of formic acid generated by the oxidation of glycerol by periodate, which can be considered a very simple method easily assessed by local producers [123].

Electrochemical approaches (cyclic voltammetric, amperometric and pulsed-amperometric detectors) were also developed and coupled with chromatographic techniques, flow-injection methods, and enzymatic reactions. Electrochemical methods present high sensitivity, selectivity, can be easily miniaturized and require portable commercially-available instrumentation. Additionally, glycerol can be electrochemically detected without the use of derivatization reagents. Amperometric detection was often applied due to the easiness of its association with flow techniques such as chromatography or flow-injection methods [106,117,125,128,130]. The selection of amperometry or pulsed-amperometry is more related to the electrochemical oxidation of glycerol at different working electrode materials. Enzymatic assays employed enzymes which specifically convert glycerol to dihydroxyacetone phosphate generating H_2O_2 and consuming oxygen. Then, H_2O generation or oxygen consumption can be easily monitored by electrochemical techniques (similarly to portable glucose-sensors) as well as using colorimetric assays (the commercial-available Greenhill assay provides the formation of a quinoneimine dye that shows maximum absorbance at 540 nm) [108]. The development of such kits for fast and low-cost monitoring of glycerol in biodiesel is an alternative to GC methods which employ bulky instrumentation and organic solvents during the derivatization step. However, enzymes require special condition of storage and limited shelf-life time. To overcome such a drawback, the development of (electro)chemical sensors based on the direct detection of glycerol is promising.

Analytical methods developed for the determination of total glycerol in biodiesel (Table 9) typically reported the conversion of mono-, di-, and triglycerides into glycerol by saponification or transesterification reactions except GC or HPLC methods which separate and directly quantify each component present in biodiesel. Other analytical methods developed for the determination of free glycerol (listed in Table 8) can also be readily adapted to perform the determination of total glycerol.

6.2. Trace metals

The presence of metals in biodiesel can be arisen from catalyst residues (Na, K, Ca, Mg) and due to corrosion of storage tanks and automotive engine parts (e.g. Al, Cu, Cr, Fe, Mn, Pb, Zn, etc.). Metal ions cause the formation of deposits of insoluble soaps, as well as catalyze polymerization reactions of biodiesel degradation. In this way the different regulatory agencies establishes upper limits as listed in Table 4 for group I (Na and K) and group II (Ca and Mg) cations. However, no limits are established for transition metals which are strong catalysts for biodiesel oxidation even at trace concentrations as previous works have reported [89-94]. Therefore, analytical methods capable of monitoring trace metals in biodiesel are required and this information can be correlated with biodiesel oxidation stability. Table 10 presents a list of analytical methods developed for the determination of metals in biodiesel including sample preparation steps when required and detection limits.

Atomic absorption spectrometry (AAS) and inductively coupled plasma optical emission spectrometry (ICP OES) are the recommended technique by Brazilian and European standards (Table 5). The first analytical methods reported for the determination not only of Na, K, Ca and Mg as well as other metals were the spectrometric methods such as the recommended techniques AAS and ICP OES. AAS coupled to graphite furnace (electrothermal atomization) provides higher sensitivity and for this reason it has been applied for metal determination in biodiesels, although relatively higher cost and longer analysis time are verified. The main advantage of ICP OES is the lower chemical and spectral interferences and multi-element determination; dozens of elements can be analyzed at once after simple sample dilution, including phosphorus and sulfur which are also regulated by European, American and Brazilian norms (Table 4 and 5). However, the overall analysis costs can be so elevated that this method is not available to all analytical laboratories of quality control. ICP MS provides higher sensitivity than ICP OES if sub-ppb levels are required. Most of the spectrometric methods require very simple preparation steps such as sample dilution or sample microemulsion with surfactants, which is an elegant strategy to avoid sample digestions using of concentrated acids. Pre-concentration steps can be applied for AAS with flame atomization which is a low-cost spectrometer but does not provide low detection limits generally required for trace metal determinations. Then pre-concentration steps supplies the low sensitivity of flame atomic absorption (or emission) spectrometers. More details on the use of spectrometric methods to determine metals and metalloids in automotive fuels (including biodiesel) can be found in a recent review [156].

Method	Analyte	Sample preparation	Detection limit	Ref.
Inductively coupled plasma optical emission spectrometry	Ca, K, Mg, Na	Sample dilution in kerosene	0.4 – 0.9 mg kg^{-1}	[135]
Inductively coupled plasma mass spectrometry	31 elements	Sample dilution in kerosene	0.0109 – 22.7 µg kg^{-1}	[136]
Inductively coupled plasma optical emission spectrometry (axial viewing)	Ca, Mg, K	Sample dilution in ethanol	0.005 – 0.1 µg g^{-1}	[137]
Inductively coupled plasma optical emission spectrometry (axial viewing)	Ca, Cu, Fe, Mg, Mn, Na	Sample emulsion with Triton X-100 and water	0.007 – 0.165 µg g^{-1}	[138]
Flame atomic absorption spectrometry	Na, K	Sample microemulsion with n-pentanol and Triton X-100	0.1 and 0.06 µg g^{-1}	[139]
Flame atomic emission spectrometry	Na, K	Sample microemulsion with n-pentanol and aqueous acid solution	0.1 µg g^{-1}	[140]
Graphite-furnace atomic spectrometry	As	Sample microemulsion with n-pentanol and aqueous acid solution	0.3 mg kg^{-1}	[141]
Graphite-furnace atomic spectrometry	Cu, Pb, Ni, Cd	Focused microwave wet digestion	Not cited	[142]
Cold-vapor atomic fluorescence spectrometry	Hg	Sample microemulsion with n-pentanol and Triton X-100	0.2 µg kg^{-1}	[143]
Inductively coupled plasma optical emission spectrometry	28 elements	Microwave acid digestion	0.1 – 136.5 µg g^{-1}	[144]
Inductively coupled plasma optical emission spectrometry and mass spectrometry	19 elements	Microwave acid digestion	0.001 – 0.4 µg g^{-1}	[145]
Graphite-furnace atomic spectrometry	Ni, Cd	Sample microemulsion with Triton X-100 and acid aqueous solution	0.9 and 0.1 µg L^{-1}	[146]
Flame atomic emission spectrometry	Cu, Ni, Zn	Pre-concentration by adsorption on chitosan microspheres	Not cited	[147]
Flame atomic emission spectrometry	Na, K	Sample dilution in ethanol	2.16 and 2.00 mg kg^{-1}	[148]
Inductively coupled plasma mass spectrometry	32 elements	Microwave acid digestion	10^{-6} mg kg^{-1}	[149]
Potentiometry	K	Not required	0.01 ppm	[150]
Potentiometry	K	Liquid-liquid extraction	1.9 x 10^{-5} mol L^{-1}	[151]
Ion chromatography	Na, K, Ca, Mg	Liquid-liquid extraction, heating, sonication	0.11 – 0.42 mg kg^{-1}	[152]
Ion chromatography	Na, K	Liquid-liquid extraction	Not cited	[153]
Capillary electrophoresis	Ca, K, Mg, Na	Liquid-liquid extraction	0.07 – 0.14 mg L^{-1}	[154]
Square-wave stripping voltammetry	Sn	Dry-ashing decomposition	0.14 µg L^{-1}	[155]
Potentiometric stripping analysis	Cu	Sample dilution in hydroethanolic electrolyte	200 ng g^{-1}	[81]

Table 10. Methods for trace metal determination in biodiesel.

Separation techniques such as ion chromatography and capillary electrophoresis were applied for simultaneous determination of cations in biodiesel after a simple liquid-liquid extraction [152-154]. The capillary electrophoretic (CE) method presented faster separation of ions, employed very low sample volumes and capillaries presents much lower cost than IC columns. Moreover, the CE method can also be used for glycerol determination after its chemical conversion by periodate as a previous method described [115].

Electroanalytical methods for metal determination in biofuels were recently reviewed [157,158]. Potentiometry is a well-known technique which allow sensitive detection of K in aqueous solutions and were also applied for biodiesel analysis after a liquid-liquid extraction [151]. An ion-selective electrode sensor associated with cellophane semi-permeable membrane was applied for K determination in biodiesel without a sample preparation step [150]. The elimination of the sample preparation step is a tendency in modern analytical methods since this step may provide sample contamination, analyte losses, and high analysis time, whose characteristics are avoided when developing an analytical method for routine analyses. Electroanalytical potentiometric stripping analysis was applied for Cu determination in biodiesel after its dilution in hydroethanolic electrolyte [81]. This is the first report on the use of electroanalysis for metal determination in biodiesel. Electroanalytical methods provide real advantages for routine analysis such as high sensitivity and selectivity employing a low-cost portable instrumentation. Similar technology applied for gluco-sensors (using disposable sensors) can be extended to on-site analysis of biodiesel, aiming not only the determination of trace metals but also other species in the biofuel [157,158].

6.3. Other trace contaminants

Other trace contaminants can be found in biodiesel such as water (moisture), residual alcohol (typically methanol and ethanol), sterols, phosphorus and sulfur. Table 4 shows the upper limits established for water and residual alcohol and Table 5 lists the recommended methods for the analysis of each parameter. The GC method developed Mittelbach et al [105] can also be applied for the determination of alcohol residues in biodiesel as well as the HPLC-PAD method [106]. A flow analysis method coupled to a membrane extraction was developed for methanol determination in biodiesel [159]. Methanol was detected by spectrophotometry (at 240 nm) after reaction with alcohol oxidase in aqueous solution [159].

Sterols are minor components found in animal fats and vegetable oils and that occur in biodiesel due to their solubility in the biofuel. Sterol glycosides can accelerate precipitate formation in biodiesel even at room temperature and block fuel filters [160-162]. HPLC methods [160,161], mass spectrometry [161], and a GC MS method [162] have been exploited for the determination of sterol glycosides. More information about the use of chromatographic techniques for biodiesel and biodiesel blends (not only including the analysis of sterols) can be found in a recent review [163].

The contents of phosphorus and sulfur in biodiesel are additional parameters regulated by European, American and Brazilian norms. Some ICP OES methods reported for metal determinations in biodiesel (Table 10) were also applied for the determination of phosphorus and sulfur [136,144,145,149]. A graphite-furnace AAS method was developed for the direct determination of phosphorus in biodiesel using a solid sampling accessory [164]. Simpler methodologies using spectrophotometry were developed [165,166]. In the first work biodiesel samples were mineralized (dry ashing) and their residue containing phosphate was reacted with 1-amino-2-naphthol-4-sulfonic acid to form a blue molybdenum complex [165]. In the second work biodiesel samples were digested using an acid mixture and the obtained solution containing phosphate was mixed with ammonium molybdate and potassium and antimony tartrate ion to form phosphomolybdic acid (yellow) [166]. An electroanalytical method was developed for phosphorus determination in biodiesel using a phosphomolybdic modified electrode [167]. X-ray fluorescence [168], and improvements on ICP OES [169] and ICP MS [170] methods for sulfur determination in biodiesel have been reported.

Analytical methods for monitoring of the transesterification reaction and for the determination of fatty mono-alkyl esters in biodiesel-diesel blends or in pure biodiesel as well as other parameters such as biodiesel oxidation and thermal stability are well-addressed in the review by Monteiro et al. [171] and is not discussed in this text.

7. Conclusion and perspective

Much effort has been dedicated to the development of new methods for biodiesel production and improvement of the traditional ones. A large variety of raw materials have been investigated for biodiesel production in the world, and Brazil has favourable environment and climate conditions in this scenario for the development of biodiesels from new oilseeds although the large Brazilian soybean biodiesel production. The production of biodiesels of new oilseeds has economic impact since small local producers can contribute for biodiesel production and the special characteristics of these biodiesels may improve the physical-chemical properties (e.g. acidity, oxidation stability, viscosity, etc.) of other biodiesels by using blends of biodiesels. Microalgae for biodiesel production is a promising sustainable source which has received incresing interest from researchers worldwide. Metallic corrosion is a real problem in storage stability of biodiesels and inside diesel engines and thus requires constant investigation. The monitoring of trace metals in the biofuel may provide information on how trace metals really affect biodiesel oxidation stability. Moreover, the presence of antioxidants and corrosion inhibitors in biodiesel plays key role on the metallic corrosion and deeper investigations would indicate the real need of additives in the biofuel and their required concentration. Modern analytical methods developed for monitoring contaminants in biodiesel have been reported and the creation of portable devices for the quality control of biofuels is a promising tendency. In the future, any local producer would have access to fast and reliable technologies to certificate the quality of its biodiesel during the production process. Additionaly, the real-time and on-site analysis of biofuels is desirable at local gas stations which receive large volumes of biodiesel and need to quickly check the real quality of the biofuel.

Acknowledgments

The authors are grateful to FAPEMIG (APQ-00236-12 and APQ-02276-09), CNPq (476667/2008-9 and 305227/2010-6) and CAPES for the financial support.

Author details

Rodrigo A. A. Munoz[1*], David M. Fernandes[1], Douglas Q. Santos[2], Tatielli G. G. Barbosa[1] and Raquel M. F. Sousa[1]

*Address all correspondence to: raamunoz@iqufu.ufu.br

1 Institute of Chemistry, Federal University of Uberlândia (UFU), Uberlândia-MG, Brazil

2 Technical School of Health Sciences, Federal University of Uberlândia (UFU), Uberlândia-MG, Brazil

References

[1] Network of Sustainability. What is Sustainability? Available in http://www.sustenta-bilidade.org.br. Accessed on September 10, 2012.

[2] Energy Information Administration (EIA). International Energy Outlook 2011 – "World Energy Demand and Economy Outlook". Available in (http://www.oecd.org). Accessed on September 10, 2012.

[3] SEBRAE – Spelling Book on Biodiesel. Available in www.biblioteca.biodiesel.se-brae.com.br Accessed on September 10, 2012.

[4] Diesel R. Die Entstehung des Dieselmotors. Berlin: Verlag von Julius Springer; 1913

[5] Knothe G, Gerpen JV, Krahl J, Ramos LP. The Biodiesel Handbook, Berlin: Ed. Edgard Blücher; 2006

[6] Fangrui M, Hanna MA. Biodiesel: A review. Bioresource Technology 1999; 70 1-15.

[7] Expedito Parente – Avalaible in http://www.inovacao.unicamp.br/report/entre-expedito.php Accessed on September 10, 2012.

[8] ANP – National Agency of Petroleum, Natural Gas and Biofuels. The ANP and the Biofuels. Available in http://www.anp.gov.br. Accessed on September 10, 2012.

[9] Secretary of Petroleum, Natural Gas and Renewable Fuels and Department of Renewable Fuels. Monthly Bulletin of Renewable Fuels, Issue No. 53, June 2012.

[10] CONAB - National Supply Company. Available in http://www.conab.gov.br. Accessed on September, 10, 2012.

[11] Jatropha Project EPAMIG /FINEP. Final Report relating 1st period finalized in March 31, 1985. Available in http://www.epamig.br/index.php?option=com_docman&task=doc_download&gid=14&Itemid=109 Accessed on September 10, 2012.

[12] Borsato D, Maia ECR, Dall'Antonia LH, da Silva, HC, Pereira JL. Kinetics of oxidation of biodiesel from soybean oil mixed with TBHQ: determination of storage time. Química Nova 2012; 35 733-737.

[13] ABRAPA – Brazilian Association of Cotton Producers. Management Report, Biennium 2008-2010, 2010, 1, 12.

[14] ANVISA – National Agency of Sanitary Surveillance. Resolution No. 482, September 23, 1999.

[15] Jorge FAS. Study of adsorption of gossypol existing in cottonseed oil. Dissertation of Master's degree, Fortaleza, 2006.

[16] EMBRAPA - Brazilian Agricultural Research Company. Agroenergy review No 2. EMBRAPA Agroenergy May, 2011.

[17] USDA – United States Department of Agriculture. Canola Production and Processing. Available in http://www.ers.usda.gov. Accessed on July 30, 2012.

[18] Georgogianni KG, Katsoulidis AK, Pomonis PJ, Manos G, Kontominas MG. Transesterification of rapeseed oil for the production of biodiesel using homogeneous and heterogeneous catalysis. Fuel Processing Technology 2009; 90 671-676.

[19] Endalew AK, Kiros Y, Zanzi R. Heterogeneous catalysis for biodiesel production from Jatropha curcas oil (JCO). Energy 2011; 36 2693–2700.

[20] Heller J. Physic Nut: jatropha curcas L. International Plant Generic Resources Institute, 1996.

[21] Bártoli JAA. Physic Nut (jatropha curcas) Cultivation in Honduras Handbook, Agricultural Communication Center of the Honduran Foundation for Agricultural Research, 2008.

[22] Oliveira JS, Leite PM, Souza LB, Mello VM, Silva EC, Rubim JC Meneghetti SMP, Suarez PAZ. Characteristics and composition of Jatropha gossypol and jatropha curcas L. oils and application for biodiesel production. Biomass and Bioenergy 2009;33 449-453.

[23] Kumar A, Sharma S. An evaluation of multipurpose oil seed crop for industrial uses (jatropha curcas L.): A review, Industrial Crops and Products 2008;28 1-10.

[24] Koris A, Vatai G. Dry degumming of vegetable oils by membrane filtration, Desalination 2002;(148) 149-153.

[25] Sarin, A.; Arora, R.; Singh, N.P.; Sarin, R.; Malhota, R.K.; Sharma, M. Synergistic effect of metal deactivator and antioxidant on oxidation stability of metal contaminated Jatropha biodiesel. Energy 2010;35 2333-2337.

[26] ABIOVE - Brazilian Association of Oil Industries Vegetables. Monthly Statistic. Reference Month – July, 2012.

[27] Kulkarni MG, Dalai AK. Waste Cooking Oil An Economical Source for Biodiesel: A Review. Industrial & Engineering Chemistry Research 2006;45 2901-2913.

[28] Zheng S, Kates M, Dubé MA, McLean DD. Acid-catalyzed production of biodiesel from waste frying oil. Biomass and Bioenergy 2006;30 267-272.

[29] Refaat AA, Attia NK, Sibak HA, El Sheltawy ST, El Diwani GI. Production optimization and quality assessment of biodiesel from waste vegetable oil. International Journal of Environmental Science and Technology 2008;5 75-82.

[30] Canakci M. The potencial of restaurant waste lipids as biodiesel feedstocks. Bioresource Tecnhology 2007;98 183-190.

[31] Tang H, Salley SO, Simon Ng KY. Fuel properties and precipitate formation at low temperature in soy-, cottonseed-, and poultry fat-based biodiesel blends. Fuel 2008;87 3006-3017.

[32] Da Silva JPV, Serra TM, Gossmann M, Wolf CR, Meneghetti MR, Meneghetti SMP. Moringa Oleira oil: Studies of characterization and biodesel production. Biomass and Bioenergy 2010;34 1527-1530

[33] Rashid, U.; Anwar, F.; Moser, B.R.; Knothe, G. Moringa oleifera oil: A possible source of biodiesel. Bioresource Technology 2008;99 8175-8179

[34] Ayerza, R. Seed yield components, oil content, and fatty acid composition of two cultivars of Moringa (Moringa oleifera Lam.) growing in the Arid Chaco of Argentina. Industrial Crops and Products 2011;33 389-394.

[35] Azam MM, Waris A, Nahar NM. Prospects and potential of fatty acid methyl esters of some non-traditional seed oils for use as biodiesel in India. Biomass and Bioenergy 2005;29 293-302.

[36] Lalas S, Tsaknis J. Characterization of Moringa Oleifera Seed oil Variety "Priyakulam". Journal of Food Composition and Analysis 2002;15 65-77.

[37] Ibrahim SS, Ismail, M, Samuel G, Kamal E, El Azhari T. Benseed: A potential oil source. Agriculture Research and Review. 1974;52 47-50.

[38] Kegl B. Effects of biodiesel on emissions of a bus diesel engine. Bioresource Technology 2008;99 863-873.

[39] Meher LC, Vidya Sagar D, Naik SN. Technical aspects of biodiesel production by transesterification--a review. Renewable and Sustainable Energy Reviews 2006;10 248-268.

[40] Srivastava A, Prasad R. Triglycerides-based diesel fuels. Renewable and Sustainable Energy Reviews 2000;4 111-133.

[41] Refaat AA, El Sheltawy ST, Sadek KU. Optimum reaction time, performance and exhaust emissions of biodiesel produced by microwave irradiation. International Journal of Environmental Science and Technology 2008;5 315-322.

[42] Encinar JM, González JF, Rodríguez-Reinares A. Ethanolysis of used frying oil. Biodiesel preparation and characterization. Fuel Processing Technology 2007;88 513-522.

[43] Ma F, Hanna MA. Biodiesel production: a review. Bioresource Technology 1999;70 1-15.

[44] Gerpen JV. Biodiesel processing and production. Fuel Processing Technology 2005;86 1097-1107.

[45] Fukuda H, Kondo A, Noda H. Biodiesel fuel production by transesterification of oils. Journal of Bioscience and Bioengineering 2001;92 405-416.

[46] Marchetti JM, Miguel VU, Errazu AF. Possible methods for biodiesel production. Renewable and Sustainable Energy Reviews 2007;11 1300-1311.

[47] Singh SP, Singh D. Biodiesel production through the use of different sources and characterization of oils and their esters as the substitute of diesel: A review. Renewable and Sustainable Energy Reviews 2010;14 200-216.

[48] Sinha S, Agarwal AK, Garg S. Biodiesel development from rice bran oil: Transesterification process optimization and fuel characterization. Energy Conversion and Management 2008;49 1248-1257.

[49] Crabbe E, Nolasco-Hipolito C, Kobayashi G, Sonomoto K, Ishizaki A. Biodiesel production from crude palm oil and evaluation of butanol extraction and fuel properties. Process Biochemistry 2001;37 65-71.

[50] Faccini CS, Da Cunha ME, Moras MAS, Krause Lc, Manique MC, Rodrigues MRA, Benvenutti EV, Caramao EB. Journal of the Brazilian Chemical Society 2011; 22 558-563.

[51] Mittelbach M. Diesel fuel derived from vegetable oils, VI: Specifications and quality control of biodiesel. Bioresource Technology 1996;56 7-11.

[52] Tomasevic AV, Siler-Marinkovic SS. Methanolysis of used frying oil. Fuel Processing Technology 2003;81 1-6.

[53] Lima AL, Lima AP, Portela FM, Santos DQ, Neto WB, Hernández-Terrones MG, Fabris JD. Parameters of transesterification reaction of corn oil with ethanol for biodiesel production. Eclética Química 2010;35 101-106.

[54] Freedman B, Pryde EH, Mounts TL. Journal of the American Oil Chemists' Society 1984;61 1638-1643.

[55] Schwab AW, Bagby MO, Freedman B. Preparation and properties of diesel fuels from vegetable oils. Fuel 1987;66 1372-1378.

[56] Aksoy HA, Becerik I, Karaosmanoglu F, Yatmaz HC, Civelekoglu H. Utilization prospects of Turkish raisin seed oil as an alternative engine fuel. Fuel 1990;69 600-603.

[57] Vicente G, Martínez M, Aracil J. Integrated biodiesel production: a comparison of different homogeneous catalysts systems. Bioresource Technology 2004;92 297-305.

[58] Stamenkovic OS, Lazic ML, Todorovic ZB, Veljkovic VB, Skala DU. The effect of agitation intensity on alkali-catalyzed methanolysis of sunflower oil. Bioresource Technology 2007;98 2688-2699.

[59] Arzamendi G, Campo I, Arguiñarena E, Sánchez M, Montes M, Gandía LM. Synthesis of biodiesel with heterogeneous NaOH/alumina catalysts: Comparison with homogeneous NaOH. Chemical Engineering Journal 2007;134 123-130.

[60] Xie W, Huang X, Li H. Soybean oil methyl esters preparation using NaX zeolites loaded with KOH as a heterogeneous catalyst. Bioresource Technology 2007;98 936-939.

[61] Al-Widyan MI, Al-Shyoukh AO. Experimental evaluation of the transesterification of waste palm oil into biodiesel. Bioresource Technology 2002;85 253-256.

[62] Zheng S, Kates M, Dubé MA, McLean DD. Acid-catalyzed production of biodiesel from waste frying oil. Biomass and Bioenergy 2006;30 267-272.

[63] Shieh CJ, Liao HF, Lee CC. Optimization of lipase-catalyzed biodiesel by response surface methodology. Bioresource Technology 2003;88 103-106.

[64] Iso M, Chen B, Eguchi M, Kudo T, Shrestha S. Production of biodiesel fuel from triglycerides and alcohol using immobilized lipase. Journal of Molecular Catalysis B: Enzymatic 2001;16 53-58.

[65] Saka S, Kusdiana D. Biodiesel fuel from rapeseed oil as prepared in supercritical methanol. Fuel 2001;80 225-231.

[66] Demirbas AH, Demirbas I. Importance of rural bioenergy for developing countries. Energy Conversion and Management 2007;48 2386-2398.

[67] He H, Wang T, Zhu S. Continuous production of biodiesel fuel from vegetable oil using supercritical methanol process. Fuel 2007;86 442-447.

[68] PETROBRAS – Brazilian Oil Company. Available in http://www.petrobras.com.br/pt/quem-somos/perfil/atividades/producao-biocombustiveis/downloads/pdf/Cartilha-Biocombustiveis-PORTUGUES.pdf. Accessed on September 10, 2012.

[69] ANP, Resolução n⁰ 42 de novembro de 2004, in, 2010.

[70] Sharma YC, Singh B, Upadhyay SN. Advancements in development and characterization of biodiesel: A review. Fuel 2008;87 2355-2373.

[71] Benjumea P, Agudelo J, Agudelo A. Basic properties of palm oil biodiesel-diesel blends. Fuel 2008;87 2069-2075.

[72] Sharon H, Karuppasamy K, Soban Kumar, Sundaresan D R A. A test on DI diesel engine fueled with methyl esters of used palm oil. Renewable Energy 2012;47 160-166.

[73] Singh B, Korstad J, Sharma YC. A critical review on corrosion of compression ignition (CI) engine parts by biodiesel and biodiesel blends and its inhibition. Renewable and Sustainable Energy Reviews 2012;16 3401-3408.

[74] Díaz-Ballote L, López-Sansores JF, Maldonado-López L, Garfias-Mesias LF. Corrosion behavior of aluminum exposed to a biodiesel. Electrochemistry Communications 2009;11 41-44.

[75] Haseeb ASMA, Fazal MA, Jahirul MI, Masjuki HH. Compatibility of automotive materials in biodiesel: A review. Fuel 2011;90 922-931.

[76] Fazal MA, Haseeb ASMA, Masjuki HH. Biodiesel feasibility study: An evaluation of material compatibility; Performance; emission and engine durability. Renewable and Sustainable Energy Reviews 2011;15 1314-1324.

[77] Ambrozin ARP, Kuri SE, Monteiro MR. Metallic corrosion related to mineral fuels and biofuels utilization. Quimica Nova 2009;32 1910-1916.

[78] Norouzi S, Eslami F, Wyszynski ML, Tsolakis A. Corrosion effects of RME in blends with ULSD on aluminium and copper. Fuel Processing Technology 2012; 104 204-210.

[79] Fazal MA, Haseeb ASMA, Masjuki HH. Comparative corrosive characteristics of petroleum diesel and palm biodiesel for automotive materials. Fuel Processing Technology 2010;91 1308-1315.

[80] Kaul S, Saxena RC, Kumar A, Negi MS, Bhatnagar AK, Goyal HB, et al. Corrosion behavior of biodiesel from seed oils of Indian origin on diesel engine parts. Fuel Processing Technology 2007;88 303-307.

[81] Almeida ES, Monteiro MANA, Montes RHO, Mosquetta R, Coelho NMM, Richter EM, Munoz RAA. Direct Determination of Copper in Biodiesel Using Stripping Analysis. Electroanalysis 2010; 22 1846-1850.

[82] Haseeb ASMA, Masjuki HH, Ann LJ, Fazal MA. Corrosion characteristics of copper and leaded bronze in palm biodiesel. Fuel Processing Technology 2010;91 329-334.

[83] Fazal MA, Haseeb ASMA, Masjuki HH. Degradation of automotive materials in palm biodiesel. Energy 2012;40 76-83.

[84] Geller DP, Adams TT, Goodrum JW, Pendergrass J. Storage stability of poultry fat and diesel fuel mixtures: Specific gravity and viscosity. Fuel 2008;87 92-102.

[85] Hu E, Xu Y, Hu X, Pan L, Jiang S. Corrosion behaviors of metals in biodiesel from rapeseed oil and methanol. Renewable Energy 2012;37 371-378.

[86] Fazal MA, Haseeb ASMA, Masjuki HH. Effect of temperature on the corrosion behavior of mild steel upon exposure to palm biodiesel. Energy 2011;36 3328-3334.

[87] Salamanca M, Mondragón F, Agudelo JR, Benjumea P, Santamaría A. Variations in the chemical composition and morphology of soot induced by the unsaturation degree of biodiesel and a biodiesel blend. Combustion and Flame 2012;159 1100-1108.

[88] McCormick RL, Ratcliff M, Moens L, Lawrence R. Several factors affecting the stability of biodiesel in standard accelerated tests. Fuel Processing Technology 2007;88 651-657.

[89] Knothe G, Dunn RO. Dependence of oil stability index of fatty compounds on their structure and concentration and presence of metals. Journal of the American Oil Chemists' Society 2003;80 1021-1026.

[90] Sarin A, Arora R, Singh NP, Sharma M, Malhotra RK. Influence of metal contaminants on oxidation stability of Jatropha biodiesel. Energy 2009;34 1271-1275.

[91] Sarin A, Arora R, Singh NP, Sarin R, Malhotra RK. Oxidation Stability of Palm Methyl Ester: Effect of Metal Contaminants and Antioxidants. Energy & Fuels 2010;24 2652-2656.

[92] Sarin A, Arora R, Singh NP, Sarin R, Sharma M, Malhotra RK. Effect of Metal Contaminants and Antioxidants on the Oxidation Stability of the Methyl Ester of Pongamia. Journal of the American Oil Chemists' Society 2010;87 567-572.

[93] Santos NA, Damasceno SS, De Araujo PHM, Marques VC, Rosenhaim R, Fernandes VJ, Queiroz N, Santos IMG, Maia AS, Souza AG, Energy & Fuels 2011; 25 4190-4194.

[94] Jain S, Sharma MP. Correlation development for effect of metal contaminants on the oxidation stability of Jatropha curcas biodiesel. Fuel 2011;90 2045-2050.

[95] Fazal MA, Haseeb ASMA, Masjuki HH. Effect of different corrosion inhibitors on the corrosion of cast iron in palm biodiesel. Fuel Processing Technology 2011;92 2154–2159.

[96] Jain S, Sharma MP. Stability of biodiesel and its blends: A review. Renewable and Sustainable Energy Reviews 2010;14 667-678.

[97] Liang YC, May CY, Foon CS, Ngan MA, Hock CC, Basiron Y. The effect of natural and synthetic antioxidants on the oxidative stability of palm diesel. Fuel 2006;85 867-870.

[98] Domingos AK, Saad EB, Vechiatto WWD, Wilhelm HM, Ramos LP. The influence of BHA, BHT and TBHQ on the oxidation stability of soybean oil ethyl esters (biodiesel). Journal of the Brazilian Chemical Society 2007;18 416–423.

[99] Tang H, De Guzman RC, Simon-Ng K, Salley SO. Effect of Antioxidants on the Storage Stability of Soybean-Oil-Based Biodiesel. Energy & Fuels 2010;24 2028–2033.

[100] Almeida ES, Portela FM, Sousa RMF, Daniel D, Terrones MGH, Richter EM, Munoz RAA. Behaviour of the antioxidant tert-butylhydroquinone on the storage stability and corrosive character of biodiesel. Fuel 2011;90 3480-3484.

[101] Bailer J, de Heuber K. Determination of saponifiable glycerol in bio-diesel. Fresenius Journal of Analytical Chemistry 1991;340 186-186.

[102] Mittelbach M. Diesel fuel derived from vegetable oils, V (1): Gas chromatographic determination of free glycerol in transesterified vegetable oils. Chromatographia. 1993;37 623-626.

[103] Fillieres R, Mlayah BB, Delmas M. Ethanolysis of rapeseed oil: quantitation of ethyl esters, mono-, di-, and triglycerides and glycerol by high performance size-exclusion chromatography. Journal of the American Oil Chemists' Society. 1995;72(4):427-432.

[104] Plank C, Lorbeer E. Simultaneous determination of glycerol, and mono-, di- and triglycerides in vegetable oil methyl esters by capillary gas chromatography. Journal of Chromatography A. 1995;697 461-468.

[105] Mittelbach M, Roth G, Bergmann A. Simultaneous gas chromatographic determination of methanol and free glycerol in biodiesel. Chromatographia. 1996;42 431-434.

[106] Lozano P, Chirat N, Graille J, Pioch D. Measurement of free glycerol in biofuels. Fresenius Journal of Analytical Chemistry 1996; 354 319-322.

[107] Darnoko D, Cheryan M, Perkins EG. Analysis of vegetable oil transesterification products by gel permiation chromatography. J Liq Chrom Rel Technol. 2000;23 2327-2335.

[108] Greenhill SJ, inventor Stepan Company, assignee. Method for determination of free and combined glycerin in biodiesel. 2004.

[109] Bondioli P, Della Bella L. An alternative spectrophotometric method for the determination of free glycerol in biodiesel. European Journal of Lipid Science and Technology. 2005;107 153-157.

[110] Gandia LM, Arzamendi G, Arguinarena E, Campo I. Monitoring of biodiesel production: Simultaneous analysis of the transesterification products using size-exclusion chromatography. Chemical Engineering Journal. 2006;122 31-40.

[111] Hajek M, Skopal F, Machek J. Determination of free glycerol in biodiesel. European Journal of Lipid Science and Technology. 2006;108 666-669.

[112] Munari F, Cavagnino D, Cadoppi A. Determination of free and total glycerin in pure biodiesel (B100) by GC in compliance with EN 14105. Thermo Fisher Scientific. 2007(10215):8.

[113] Nguyen N, Kelly K, Countryman S. Reliable analysis of glycerin in biodiesel using a high temperature non-metal GC column. LC/GC North America. 2007;26(2) 60.

[114] McCurry JD, Wang C-X. Improving the analysis of biodiesel using capillary flow technology: Agilent Technologies; 2007 Contract No.: Document Number I.

[115] Goncalves-Filho LC, Micke GA. Development and validation of a fast method for determination of free glycerol in biodiesel by capillary electrophoresis. Journal of Chromatography A. 2007;1154 477-480.

[116] Cole J, Lefler J, Rui C. Fast separation of FFA, FAME and glycerol for biodiesel analysis by supercritical fluid chromatography. LC-GC Europe. 2008; 40-41.

[117] Gandhi J, Wille A, Steinbach A. Ion Chromatographic Determination of Free and Total Glycerol in Biodiesel and Biodiesel Blends. Lc Gc Europe. 2009: 8-14.

[118] de Castro MDL, Pinzi S, Capote FP, Jimenez JR, Dorado MP. Flow injection analysis-based methodology for automatic on-line monitoring and quality control for biodiesel production. Bioresource Technology 2009;100 421-427.

[119] Lourenco LM, Stradiotto NR. Determination of free glycerol in biodiesel at a platinum oxide surface using potential cycling technique. Talanta. 2009;79 92-96.

[120] Rocha FRP, Silva SG. A flow injection procedure based on solenoid micro-pumps for spectrophotometric determination of free glycerol in biodiesel. Talanta. 2010;83 559-564.

[121] D'Elia E, Luetkmeyer T, dos Santos RM, da Silva AB, Amado RS, Vieira ED. Analysis of Free and Total Glycerol in Biodiesel Using an Electrochemical Assay Based on a Two-Enzyme Oxygen-Electrode System. Electroanalysis. 2010;22 995-999.

[122] D'Elia E, Pêgas MM, Amado RS, de Castro EV. Analysis of free glycerol in biodiesel using an electrochemical assay based on a two-enzyme platinum microelectrode system. Journal of Applied Electrochemistry. 2010;40 2061-2063.

[123] Pisarello ML, Dalla Costa BO, Veizaga NS, Querini CA. Volumetric method for free and total glycerin determination in biodiesel. Industrial & Engineering Chemistry Research 2010;49 8935-8941.

[124] Hollebone BP, Yang ZY, Wang ZD, Yang C, Landriault M. Determination of polar impurities in biodiesels using solid-phase extraction and gas chromatography-mass spectrometry. Journal of Separation Science 2011;34 409-421.

[125] Christison TT, De Borba BM, Rohrer JS. Determination of Free and Total Glycerol in Biodiesel. Lc Gc North America. 2011:39.

[126] Morales-Rubio A, Silva SG, de La Guardia M, Rocha FRP. Sequential spectrofluorimetric determination of free and total glycerol in biodiesel in a multicommuted flow system. Analytical and Bioanalytical Chemistry.2011;401 365-371.

[127] Halaweish F, Mercer EJ. Determination of Free Glycerol in Biodiesel via Solid-Phase Extraction and Spectrophotometric Analysis. Journal of the American Oil Chemists' Society 2011;88 655-659.

[128] Paixão TRLC, Maruta AH. Flow injection analysis of free glycerol in biodiesel using a cooper electrode as an amperometric detector. Fuel 2012;91 187-191.

[129] Valdez HdC, Amado RS, Souza FCd, D'Elia E, Vieira EdC. Determinação de glicerol livre e total em amostras de biodiesel por método enzimático com detecção colori-métrica. Química Nova 2012;35 601-607.

[130] Barbosa TGG, Richter EM, Muñoz RAA. Flow-injection pulsed-amperometric deter-mination of free glycerol in biodiesel at a gold electrode. Electroanalysis 2012 May 2012;24 1160-1163.

[131] Tehrani RMA, Ghani SA. Electrocatalysis of free glycerol at a nanonickel modified graphite electrode and its determination in biodiesel. Electrochimica Acta 2012;70 153-157.

[132] Dias AN, Cerqueira MBR, de Moura RR, Kurz MHS, Clementin RM, D'Oca MGM, et al. Optimization of a method for the simultaneous determination of glycerides, free and total glycerol in biodiesel ethyl esters from castor oil using gas chromatography. Fuel 2012;94 178-183.

[133] Prados CP, Rezende DR, Batista LR, Alves MIR, Antoniosi Filho NR. Simultaneous gas chromatographic analysis of total esters, mono-, di- and triacylglycerides and free and total glycerol in methyl or ethyl biodiesel. Fuel 2012;96 476-481.

[134] Chadha A, Reddy SR, Titu D. A Novel Method for Monitoring the Transesterification Reaction of Oil in Biodiesel Production by Estimation of Glycerol. Journal of the American Oil Chemists' Society 2010 ;87 747-754.

[135] Edlund M, Visser H, Heitland P. Analysis of biodiesel by argon–oxygen mixed-gas inductively coupled plasma optical emission spectrometry. Journal of Analytical Atomic Spectrometry 2002; 17 232-235.

[136] Woods GD, Fryer FI. Direct elemental analysis of biodiesel by inductively coupled plasma–mass spectrometry. Analytical and Bioanalytical Chemistry 2007; 389 753-761.

[137] Santos EJ, Herrmann AB, Chaves ES, Vecchiatto WWD, Schoemberger AC, Frescura VLA, Curtius AJ. Simultaneous determination of Ca, P, Mg, K and Na in biodiesel by axial view inductively coupled plasma optical emission spectrometry with internal standardization after multivariate optimization. Journal of Analytical Atomic Spec-trometry 2007; 22 1300-1303.

[138] Sousa RM, Leocadio LG, Silveira CLP. ICP OES Simultaneous Determination of Ca, Cu, Fe, Mg, Mn, Na, and P in Biodiesel by Axial and Radial Inductively Coupled Plasma-Optical Emission Spectrometry. Analytical Letters 2008; 41 1615-1622.

[139] Jesus A, Silva MM, Vale MGR. The use of microemulsion for determination of sodium and potassium in biodiesel by flame atomic absorption spectrometry. Talanta 2008; 74 1378-1384.

[140] Chaves ES, Saint Pierre TD, Santos EJ, Tormen L, Bascunan VLAF, Curtius AJ. Determination of Na and K in Biodiesel by Flame Atomic Emission Spectrometry and Microemulsion Sample Preparation. Journal of the Brazilian Chemical Society 2008; 19(5) 856-861.

[141] Vieira MA, Oliveira LCC, Gonçalves RA, Souza V, Campos RC. Determination of As in Vegetable Oil and Biodiesel by Graphite Furnace Atomic Absorption SpectrometryEnergy Fuels 2009; 23 5942-5946.

[142] Lobo FA, Gouveia D, Oliveira AP, Pereira-Filho E, Fraceto LF, Dias-Filho NL, Rosa AH. Comparison of the univariate and multivariate methods in the optimization of experimental conditions for determining Cu, Pb, Ni and Cd in biodiesel by GFAAS. Fuel 2009; 88 1907-1914.

[143] Aranda PR, Pacheco PH, Olsina RA, Martinez LD, Gil RA. Total and inorganic mercury determination in biodiesel by emulsion sample introduction and FI-CV-AFS after multivariate optimization. Journal of Analytical Atomic Spectrometry 2009; 24 1441-1445.

[144] Iqbal J, Carney WA, LaCaze S, Theegala CS. Metals Determination in Biodiesel (B100) by ICP-OES with Microwave Assisted Acid Digestion. The Open Analytical Chemistry Journal 2010, 4 18-26.

[145] Chaves ES, Santos EJ, Araujo RGO, Oliveira JV, Frescura VLA, Curtius AJ. Metals and phosphorus determination in vegetable seeds used in the production of biodiesel by ICP OES and ICP-MS. Microchemical Journal 2010; 96 71-76.

[146] Lobo FA, Gouveia D, Oliveira AP, Romao LPC, Fraceto LF, Dias-Filho NL, Rosa AH. Development of a method to determine Ni and Cd in biodiesel by graphite furnace atomic absorption spectrometry. Fuel 2011; 90 142-146.

[147] Prado AGS, Pescara IC, Evangelista SM, Holanda MS, Andrade RD, Suarez PAZ, Zara LF. Adsorption and preconcentration of divalent metal ions in fossil fuels and biofuels: Gasoline, diesel, biodiesel, diesel-like and ethanol by using chitosan microspheres and thermodynamic approach. Talanta 2011; 84 759-765.

[148] Barros AI, Oliveira AP, Magalhães MRL, Villa RD. Determination of sodium and potassium in biodiesel by flame atomic emission spectrometry, with dissolution in ethanol as a single sample preparation step. Fuel 2012; 93 381-384.

[149] Pillay AE, Elkadi M, Fok SC, Stephen S, Manuel J, Khan MZ, Unnithan S. A comparison of trace metal profiles of neem biodiesel and commercial biofuels using high performance ICP-MS. Fuel 2012; 97 385-389.

[150]   Rapta P, Paligova J, Rotheneder H, Cvengros J. Determination of potassium in fatty acid methyl esters applying an ion-selective potassium electrode. Chemical Papers 2007; 61, 337-341.

[151]   Castilho MS, Stradiotto NR. Determination of potassium ions in biodiesel using a nickel(II) hexacyanoferrate-modified electrode. Talanta 2008; 74 1630-1634.

[152]   Steinbach A, Willie A, Subramanian NH. Biofuel analysis by ion chromatography. LG GC Northe America 2008, S 32-35.

[153]   Caland LB, Silveira ELC, Tubino M. Determination of sodium, potassium, calcium and magnesium cations in biodiesel by ion chromatography. Analytica Chimica Acta 2012; 718 116-120.

[154]   Nogueira T, Lago CL, Microchemical Journal 2011, 99, 267.

[155]   Frena M, Campestrini I, Braga OC, Spinelli A. In situ bismuth-film electrode for square-wave anodic stripping voltammetric determination of tin in biodiesel. Electrochimica Acta 2011; 56, 4678-4684.

[156]   Korn MGA, Dos Santos DSSS, Welz B, Vale MGR, Teixeira AP, Lima DC, Ferreira SLC. Atomic spectrometric methods for the determination of metals and metalloids in automotive fuels – A review. Talanta 2007; 73 1-11.

[157]   Santos AL, Takeuchi RM, Fenga PG, Stradiotto NR. Electrochemical Methods in Analysis of Biofuels. Rijeka: In Tech; 2011; 451-494.

[158]   Santos AL, Takeuchi RM, Munoz RAA, Angnes L, Stradiotto NR. Electrochemical Determination of Inorganic Contaminants in Automotive Fuels. Electroanalysis 2012; 24 1681-1691.

[159]   Araujo ARTS, Saraiva LMFS, Lima JLFC, Korn MGA. Flow methodology for methanol determination in biodiesel exploiting membrane-based extraction. Analytica Chimca Acta 2008; 613 177-183.

[160]   Wang H, Tang H, Salley S, Simon Ng KY. Analysis of Sterol Glycosides in Biodiesel and Biodiesel Precipitates. Journal of the American Oil Chemists' Society 2010; 87 215-221.

[161]   Moreau RA, Scott KM, Haas MJ. The Identification and Quantification of Steryl Glucosides in Precipitates from Commercial Biodiesel. Journal of the American Oil Chemists' Society 2008; 85 761-770.

[162]   Bondioli P, Cortesi N, Mariani C. Identification and quantification of steryl glucosides in biodiesel. European Journal of Lipid Science Technology 2008; 110 120-126.

[163]   Pauls RE. A Review of Chromatographic Characterization Techniques for Biodiesel and Biodiesel Blends. Journal of Chromatographic Science 2011; 49 384-396.

[164]   Lyra FH, Carneiro MTWD, Brandao GP, Pessoa HM, De Castro EVR. Direct determination of phosphorus in biodiesel samples by graphite furnace atomic absorption

spectrometry using a solid sampling accessory. Journal of Analytical Atomic Spectrometry 2009; 24 1262-1266.

[165] Silveira ELC, De Caland LB, Tubino M. Molecular absorption spectrophotometric method for the determination of phosphorus in biodiesel. Fuel 2011; 90 3485-3488.

[166] Lira LFB, Dos Santos DCMB, Guida MAB, Stragevitch L, Korn MDA, Pimentel MF, Paim APS. Fuel 2011; 90 3254-3258.

[167] Zezza TRC, Castilho MD, Stradiotto NR. Determination of phosphorus in biodiesel using 1:12 phosphomolybdic modified electrode by cyclic voltammetryFuel 2012; 95 15-18.

[168] Barker LR, Kelly R, Guthrie WF. Determination of Sulfur in Biodiesel and Petroleum Diesel by X-ray Fluorescence (XRF) Using the Gravimetric Standard Addition Method-II. Energy & Fuels 2008; 22 2488-2490.

[169] Young CG, Amais RS, Schiavo D, Garcia EE, Nobrega JA. Determination of sulfur in biodiesel microemulsions using the summation of the intensities of multiple emission lines. Talanta 2011; 84 995-999.

[170] Amais RS, Donati GL, Nobrega JA. Interference Standard Applied to Sulfur Determination in Biodiesel Microemulsions by ICP-QMS. Journal of the Brazilian Chemical Society 2012; 12 797-803.

[171] Monteiro MR, Ambrozin ARP, Liao LM, Ferreira AG. Critical review on analytical methods for biodiesel characterization. Talanta 2008; 77 593-605.

Chapter 7

Non-Catalytic Production of Ethyl Esters Using Supercritical Ethanol in Continuous Mode

Camila da Silva, Ignácio Vieitez, Ivan Jachmanián,
Fernanda de Castilhos, Lúcio Cardozo Filho and
José Vladimir de Oliveira

Additional information is available at the end of the chapter

1. Introduction

Development of alternative renewable energy has become necessary because, among other factors, the possible shortage of fossil fuels and environmental problems. Among the renewable resources available for alternative fuel production, the conversion of fats and oils to biodiesel has been investigated and well documented in the literature [1-4].

The merits of biodiesel as an alternative to mineral diesel comprise a nontoxic, biodegradable, domestically produced, and renewable resource. Besides, biodiesel possesses a higher cetane number compared to diesel from petroleum and a favorable combustion emissions profile, such as reduced levels of particulate matter, carbon monoxide, and, under some conditions, nitrogen oxides [5,6]. Because of these environmental benefits, which means reduction of environmental investments, and also due to the relief from reliance on import needs, biodiesel fuel can be expected to become a good alternative to petroleum-based fuel.

The establishment of the Brazilian national program on biodiesel has prompted several studies on biodiesel production using different techniques and a variety of vegetable and animal sources. Methanol has been the most commonly used alcohol to perform transesterification reactions. However, in the Brazilian context, ethanol has been the natural choice since Brazil is one of the world's biggest ethanol producers, with a well-established technology of production and large industrial plant capacity installed throughout the country. Due to the fact that ethanol also comes from a renewable resource, thus, ethanol biodiesel appears as a 100% renewable alternative additionally enabling the replacement of traditionally used methanol by an innocuous reagent [7].

Typical raw materials investigated for the production of biodiesel, include soybean, sunflower, castor, corn, canola, cottonseed, palm, peanuts [1] and more recent studies highlight the use of *Jatropha curcas* oil [8,9]. A fact to be also considered to lower manufacturing costs and make biodiesel competitive, is the use of degummed oils that have lower cost than refined oils, besides the possibility of recycling the waste oils [10,11]. However, the choice of the oilseed to be used must consider the content in vegetable oil, yield and territorial adaptation.

Among other processes used for the production of biofuels from vegetable oils, such as pyrolysis and microemulsification, transesterification is the most common way to produce biodiesel [1,3]. Transesterification, also called alcoholysis, refers to the reaction of a triglyceride (from animal or vegetable source) with an alcohol in the presence or absence of catalyst to form fatty acid alkyl esters (i.e., biodiesel) and glycerol as a byproduct.

The complete transesterification is the reaction of one mole of triglyceride with three moles of alcohol, resulting in the production of 3 moles of esters and 1 mol of glycerol as shown in Figure 1. Transesterification is a reversible reaction which occurs in three steps with formation of intermediate products: diglycerides and monoglycerides.

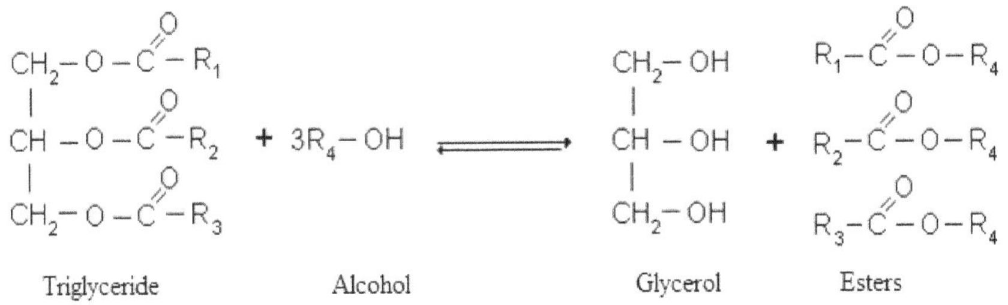

Figure 1. Transesterification reaction of a triglyceride with an alcohol.

The transesterification process reduces the average molar mass to approximately 1/3 compared to triglycerides, hence decreasing the viscosity and enhancing the mixture volatility. Unlike the original oil, biodiesel has similar properties and full compatibility with petroleum diesel, accordingly conventional diesel engines can be powered on biodiesel without requiring substantial mechanical modification [12]. After the reaction, the products consist of a mixture of fatty acid esters, glycerol, remainder alcohol, catalyst and a low percentage of tri-, di-and monoglycerides [13].

Among the factors affecting the yield of the transesterification reaction, one can cite: type and amount of catalyst, reaction time, temperature, molar ratio of oil to alcohol, content of free fatty acids and water in the substrates, agitation power, solubility between the phases and nature of the alcohol [10]. However, the extent of variables effect will necessarily depend on the method used [14].

The homogeneous chemical catalysis (acid or basic) is the most used technique in the trans-esterification reaction at industrial scale, since it allows, in the case of alkaline catalysis, reaching high conversions at shorter reaction times [15-23].

The chemical method using homogeneous alkali catalysts, although simple, fast and with high yields, presents several drawbacks, such as costs of catalyst separation and difficulty of purification and separation of reaction products, which involves high production costs and energy consumption [24]. Because alkali catalyzed systems are very sensitive to both water and free fatty acids contents, the glycerides and alcohol must be substantially anhydrous. Water makes the reaction partially change to saponification, which produces soaps, thus consuming the catalyst and reducing the catalytic efficiency, as well as causing an increase in viscosity, formation of gels, and difficulty in separations [1,3,25]. As a consequence, the water and free fatty acids content should be less than 0.06% (w/w) and 0.5% (w/w) for trans-esterification reaction with alkali catalysts, respectively [1,26].

The transesterification reaction using homogeneous acid catalysts is preferred for the con-version of raw materials containing high levels of free fatty acids, because the acid catalyst can promote simultaneously the transesterification of the triglycerides and esterification of the free fatty acids to alkyl esters [27]. Although esterification of free fatty acids may pro-ceed with a relatively high rate and high yields can be achieved, the kinetics of triglycerides transesterification is much slower, requiring high temperatures (above 373 K) and 24 hours of reaction for completion [12].

Thus obtaining of esters in two reaction steps for substrates with high acidity has been pro-posed, consisting of two approaches: (a) the acid esterification of free fatty acids and subse-quent the alkaline transesterification of triglycerides [28-31] or (2) enzymatic hydrolysis of triglycerides, followed by the acid esterification of the fatty acids produced [32-34].

The use of heterogeneous chemical catalysts in alcoholysis of vegetable oils reduces the diffi-culties of separation of products and catalyst, resulting in the generation of lower effluents volume. The literature suggests the use of various acid and basic catalysts [35-37], with cata-lysts reuse in the process. However, heterogeneous chemical catalysis generally shows low yields compared to homogeneous alkaline catalysis.

The reaction catalyzed by enzymes (lipases) provides easy separation of catalyst from the re-action medium, catalyst reusability and higher purity of the reaction products. However, to date, the main disadvantages of this method refers to the long reaction times needed and the high cost of the enzymes [14], that progressively are deactivated during reaction course. The enzyme method can be conducted in the presence of organic solvents in order to minimize mass transfer limitations, immiscibility between phases and catalyst deactivation, requiring the use of higher ratios of solvent/vegetable oil (in the order of 40/1) to provide satisfactory reaction rates [38]. For the production of biodiesel in enzyme systems using pressurized sol-vents, smaller amounts of solvent can be used and the solvent can be easily separated from the reaction medium by system decompression [38-41]. High conversions have been report-ed for both systems but the use of high enzyme to substrates ratios has hindered large-scale implementation of such technique.

The efficiency of microwave irradiation [42-44] and the use of ultrasonic technology [45-47] in the transesterification of vegetable oils using different catalysts has been reported with the advantage of high reaction rates compared to conventional processes.

Recently, a catalyst-free technique for the transesterification of vegetable oils using an alcohol at supercritical conditions has been proposed, keeping the benefits of fuel quality and taking into account environmental concerns [48-53]. According to the current literature, catalyst-free alcoholysis reactions at high temperature and pressure conditions provide improved phase solubility, decreased mass-transfer limitations, afford higher reaction rates and simpler separation and purification steps [24]. Besides, it has been shown that the so-called supercritical method is more tolerant to the presence of water [54] and free fatty acids [54,55] than the conventional alkali-catalyzed technique, and hence more tolerant to various types of vegetable oils, even for fried and waste oils.

The reaction for biodiesel production at supercritical conditions requires high alcohol to oil molar ratios and the adoption of high temperatures and pressures for the reaction to present satisfactory conversion levels, leading to high processing costs and causing in many cases the degradation of the fatty acid esters formed [56-60] and reaction of glycerol formed with other components of the reaction medium [61-64], hence decreasing the reaction conversion [65-68,57,58]. Current literature shows some alternatives to reduce the expected high operating costs and product degradation, and such strategies usually involve: (i) addition of co-solvents [69-74]; (ii) two-step process with glycerol removal in the first step [75-77]; (iii) two-step process comprising hydrolysis of triglycerides in subcritical water and subsequent esterification of fatty acids [65,66,78]; (iv) use of microreator systems operating in continuous mode [74,79] and use of packed bed reactor [80].

The aim of this work is to provide a brief review on the continuous production of fatty acid ethyl esters (FAEE) by non-catalytic process using ethanol at supercritical conditions. These results are part of a broader project aimed at building a platform to allow the development of a new process for the production of biodiesel from vegetable oils. A section of this chapter will be dedicated to reviewing the characteristics of the supercritical method, comprising the research in the production of FAEE in continuous mode evaluating the role of process variables such as temperature, pressure, molar ratio of oil:ethanol and residence time. This review also focuses on the different configurations of reaction systems, like tubular reactor, microtube reactor, packed bed tubular reactor, as well as the experimental simulation of reactors in series and reactor with recycle. The effect of addition of co-solvent (carbon dioxide), water and free fatty acids to the reaction medium on the FAEE yield are evaluated and decomposition of FAEE produced and conversion of oil to FAEE are also considered.

2. Characteristics of the non-catalytic supercritical method for biodiesel production

The transesterification reaction using a solvent at pressurized conditions is one of the methods used for the synthesis of biodiesel [48]. This can be a secure way, without causing environmental damage, and requires less investment in the overall process, since the equipment cost is offset by the high reaction rates, better efficiency and lower cost of products purification.

Glisic & Skala [81] reported the economic analysis of the processes for biodiesel production using homogeneous alkaline catalysis and supercritical method, noting that energy consumption is extremely similar in both cases. Since in the supercritical method the heating step involves high energy consumption, costs are compensated by the simpler purification step of the products (esters and glycerol), requiring lower power consumption, which leads to a high costs step of the conventional process. Deshpande et al. [82] reported an economic analysis of the proposed supercritical process and found that the biodiesel processing cost through the proposed technology could be half of that of the actual conventional methods.

The production costs of biodiesel can be minimized by the sale of by-products generated by the transesterification process, such as glycerin. However, when using the conventional method by alkaline catalysis, traces of catalyst can be found in the glycerin, which limits the use of this product. Thus subsequent purification steps are required [83,84], a fact that is not needed in the supercritical method, which proceeds with simple purification and separation of the biofuel produced and generates a high-pure glycerin [48,49,85].

Marchetti & Errazu [85] evaluated different processes for biodiesel production using vegetable oils with high content of free fatty acids, including the supercritical method and stated that the supercritical method is an attractive alternative from a technological point of view. Additionally, from the economic point of view, less wastewater is produced and a high quality glycerin is generated as a byproduct, however higher energy is required by the reaction step.

The reactivity in the supercritical state is higher than in the liquid or gas, which facilitates the transesterification reaction [86]. The supercritical point of ethanol and methanol are 514 K and 6.14 Mpa [27,51] and 513 K and 8.09 Mpa [48], respectively. The non-catalytic production of biodiesel with supercritical alcohol provides high reaction yields, since it promotes the simultaneous hydrolysis and transesterification of triglycerides and esterification of free fatty acids present in vegetable oil [50].

The supercritical method has the following advantages over other methods used for biodiesel production [67]:

a.   Catalyst is not used in the reaction and purification procedures are much simpler, since the separation process of the catalyst and the saponified product is not required;

b.   The supercritical reaction requires shorter reaction time than the traditional catalytic transesterification and the conversion rate is high. The catalytic transesterification requires, in some cases, hours to reach the reaction equilibrium, while supercritical method only minutes;

c.   Low quality substrates of can be used in the supercritical method, since high levels of free fatty acids and water do not have a negative effect on the reaction.

The alcohol in the supercritical state solves or reduces the possible formation of two phases to form a single homogeneous phase, by decreasing the dielectric constant of alcohol in the supercritical state, which results in increased solubility of the oil [24]. Ma & Hanna [1] reported that the solubility of triglycerides in methanol increases at a rate of 2 to 3% (w/w) of 10 K increase in temperature.

Some disadvantages of supercritical method are nevertheless pointed out: high alcohol to oil ratios are required (in the order of 40:1), best results are obtained at temperatures above 573 K and high pressures, typically 20 MPa, which leads to high processing costs and energy consumption. In addition, the quality of biodiesel may be compromised by the low stability of certain fatty acid esters exposed to the drastic reaction conditions required. Thus, due to drastic increase in costs associated with the use of excess alcohol and equipment due to operation at high temperatures and pressures, improvements to the supercritical method for producing biodiesel are required [87].

Kiwjaroun et al. [88] investigated the biodiesel production processes by supercritical methanol combined with and alkaline catalyst and the impacts generated by each process on the environment, using LCA (life cycle analysis) as a tool. It was observed by these researchers that the supercritical method is advantageous compared to conventional method due to the less amount of wastewater generated, however, creates a high impact on the environment, mainly due to the large amount of alcohol used in the process, emphasizing the need for research regarding the reduction in operating conditions (temperature, pressure) and the amount of alcohol used in the process. Marulanda [89] evaluated the potential environmental impact assessment of the process for biodiesel production by non-catalyst supercritical method and conventional base-catalyzed process. The environmental assessment results indicated the supercritical process, even when working at a 42:1 molar ratio, has a lower impact than the conventional base-catalyzed process.

2.1. Decomposition

During supercritical transesterification, the high temperatures (above 573 K) employed and long reaction periods, a decrease in the conversion can be observed [7,57,65-68,73].

He et al. [67] evaluated the results obtained for the transesterification of soybean oil in supercritical methanol and concluded that the reason for the decrease in reaction yield is the decrease in the content of unsaturated esters, caused by isomerization, hydrogenation and thermal decomposition that would consume such esters, especially C18:2 (linoleic) and C18:3 (linolenate). Imahara et al. [56] evaluated the thermal stability of different samples of biodiesel and fatty acid esters in different conditions of temperature and pressure. The authors found that thermal degradation is more pronounced for the unsaturated esters above 573 K and 19 MPa and thermal stability of saturated esters is also affected. Kasim et al. [63] report that the percentage of trans isomers can reach levels up to 16% under certain reaction conditions (30 MPa, 573 K) for the transesterification of rice bran oil in methanol.

At the supercritical reaction conditions, side reactions with the participation of the glycerol formed as byproduct can cause the degradation of other components present in the reaction medium. For instance, Anistescu et al. [61] performed the alcoholysis reactions using supercritical methanol at temperatures around 623-673 K and reported the absence of glycerol in the reaction products, the authors cogitated that reaction of glycerol with other compounds may have occurred. Aimaretti et al. [62] evaluated the reaction of refined soybean oil with supercritical methanol at different reaction conditions and at the conditions studied by the authors, glycerol was not formed. It is reported that glycerol is converted into lower molecular weight products and water at the beginning of the reaction and that water reacts with triglyceride to form free fatty acid, thus increasing the acidity of the product. In the course of the reaction, these fatty acids are converted into methyl esters. Also, the glycerol may react in different ways: (i) decomposition to produce products of lower molecular weight, such as acrolein, acetaldehyde, acetic acid, among others, (ii) polymerization to form polyglycerols, which occur at high temperature conditions and (iii) etherification with methanol to produce ethers of glycerol, thus consuming the alcohol in the reaction medium. Lee et al. [90], in the synthesis of biodiesel from waste canola oil, reported that side reaction was obtained by reacting glycerol and supercritical methanol at 543 K/10 MPa for 15, 30 and 45 minutes. The experimental results showed that these reactions could positively affect the overall biodiesel yield by providing oxygenated compounds such as 3-methoxy-1,2-propanediol, dimethoxymethane, and 2,2-dimethoxypropane as well methyl palmitate and methyl oleate.

In Vieitez et al. [57] a novel and simple GC method was proposed to evaluate de percentage of overall decomposition. Samples were treated with $BF_3/MeOH$ [91] to derivatize all of the fatty acids (mono-, di-, and triglycerides, free fatty acids, and also ethyl esters) to the corresponding methyl esters, and then analyzed by GC. For the evaluation of the degradation percentage, palmitic acid was assumed not liable to degradation, considering its high stability, and was taken as reference (as an internal standard "native"). Thus, degradation was estimated as:

$$Decomposition\ (\%) = 100 \times \left[1 - \left(\frac{\sum P_i}{P_{16:0}} \right)_S \times \left(\frac{P_{16:0}}{\sum P_i} \right)_O \right]$$

where $\sum P_i$ was the summation of all fatty acid methyl ester percentages, $P_{16:0}$ was the percentage of 16:0 ethyl ester, and subscripts "s" and "o" indicate that the expressions between brackets were evaluated considering the composition of the sample product and the original oil, respectively [57].

The use of the term "decomposition" of fatty acids referred to the decrease in its percentage (determined by gas chromatography) due to the formation of other compounds (not necessarily imply that they have "broken" but have suffered some type of alteration). Since there is no information about the determination of this parameter type, the method described below can be considered a new contribution to the area of the synthesis of biodiesel in supercritical alcohols.

2.2. Addition of co-solvent

A question to be considered is the addition of co-solvents to the reaction medium that can provide milder operation conditions, since the use of co-solvents reduces the limitations of mass transfer between phases involved [92] and increases the reaction rate offering an homogeneous reaction media [69,70].

As co-solvents in supercritical transesterification it can be used non-polar compressed gases, for example, carbon dioxide, methane, ethane, propane, n-butane and their mixtures [92]. Some studies have reported the use of heptane/hexane as co-solvent [93-95]. Among these the use of CO_2 at supercritical conditions has shown a promising future for environmentally friendly chemical processes, because it comprises a nonflammable solvent, nontoxic, inexpensive and readily available in high purity. Indeed, besides being a good solvent for extraction, carbon dioxide has also proved useful as solvent reaction medium [96]. However, a limiting factor for the use of carbon dioxide is low mutual solubility CO_2-triglycerides, which means that high pressures are required to solubilize the reagents [97].

The use of propane and n-butane as compressed solvent or even in the supercritical state seems to be a nice substitute for a variety of solvent in reactive systems. These gases offer as the main advantage the low pressure transitions systems found mainly in vegetable oils due to the higher solubility exhibited compared to that the use of CO_2 [97,98]. Pereda et al. [99] reported that the use of propane in the hydrogenation of triglycerides increases the miscibility of the components of the mixture, allowing the reaction to occur under conditions of a single homogeneous phase.

Yin et al. [72] reported that esters yield for the reaction using supercritical methanol increased when using carbon dioxide as cosolvent. Imahara et al. [93], in the alcoholysis of canola oil in methanol with the addition of supercritical CO_2, found that the addition of co-solvent increases the reaction yield, however, high molar percentage of CO_2 (above 0.1 CO_2/ methanol) led to a decrease in reaction conversion.

2.3. Two-step reaction

Based on the reports available in the literature it is suggested that the transesterification of vegetable oils at supercritical conditions can be conducted on alternative systems in order to reduce raw material costs and operating costs. There is a growing emphasis on the proposed system with a two-step reaction using reactors in series, with higher conversions to the system in one step [66] at mild operating temperatures and pressures and decreasing the amount of alcohol used in the process [87].

Kusdiana & Saka [65] and Minami & Saka [66] proposed the continuous synthesis of biodiesel from canola oil in two reaction steps, which consists primarily in the hydrolysis of triglycerides in pressurized water and subsequent esterification of fatty acids in supercritical methanol, with glycerol removed prior to FFA methyl esterification. This process is carried out under more moderate temperature and pressure compared to the process in one step.

Busto et al. [100] reported that tubular reactors for supercritical transesterification must operate in order to minimize the axial dispersion, and as suggested by the authors, to satisfy this condition: reactions in a tubular reactor with separation step of unreacted products, with recycle the same or two or more reactors in series with intermediate separation of glycerol generated. One advantage of removing the glycerol formed in the reaction mixture is to allow the reaction to occur at lower ratios of alcohol to oil increasing the reaction rate for the production of biodiesel [87]. As cited by Aimaretti et al. [62], along the reaction, the alcohol used in the process is required by secondary reactions, which occur with glycerol.

D'Ippolito et al. [75] evaluated theoretically the non-catalytic process for producing biodiesel from experimental data and information available in the literature to determine an operating mode and operating conditions that reduce energy consumption and increase product quality. Results obtained suggest that the two-step process with intermediate removal of glycerol decreases the ratios of methanol to oil to about 10-15 times. Furthermore, not only the system pressure can be reduced as energy costs. In the process proposed by Crawford et al. [87], it is suggested that the obtained esteres by supercritical route can be made by transesterification of triglycerides with continuous removal of glycerol formed in the process, periodically or continuously, increasing the rate of ester formation. These authors argued that the reaction proceeded in this way can greatly decrease the amount of alcohol to be used in the process.

2.4. Intensification technologies in continuous biodiesel production

In the transesterification of vegetable oils, reaction rate can be limited by mass transfer between oil and alcohol because the very poor mutual miscibility. Hence, some process intensification technologies have been developed and applied to improve mixing and mass/heat transfer between the two liquid phases in recent years. Reaction rate is greatly enhanced and thus residence time may be reduced. Some of the technologies have been applied successfully in commercial production [101]. To reduce the limitations of mass and heat transfer in chemical reactions, literature indicates to conduct these reactions in microreactors [102-105] and in packed bed reactors [106-109].

In microreactors, mass and heat transfer increase due to the small size and large contact area [110] and the lowest internal diameters promote interaction with the reagents at the molecular level [111]. The internal diameter of microreactors, are typically10-300 μm [102,103]. Sun et al. [109] used reactors with 0.025 to 0.053 cm inner diameter and Guan et al. [112] used reactors with different inner diameters: 0.04, 0.06, 0.08 and 0.1 cm, calling them as microtube reactors. Furthermore, higher conversion and selectivity are obtained in a shorter reaction time as compared to batch system [102,113].

The rates of transesterification for biodiesel production are controlled by the rate of mass transfer between phases [112], being applied high rates of agitation for the batch system. Sun et al. [109] studied the production of biodiesel using alkaline catalysis with capillaries microreactors, and reported that the residence time is significantly reduced by the use of these reactors compared to the conventional process in batch mode. Guan et al. [112] investigated the synthesis of biodiesel using microtube reactors for the alcoholysis of sunflower oil by basic catalysis, evaluating the influence of the length and internal diameter of the reactor. The conversion of the oil was strongly influenced by reactor geometry and the best results were obtained for the reactor with smaller diameter and greater length.

Although the phenomenon related to mass transfer is a key parameter to obtain better yields in biodiesel by the supercritical method and one approach suitable is the use of packed bed reactor. The packed bed system maximizes the interfacial surface area between the two phases (oil and alcohol) and the contact of the immiscible liquid-liquid two phases are improved towards achieving excellent mass transfer performance, which is obtained by extruding one phase into another, as the two phases flow through the particles openings, as commonly found in a packed bed reactor [106,114, 115].

Ataya et al. [106] reported the acid-catalyzed transesterification of canola oil with methanol using a packed bed reactor and showed that the mass-transfer limitations for two-phase experiments can be effectively overcome using a liquid-liquid packed bed reactor. Santacesaria et al. [108] performed the transesterification reactions in a simple tubular reactor filled with stainless steel spheres of different sizes and obtained that the reactions like methanol–soybean oil transesterification, mass transfer rate can greatly be increased also by favoring an intense local turbulence. The effects of packed bed reactor can be observed in other chemical reactions, for instance, Su et al. [115] evaluated the effect of packed microchannel reactors to perform the nitration of o-nitrotoluene with mixed acid and reported that the yield of this liquid-liquid multiphase reaction is increased by conducting the reaction using the packed reactor.

3. Configuration of reactors in continuous mode for supercritical ethanolysis

The following sections are dedicated to provide an overview of results obtained in supercritical ethanolysis in different reactor configurations. The schematic diagram of the experimental setup, developed by our research group, is shown in Figure 2. In these experiments, the residence time was simply computed dividing the volume of the reactor (mL) by the flow rate of substrates (mL/min) set in the liquid pump.

Results reported are in relation to content of esters in the sample determined by gas chromatography, following the European normative EN 14103 [116]. The data related to decomposition refer to derivatization of the samples with BF_3/methanol [91] to derivatize all of the fatty acids (mono-, di-, and triglycerydes, free fatty acids, and also ethyl esters) to the corresponding methyl esters and then analyzed by gas chromatography. For the evaluation of the decomposition percentage, palmitic acid was assumed not liable to degradation, considering its high stability [56,67]. These experimental procedures as well as analytical methods used are described in detail in the work of Vieitez et al. [57] and Silva et al. [79].

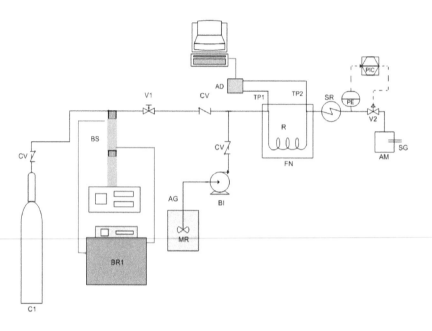

Figure 2. Schematic diagram of the experimental apparatus. RM - reactional mixture; MS - mechanical stirring device; LP - high-pressure liquid pump; CV - check-valve; A - solvent reservoir; B - thermostatic baths; SP - syringe pump; F - furnace; R - reactor; T1 - temperature indicator at the reactor inlet; T2 - temperature indicator at the reactor outlet; DA - data acquisition system; CS - cooling system; V1 - feed valve; PI - pressure indicator; PIC - controller; V2 - pressure control valve; S - glass collector; G - gas output. Taken with permission from Silva et al. [79].

3.1. Tubular reactor

The tubular reactor utilized was made of stainless steel tubing (316L 1/4 in. OD internal diameter of 3.2 mm HIP), being used in the work of Silva et al. [68], Vieitez et al. [57], Vieitez et al. [7], Bertoldi et al. [73], Vieitez et al. [117], Vieitez et al. [58], Vieitez et al. [118], Silva et al. [77], Vieitez et al. [119] and Vieitez et al. [120]. In these works, several approaches were made in order to optimize transesterification reactions for biodiesel production in supercritical ethanol in continuous tubular reactor and the better yields achieved for each study are presented in Table 1.

Silva et al. [68] investigated the effect of the variables temperature, pressure, oil to ethanol molar ratio and residence time on the yield of ethyl esters in the transesterification reaction of refined soybean oil. In that work, it was observed that an increase in temperature led to a sharp enhancement of reaction conversions and faster initial reaction rates. Also, as reaction time develops, a decline in the conversion reaction was observed for the temperature of 648 K. The reaction pressure had influence on the FAEE yields, with better yields obtained at 20 MPa. Regarding the effect of oil to ethanol molar ratio, results obtained by that study demonstrated that after a certain period of time higher values of molar ratio of ethanol to oil afford better conversions in shorter reaction times. This fact could be expected to a certain extent because in catalyst-free reactions an increase in the alcohol-to-oil molar ratio should provide greater contact between substrates, thus favoring reaction conversion. Besides, an excess of reactant could also shift the reaction to ethyl esters formation. In the experimental range investigated the authors reported ~80% in ethyl esters at the operating conditions shown in Table 1.

Vegetable oil	Conditions and additional information	FAEE yield [%]	Decomposition [%]	Reference
Refined soybean oil	1:40 oil to ethanol molar ratio; 623 K; 20 MPa; 35 min	~80.0	NR	[68]
Refined soybean oil	1:40 oil to ethanol molar ratio; 623 K; 20 MPa; 28 min and water content of 2.5 wt%	70.0	~ 14.0	[57]
Refined soybean oil	1:40 oil to ethanol molar ratio; 623 K; 20 MPa; 28 min and water content of 2.5 wt%	70.0	~ 14.0	
Refined soybean oil	1:40 oil to ethanol molar ratio; 573 K; 20 MPa; 52.5 min and water content of 5 wt%	70.0	3.0	[7]
Degummed soybean oil	1:40 oil to ethanol molar ratio; 623 K; 20 MPa; 28 min and water content of 10 wt%	55.0	NR	[117]
Castor oil	1:40 oil to ethanol molar ratio; 573 K; 20 MPa; 28 min and water content of 5 wt%	75.0	~11.0	[58]
Sunflower oil	1:40 oil to ethanol molar ratio; 623 K; 20 MPa; 42 min and water content of 5 wt%	~69.0	~14.0	[119]
High oleic sunflower oil	1:40 oil to ethanol molar ratio; 623 K; 20 MPa; 42 min and water content of 5 wt%	~75.0	<5.0	
Refined soybean oil	1:40 oil to ethanol molar ratio; 573 K; 20 MPa; ~48 min and addition of 10% of free fatty acids to oil	90.0	<5.0	[120]
High oleic sunflower oil	1:40 oil to ethanol molar ratio; 623 K; 20 MPa; ~48 min and addition of 10% of free fatty acids to oil	85.0	~8.0	
Rice bran oil	1:40 oil to ethanol; 573 K; 20 MPa; 26 min and addition of 10% of free fatty acids to oil	82.0	<5.0	
Refined soybean oil	1:40 oil to ethanol molar ratio; 598 K; 20 MPa; 110 min and CO$_2$ to substrates mass ratio of 0.05:1	76.0	NR	[73]
Refined soybean oil	1:1 oil to ethanol mass ratio; 598 K; 20 MPa; 30 min and operated with two reactors in series	74.0	~5.0	[77]
	1:1 oil to ethanol mass ratio; 598 K; 20 MPa; 30 min and operated with recycle of 40 wt%	75.0	~4.0	

NR = not reported

Table 1. Comparison of results obtained for transesterification reactions with supercritical ethanol in tubular reactor.

Industrial scale synthesis of biodiesel generally relies on the transesterification of vegetable oils with a short-chain alcohol, mainly methanol, using chemical catalysts [12]. Because ethanol is readily available from fermentative processes using biomass from a varied source, ethanol biodiesel appears as a 100% renewable alternative, additionally enabling the replacement of traditionally used methanol by an innocuous reagent. Besides, in the Brazilian context, ethanol has been the natural choice because Brazil is one of the biggest ethanol producers in the world, with a well established technology of production and large industrial plant capacity installed throughout the country. However, the cost of ethanol is still higher than that of methanol, in particular where absolute (dry) ethanol is used in processes based on conventional catalytic methods [1,3].

Adopting the best experimental conditions (soybean oil to ethanol molar ratio of 1:40, 623 K and 20 MPa) reported by Silva et al. [68], Vieitez et al. [57] evaluated the effect of water content (2.5 wt% to 10 wt%) on the reaction yield. Results showed that the presence of water in the reaction medium seems to have a positive effect on the FAEE production. A significant increase in the ester content was observed for 598, 573, and 548 K for all residence time studied, suggesting that reaction conversions should be improved by the presence of water in the reaction medium. No relevant changes were observed corresponding to 623 K, probably due to the persistence of side degradation reactions. A moderate increase in the ester content also was found for the reaction performed at 523 K, which seemed to be the minimum temperature value that should be considered for conducting catalyst-free transesterification reactions under supercritical conditions. For all values of water content in the reaction medium, a point of maximum of ester yield was found within the residence time range investigated. The maximum FAEE concentration was found at 28 minutes of residence time for water content values of 0, 2.5, and 5%, while higher values of water content (7.5 and 10%) showed a maximum ester content for 42 minutes of residence time. The maximum point of ester content was positively affected by the presence of water in the reaction medium; i.e., at 300ºC and 52.5 min, an increase in water content from 0 to 5% led to an increase in FAEE concentration from 29.7 to 70.0%, respectively. Therefore, the presence of water in the reaction medium showed a favorable effect on the ester synthesis, due to its possible catalytic role for the transesterification process and reduction of fatty acids degradation [7].

As observed by Silva et al. [68], a decrease in reaction yield by increasing the reaction time was found. As shown in Figure 3 [57], significant differences were noticed between the fatty acid composition of the starting soybean oil compared to that of the original product, involving the reduction in the polyunsaturated fatty acid ethyl ester percentage (C18:1, C18:2 e C18:3) and the production of trans isomers, originally absent (Figure 3). Also, the authors reported high percentage of fatty acid decomposition in the temperature of study (623 K). For yields > 80% about 12 wt % of decomposition was observed and about 4.0% of triglycerides for system without addition of water.

Considering the occurrence of fatty acids decomposition at high residence times and the formation of isomers of ethyl esters formed, Vieitez et al. [7] reported the effect of temperature (523 K to 623 K) and water content (5 wt% to 10 wt%) on these factors and yield of esters. It was observed that temperature strongly affected the degree of degradation with values of about 12 wt% and 28 wt% at 598 K and 623 K, respectively, for addition of 5 wt% of water in the reaction medium and high residence times. Moreover, the degradation phenomenon decreased as water concentration increased from 0 wt% to 10 wt%.

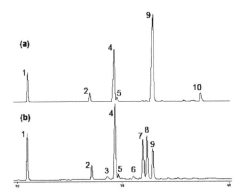

Figure 3. GC analysis of (a) soybean oil and (b) the product of the reaction (processed oil) performed at 350 °C, 20 MPa, 0% water, oil to ethanol molar ratio of 1:40 and 28 min of residence time. Peaks identification: C16:0 (1), C18:0 (2), *trans*-C18:1 (3), *cis*-9-C18:1 (4), *cis*-11-C18:1 (5), *trans*-6,12-C18:2 (6), *cis*-6,*trans*-12-C18:2 (7), *trans*-6,*cis*-12-C18:2 (8), *cis*-6,*cis*-12-C18:2 (9), and *cis*-9,12,15-C18:3 (10). Taken with permission from Vieitez et al. [57].

With respect to the effect of water concentration in the reaction medium on the degradation level, it was observed that the degradation phenomenon decreased as water concentration increased from 0 to 10 wt %. This reduction is in agreement with results showing that the addition of water may provide lower degradation levels and, accordingly, higher reaction conversions. Although no previous studies under similar conditions were found, these results are in agreement with some available references concerning the well-known favorable effect of the relatively low water activity on the oxidative stability of methyl linoleate or of vegetable oils. This phenomenon was attributed to different mechanisms, like the bonding of hydroperoxides, which decreases their reactivity, and an antioxidant effect due to hydration of traces of metals, which reduces their catalytic action [7].

The feedstock flexibility is the most important advantage to consider for biodiesel production methods because the resultant biodiesel price strongly depends on the feedstock price [121,122]. The cost of the raw materials currently represents about 70% of the total production cost [11]. The free fatty acids and water content in low grade feedstocks and hydrated ethanol pose a negative effect on the conventional homogeneous alkali-catalyzed process and heterogeneous catalytic methods, but can be successfully used in the transesterification reaction using an alcohol at its supercritical conditions. The evaluation of the effect of these variables on the efficiency of the transesterification reaction is highlighted. As shown in previous studies, the water content promotes the conversion of esters and decreases the degree of decomposition. Regarding the quality of the vegetable oil, studies concerning the effect of the vegetable oil type and free fatty acid content should be performed.

While growth within the biodiesel sector can contribute to increase the price of soybean oil and other biodiesel source materials, the competitiveness of the sector can be adversely affected by these very same prices changes, as well as other economic factors. These emerging trends suggest that food and energy markets are likely to be more strongly linked in future - such that spikes and fluctuations in the prices of energy lead to corresponding changes in food prices [123]. Currently, the main resource for biodiesel production in Brazil is soybean oil, comprising about 80% of total feedstock [124], however, recently raw material price increases has motivated the use other raw materials towards a future global leadership of the country in biodiesel production and use of non-edible and waste oils with low-added value.

The ethanolysis of degummed soybean oil was reported by Vieitez et al. [117] to evaluate the use of alternative raw materials in order to reduce the production costs. The experiments were performed at 20 MPa, 623 K and oil to ethanol molar ratio of 1:40 and lower ester contents were obtained with degummed oil than from using refined oil. At 28 minutes of residence time about ~ 80% and 40% of esters were obtained for refined and degummed oil, respectively. Many possible reasons for these results are mentioned by the authors, like the possible adverse effect of some minor components with a higher concentration in the degummed oil, e.g. pigments or hydroperoxides, with a known pro-oxidant effect on the fatty acids.

In the same context, to in the search of low-cost raw materials, alternative to refined soybean oil, Vieitez et al. [58] evaluated the possibility of producing ethyl esters from castor oil, a plant considered interesting as a potential raw material for biodiesel production. The effect of temperature (523 K to 623 K) and water content (5 wt% to 10 wt%) was evaluated by keeping the pressure fixed at 20 MPa and oil to ethanol molar ratio of 1:40. The authors reported FAEE yields in the order of 75% at 573 K, 5 wt% of water content and 28 minutes of residence time. The authors emphasized that special care should be taken into account concerning reaction temperature, which could favor the occurrence of side reaction involving the consumption of high percentage do fatty acids when increased over 573K. For example, it was related >70.0 % of decomposition at 623 K for high residence times (> 28 minutes).

In a later study, Vieitez et al. [119] focused on the dependence of esters yield and decomposition as a function of vegetable oil composition (Table 2). The results obtained show a relation between the composition of vegetable oil and content of esters. Note that the content of esters, regardless of residence time considered, decreases in the following order: high oleic sunflower oil> sunflower oil> soybean oil> castor oil. This order, except by castor oil, is inversely with the degree of unsaturation of each oil, which confirms that the efficiency of the process dependency of the stability of the oil used. The castor oil has a high percentage of decomposition. This percentage increases in the following order for the vegetable oils studied: high oleic sunflower oil <sunflower oil <soybean oil << castor oil.

Considering that decomposition phenomenon may strongly affect the ester yield and that the chemical stability is mainly determined by the insaturation degree of the fatty material, it is of major interest to study the behavior of oils with different fatty acid compositions in this process. Table 2 shows the composition of the different oils studied and their corresponding iodine value (IV), which indicates concerning solely the fatty acid composition, HO-SFO should be the oil with the higher stability (lower IV), followed by SFO and SBO.

Fatty acid	Soybean oil	Castor oil	Sunflower oil	High oleic sunflower oil
16:0	10.9	1.0	6.2	3.5
18:0	3.5	0.9	3.3	2.5
18:1	26.0	3.4	32.0	87.4
18:2	52.7	4.6	56.3	4.7
18:3	5.0	0.4	0.4	0.2
18:1-OH	---	88.7	---	---
IV[(a)]	129.2	---	128.5	85.3

Table 2. Fatty acid composition (wt%) of de vegetable oils studied. ((a)IV was calculated according method AOCS Cd 1c-85 [125])

The decomposition phenomenon was also studied in the work of Vieitez et al. [118], in which the stability of ethyl esters from soybean oil (SBOEE) exposed to high temperatures in supercritical ethanol was determined. In order to separately study the effect of such phenomenon, pure SBOEE were mixed with ethanol at a molar ratio 40:3 (ethanol:SBOEE) and exposed for different periods to supercritical conditions in a continuous system, at 20MPa and different temperatures. It was experimentally observed that the ester content of the processed samples were lower than that corresponding to the original SBOEE, indicating the occurrence of decomposition processes, which were more important as the temperature and residence time increased. The content of polyunsaturated esters of the treated SBOEE was lower than that of the starting mixture, showing that the decomposition rate was highly dependent on the nature and instauration degree of the alkyl chain. Therefore, results show that the exposure of the SBOEE to severe conditions required for efficiently performing the ethanolysis of vegetable oils by the supercritical method could cause the occurrence of important degradation processes of the lipid material. Such phenomenon could be identified as the main reason why the products from the supercritical transesterification of oils are less unsaturated than the raw materials. According to the results, the decomposition phenomenon is "selective" towards the polyunsaturated fatty esters, and there are no reasons to attribute such selectivity to the transesterification itself. Results also suggest that, in terms of the preservation of the integrity of the fatty acid chain, a supercritical transesterification process should not be performed at temperatures above 573 K, due to the high increase in the decomposition rate.

Recently, Vieitez et al. [120] evaluated the effect of the concentration of free fatty acids (FFAs) and type of vegetable oil on the yield of the reaction and decomposition of fatty acids. That work studied the effect of the addition of FFAs at various proportions to different vegetable oils (soybean oil, rice bran oil, and high oleic sunflower oil) on the efficiency of their conversion to ethyl esters by a continuous supercritical ethanolysis. When the reactor was operated at 573 K and 20 MPa with soybean oil using an alcohol/oil molar ratio of 40:1, an ester content of 53% was obtained. Under identical conditions but processing soybean oil with 10% of FFAs, the ester content rose to 91%. A similar favorable effect of the addition of

FFAs on the efficiency of the process was observed when processing rice bran oil and high oleic sunflower oil. Processing oils from different origins may lead to different ester contents in the final product because of the occurrence of decomposition phenomenon at different extents depending upon oil composition and stability. Results showed that the addition of FFAs is a useful tool for favoring alcoholysis against decomposition, with the consequence of a substantial increase in process efficiency. Therefore, the addition of FFAs could be a useful for improving the supercritical transesterification of oils with a low initial acidity and low-quality fats, such as highly hydrolyzed RBO, which could be efficiently converted to biodiesel using this technology. Several favorable effects on the process can be attributed to the presence of high levels of FFAs in the raw material: a catalytic role in the transesterification of triacylglycerides, a high esterification rate of FFA themselves, and a dilutive effect on the glycerol in the reaction medium (thus avoiding several unwanted side reactions). The contribution of all of these factors permitted us to achieve high efficiencies even at milder reaction conditions, thus minimizing the decomposition phenomenon, which has been pointed out as one of the main drawbacks of the supercritical method [119].

As observed in the studies presented in Table 1, the high transesterification conversion requires high temperature, high pressure and high alcohol to oil molar ratio. Indeed, the high temperature and pressure require high initial investments (equipment costs) for the implementation of such process operated and safety management policy. As a result of the high alcohol to oil molar ratio greater energy consumption in the reactants pre-heating and recycling steps is unavoidable. Moreover, the high amount of alcohol in the biodiesel product retards the biodiesel-glycerol phase separation. Therefore, the use of those original parameters results in high capital costs, especially for the reactor and pump, being somewhat higher than the novel catalytic methods [126]. To increase the technical and economical feasibility of supercritical method, further studies are required to reduce the energy consumption and operating parameters of this process.

In an attempt to reduce the operating conditions of the transesterification reaction, Bertoldi et al. [73] proposed for the first time the addition of carbon dioxide as a co-solvent in the reaction medium for reactions in continuous mode. The experiments were performed in the temperature range of 573-623 K, from 7.5 to 20 MPa, oil to ethanol molar ratio of 1:10 to 1:40 and co-solvent to substrates mass ratio from 0:1 to 0.5:1. Results showed that the yield of ethyl esters decreased with increasing addition of carbon dioxide to the system. At 623 K; 20 MPa; oil to ethanol molar ratio of 1:40 and 35 min it was observed about 80% of esters yield for system without co-solvent [68] and about 40 % for addition of CO_2 to substrates mass ratio of 0.05:1. Phase equilibrium data for the binary system ethanol-CO_2 shows the existence of high mutual solubility for these compounds [127,129]. On the other hand, very poor solubility of carbon dioxide in soybean oil has been reported in the literature [97]. Thus, it is possible that the co-solvent is dragging some amount of ethanol from the oil phase, causing the occurrence of a two-phase flowing system, decreasing the content of ethanol in contact with the vegetable oil with a consequent reduction in reaction conversion.

Another proposal considered was the non-catalytic production of fatty acid ethyl esters from soybean oil in a two-step process with experimental simulation of two reactors operated in series and a reactor with recycle, reported by Silva et al. [77]. The justification of the authors refers to the reaction conducted in two steps with reactors in series and/or recycling the leaving stream with intermediate removal of glycerol can increase the yield of the reaction, since the reaction may take place at lower alcohol to oil ratios, increasing the reaction rates of ester production [87]. The reaction of glycerol formed during the process with other components of the reaction medium may lead to a decrease in ester yield [66] and the undesirable consumption of alcohol [62,90]. Another important point of conduction of reactions in two steps is that the non-reacted products, diglycerides and monoglycerides, and also the esters formed may act as co-solvents in the reaction medium, increasing the solubility between the phases [100,130]. For the reactor in series it was reported 74% in esters at 598 K, 20 MPa, oil to ethanol mass ratio of 1:1 and 30 minutes of residence time for the second reaction step. For the system with recycle of 40 wt% at similar conditions it was obtained 75%. In both cases the degree of decomposition was lower than 5.0%.

3.2. Microtube reactor

Microreactor systems designed for continuous production have been studied in recent years for the transesterification of vegetable oils [109,112]. In the microreactor system, mass and heat transfer could be greatly intensified due to its small space with a large surface area-to-volume ratio [112], providing high process yields in low reaction times [109].

In this context, Silva et al. [79] developed a microtube reactor of stainless steel tubing (316L 1/16 in. OD internal diameter of 0.76 mm HIP) to evaluate the effects of inner diameter on the FAEE yield and compare the results with those reported by Silva et al. [68] for the same conditions using a tubular reactor. At lowest temperature (523 K) only 3.12 % FAEE yield is obtained in the tubular reactor, while 19% is reached using the microtube reactor. At 598 K this yield is increased from 38% to 53% when changing from the tubular to the microtube reactor at the same residence time. Such results demonstrate that higher ethyl esters yields can be achieved at lower temperatures, short reaction times with a smaller reactor inner diameter, hence minimizing the total decomposition of fatty acids.

In the work of Silva et al. [79] it was evaluated the effect of process variables (temperature, pressure and oil:ethanol molar ratio) on the yield of esters and decomposition. It was found that this variable had a positive effect on FAEE yield. In that work, it was noticed that an increase pressure and lowest ratios of ethanol to oil led to higher degrees of decomposition. It was also observed higher decomposition rates for oil:ethanol molar ratio of 1:10 and pressure of 20 MPa. In the experimental range investigated, appreciable yields were obtained (70%) at 598 K, 20 MPa and oil to ethanol molar ratio of 1:20, with low total decomposition of fatty acid (<5.0 wt%).

Considering the increasing reaction rates and improved mass transfer between phases in the conduction of reactions in a microtube reactor and the results obtained by Bertodi et al. [73] when using a cosolvent for the continuous tubular reactor, Trentin et al. [74] evaluated the addition of carbon dioxide on the reaction medium of soybean oil transesterification carried out in a microtube reactor. Results showed that ethyl esters yields obtained increased with increasing addition of carbon dioxide to the system and the highest yields were obtained with addition of co-solvent to substrate mass ratio of 0.2:1 to the reaction medium. The authors reported that the differences found in relation the conduct of the reactions in the tubular reactor [73] can be attributed to the problems of mass transfer in the tubular reactor and due to the fact that the mass and heat transfer may be greatly enhanced due to the smaller internal space (which means higher fluid velocity at the same flow rate), and the higher surface area-to-volume ratio, leading to higher process yields.

Silva et al. [76], conducted reactions in two steps in a microtube reactor: two-series reactors and reactor with recycle, conducted. It was obtained about 78% of ethyl esters yields and <2.0 wt% of decomposition for 45 min in the simulation of two reactors operated in series at 573 K, 20 MPa, oil to ethanol mass ratio of 1:1 (for the one-step process the authors shows 40 % of ethyl esters in the same conditions at 25 min). These results are higher than those reported by Silva et al. [77] at lower temperature and lower decomposition degree, as was also observed for reactions with recycle. Furthermore, in that work, glycerol was obtained with ~90 wt% of purity (after evaporation of ethanol and simple decantation) for the system with recycle and this fact of course should be taken into account for the purpose of implementation of a cost-effective transesterification process.

3.3. Packed-bed reactor

Results presented for the transesterification in microtube reactors are undoubtedly significant. However, production capacities of the above microreactors are considerably lower than those of conventional reactors by reason of their specific structures. Fulfilling the volume requirements of small-fuel biodiesel processing plants for distributive applications seems difficult. It is thus a challenge to identify a method for maximizing high synthesis efficiency by mixing at the microscale as well as for increasing biodiesel production remarkably [107]. An alternative to these problems would be to conduct the reactions in packed-bed reactors filled with different materials in different diameters, such as stainless steel spheres [108], metal foams [107] and glass beads [106].

In the work of Andrade et al. [80] a packed-bed tubular reactor was developed, which was made of stainless steel tubing (316 L 1/4 in OD inner diameter 3.2 mm) and stainless steel tubing (304 L 30.5 mm OD inner diameter 13 mm HIP) packed with glass beads (4.5 mm diameter). The results obtained by authors demonstrate that much higher ethyl esters yields can be achieved with this configuration. It can be seen observed in the results that at 548 K only 11.5% FAEE yield was obtained in the tubular reactor (TR), while 35% is reached using packed-bed tubular reactor (PBTR). At 573 K this yield is increased from 16% to 55% from the use TR to the PBTR at the same residence time. Such results demonstrate that much higher ethyl esters yields can be achieved at lower temperatures, small reaction times, also minimizing the total decomposition of fatty acids with the use of packed-bed tubular reactor. The increased performance of the reaction in the PBTR may be possibly due to the maximized interfacial surface area between the two flowing phases.

Silva et al. [79] proposed the use of microtube reactor for continuous synthesis of FAEE and reported yields of about 53% at 598 K, 20MPa, oil to ethanol molar ratio of 1:20 and residence time of 25 min. At similar conditions with addition of carbon dioxide as co-solvent

(CO$_2$ to substrate mass ratio of 0.2:1) in the microtube reactor, Trentin et al. [74] reported 58% of FAEE yield. At this same condition, the reaction conducted in the work of Andrade et al. (2012) in the PBTR, resulted in FAEE yields about 60%.

With the use of PBTR it can be obtained yields as high as ~ 83% at 598 K, 20 MPa, oil to ethanol molar ratio of 1:40 and 42 minutes of residence time. In such condition, it was observed 6.0 wt% of decomposition. In the evaluation of the effect of water content on the conversion, the authors reported 90% yield of ethyl esters and <5.0 wt% of decomposition at similar conditions with addition of 10 wt% of water to the reaction medium.

4. Conclusion

The non-catalytic transesterification at supercritical conditions is a promising method for esters production and has strong advantages, such as fast reaction time, feedstock flexibility, production efficiency and environmentally friendly benefits, but as observed in this manuscript the application of this methodology has some limitations, such as the operation conditions of elevated temperature and pressure and the use of higher amounts of alcohol in the reaction medium, which results in high energy costs for the process and degradation of the products generated. The analysis of these facts generate critical of the use of supercritical technology in transesterification reactions making them an open problem. Furthermore, prospective research is reducing the operating parameters and the decomposition of the reaction components are required to industrial scale application of the supercritical method. Acknowledgements

The authors thank CNPq, PROCAD/Pro-Engenharia - CAPES, BIOEN FAPESP, Fundação Araucária and Maringa State University (UEM).

Author details

Camila da Silva[1], Ignácio Vieitez[2], Ivan Jachmanián[2], Fernanda de Castilhos[3], Lúcio Cardozo Filho[1] and José Vladimir de Oliveira[4]

1 Program of Post-Graduation in Chemical Engineering, State University of Maringa, Maringá, Brazil

2 Departamento de Ciencia y Tecnología de los Alimentos, Facultad de Química, UDELAR, Montevideo, Uruguay

3 Department of Chemical Engineering, Paraná Federal University, Curitiba, Brazi

4 Department of Chemical and Food Engineering, Federal University of Santa Catarina, Florianópolis, Brazil

References

[1] Ma F, Hanna M. Biodiesel production: a review. Bioresource Technology 1999; 70(1), 1-15.

[2] Srivastava A, Prasad R. Triglycerides-based diesel fuels. Renewable and Sustainable Energy Reviews 2000; 4(2), 111-133.

[3] Fukuda H, Kondo A, Noda H. Biodiesel Fuel Production by Transesterification of Oils. Journal of Bioscience and Bioengineering 2001; 92(5), 405-416.

[4] Meher LC, Vidya SD, Naik SN. Technical aspects of biodiesel production by transesterification: a review. Renewable and Sustainable Energy Reviews 2006; 10(3), 248–268.

[5] Altin R, Çetinkaya S, Yucesu H S. The potential of using vegetable oil fuels as fuel for diesel engines. Energy Conversion and Management 2001; 42(5), 529-538.

[6] McCormick RL, Graboski MS, Alleman TL, Herring AM. Impact of biodiesel source material and chemical structure on emissions of criteria pollutants from a heavy-duty engine. Environmental Science & Technology 2001; 35(9), 1742-1747.

[7] Vieitez I, Silva C, Alkimim I, Borges GR, Corazza FC, Oliveira JV, Grompone MA, Jachmanián I. Effect of temperature on the continuous synthesis of soybean esters under supercritical ethanol. Energy & Fuels 2009; 23(1), 558-563.

[8] Achten WMJ, Verchot L, Franken YJ, Mathijs E, Singhe VP, Aertsa R, Muysa B. Jatropha bio-diesel production and use. Biomass and Bioenergy 2008; 32(12), 1063-1084.

[9] Berchmans HJ, Hirata S. Biodiesel production from crude Jatropha curcas L. seed oil with a high content of free fatty acids. Bioresource Technology 2008; 99(6), 1716–1721.

[10] Akgun, N., Iyscan E. Effects of process variables for biodiesel production by transesterification. European Journal of Lipid Science and Technology 2007; 109(5), 486-492.

[11] Robles-Medina A, González-Moreno PA, Esteban-Cerdán L, Molina-Grima E. Biocatalysis: towards ever greener biodiesel production. Biotechnology Advances 2009; 27(4), 398-408.

[12] Knothe G, Gerpen JV, Krahl J. The Biodiesel Handbook. AOCS Press: Champaign, USA; 2005.

[13] Pinto AC, Guarieiro LLN, RezendeMJC, Ribeiro NM, Torres EA, Lopes WA, Pereira PA P, Andrade JB. Biodiesel: An overview. Journal of Brazilian Chemical Society 2005; 16(6), 1313-1330.

[14] Marchetti JM, Errazu AF. Technoeconomic study of supercritical biodiesel production plant. Energy Conversion and Management 2008; 49(8), 2160–2164.

[15] Freedman B, Butterfield RO, Pryde EH. Transesterification kinetics of soybean oil. Journal of the American Oil Chemists' Society 1986; 63(10), 1375-1380.

[16] Noureddini H, Zhu D. Kinetics of transesterification of soybean oil. Journal of the American Oil Chemists' Society 1997; 74(11), 1457-1461.

[17] Darnoko D, Cheryan M. Kinetics of palm oil: Transesterification in a batch reactor. Journal of the American Oil Chemists' Society 2000; 77(12), 1263-1266.

[18] Ferrari RA, Silva V, Oliveira EAS. Biodiesel de soja: Taxa de conversão em ésteres etílicos, caracterização físicoquímica e consumo em gerador de energia. Química Nova 2005; 28(1), 19-23.

[19] Martinez M, Vicente G, Aracil J, Esteban A. Kinetics of sunflower oil methanolysis. Industrial & Engineering Chemistry Research 2005; 44(15), 5447-5454.

[20] Martinez M, Vicente G, Aracil J. Kinetics of Brassica carinata oil methanolysis. Energy & Fuels 2006; 20(4), 1722-1726.

[21] Oliveira D, Di Luccio M, Faccio C, Rosa CD, Bender JP, Lipke N, Amroginski C, Dariva C, Oliveira JV. Optimization of alkaline transesterificação of soybean oil and castor oil for biodiesel production. Applied Biochemistry and Biotechnology 2005; 121(1), 553-559.

[22] Meneghetti SMP, Meneghetti MR, Wolf CR, Silva EC, Lima GES, Silva LL, Serra TM, Cauduro F, Oliveira LG. Biodiesel from castor oil: A comparison of ethanolysis versus methanolysis. Energy & Fuels 2006; 20(5), 2262-2265.

[23] Lima JRO, Silva RB, Silva CCM, Santos LSS, Santos JR, Moura EM, Moura CVR. Biodiesel de babaçu (Orbignya sp.) obtido por via etanólica. Química Nova 2007; 30(3), 600-603.

[24] Kusdiana D, Saka S. Biodiesel fuel from rapeseed oil as prepared in supercritical methanol. Fuel 2001; 80(2), 225-231.

[25] Zhang Y, Dubé M A, Mclean D D, Kates M. Biodiesel production from waste cooking oil. Bioresource Technology 2003; 89(1), 1-16.

[26] Vyas AP, Verma JL, Subrahmanyam N. A review on FAME production processes. Energy & Fuel 2010; 89(1), 1–9.

[27] Pinnarat T, Savage P. Assessment of non-catalytic biodiesel synthesis using supercritical reaction conditions. Industrial & Engineering Chemistry Research 2008; 47(18), 6801-6808.

[28] Berrios M, Martin MA, Chica AF, Martin A. Study of esterification and transesterification in biodiesel production from used frying oils in a closed system. Chemical Engineering Journal 2010; 160(2), 473–479.

[29] Srilatha K, Issariyakul T, Lingaiah N, Sai Prasad PS, Kozinski J, Dalai AK. Efficient Esterification and Transesterification of Used Cooking Oil Using 12-Tungstophosphoric Acid (TPA)/Nb$_2$O$_5$ Catalyst. Energy & Fuels 2010; 24(9), 4748–4755.

[30] Thanh LT, Okitsu K, Sadanaga Y, Takenaka N, Maeda Y, Bandow H. A two-step continuous ultrasound assisted production of biodiesel fuel from waste cooking oils: A practical and economical approach to produce high quality biodiesel fuel. Bioresource Technology 2010; 101(14), 5394–5401.

[31] Corro G, Tellez N, Jimenez T, Tapia A, Banuelos F, Vazquez-Cuchillo O. Biodiesel from waste frying oil. Two step process using acidified SiO$_2$ for esterification step. Catalysis Today 2011; 166(1), 116–122.

[32] Gan S, Ng HK, Ooi CW, Motala NO, Anas M, Ismail F. Ferric sulphate catalysed esterification of free fatty acids in waste cooking oil. Bioresource Technology 2010; 101(19), 7338–7343.

[33] Rocha LLL, Ramos ALD, Filho NRA, Furtado NC, Taft CA, Aranda D. Production of biodiesel by a two-step niobium oxide catalyzed hydrolysis and esterification. Letters in Organic Chemistry 2010; 7(8), 571-578.

[34] Sousa JS, Oliveira EAC, Freire DMG, Aranda D. Application of lipase from the physic nut (Jatropha curcas L.) to a new hybrid (enzyme/chemical) hydroesterification process for biodiesel production. Journal of Molecular Catalysis B, Enzymatic 2010; 65(1), 133-137.

[35] Antunes WM, Veloso CO, Henriques CA. Transesterification of soybean oil with methanol catalyzed by basic solids. Catalysis Today 2008; 133(1), 548–554.

[36] Park Y, Lee D, Kim D, Lee J, Lee K. The heterogeneous catalyst system for the continuous conversion of free fatty acids in used vegetable oils for the production of biodiesel. Catalysis Today 2008; 131(1), 238–243.

[37] Trakarnpruk W, Porntangjitlikit S. Palm oil biodiesel synthesized with potassium loaded calcined hydrotalcite and effect of biodiesel blend on elastomers properties. Renewable Energy 2008; 33(7), 1558-1563.

[38] Oliveira D, Oliveira JV. Kinetics of the enzymatic alcoholysis of palm kernel oil in supercritical CO$_2$. Industrial & Engineering Chemistry Research 2000; 39(12), 4450.

[39] Oliveira D, Oliveira JV. Enzymatic alcoholysis of palm kernel oil in n-hexane and SCCO$_2$. The Journal of Supercritical Fluids 2001; 19(2), 141–148.

[40] Dalla Rosa C, Morandim MB, Ninow JL, Oliveira D, Treichel H, Oliveira JV. Lipase-catalyzed production of fatty acid ethyl esters from soybean oil in compressed propane. The Journal of Supercritical Fluids 2008; 47(1), 49-53.

[41] Dalla Rosa C, Morandim MB, Ninow JL, Oliveira D, Treichel H, Oliveira JV. Continuous lipase-catalyzed production of fatty acid ethyl esters from soybean oil in compressed fluids. Bioresource Technology 2009; 100(23), 5818-5826.

[42] Leadbeater NE, Stencel LM. Fast, Easy Preparation of biodiesel using microwave heating. Energy & Fuels 2006; 20(5), 2281–2283.

[43] Barnard TM, Leadbeater NE, Boucher MB, Stencel LM, Wilhite BA. Continuous-flow preparation of biodiesel using microwave heating. Energy & Fuels 2007; 21(3), 1777–1781.

[44] Melo CAR, Albuquerque CER, Fortuny M, Dariva C, Egues S, Santos AF, Ramos ALD. Use of microwave irradiation in the non-catalytic esterification of C18 fatty acids. Energy & Fuels 2009; 23(1), 580–585.

[45] Teixeira LSG, Assis JCR, Mendonça DR, Santos ITV, Guimarães, PRB, Pontes LAM, Teixeira JSR. Comparison between conventional and ultrasonic preparation of beef tallow biodiesel. Fuel Processing Technology 2009; 90(9), 1164-1166.

[46] Yu D, Tian L, Wu H, Wang S, Wang Y, Ma D, Fang X. Ultrasonic irradiation with vibration for biodiesel production from soybean oil by Novozym 435. Process Biochemistry 2010; 45(4), 519–525.

[47] Badday AS, Abdullah AZ, Lee KT, Khayoon MS. Intensification of biodiesel production via ultrasonic-assisted process: A critical review on fundamentals and recent development. Renewable and Sustainable Energy Reviews 2012; 16(7), 4574-4587

[48] Kusdiana D, Saka S. Kinetics of transesterification in rapeseed oil to Biodiesel fuel as treated in supercritical methanol. Fuel 2001; 80(5), 693-698.

[49] Kusdiana D, Saka S. Methyl esterification of free fatty acids of rapeseed oil as treated in supercritical methanol. Journal of chemical Engineering of Japan 2001; 34(3), 383-387.

[50] Demirbas A. Biodiesel from vegetable oils via transesterification in supercritical methanol. Energy: Conversion & Management 2002; 43(17), 2349-2356.

[51] Madras G, Kolluru C, Kumar R. Synthesis of biodiesel in supercritical fluids. Fuel 2004; 83(14), 2029-2033.

[52] Warabi Y, Kusdiana D, Saka S. Reactivity of triglycerides and fatty acids of rapeseed oil in supercritical alcohols. Bioresource Technology 2004; 91(3), 283-287.

[53] Demirbas A. Biodiesel production via non-catalytic SCF method and Biodiesel fuel characteristics. Energy Conversion & Management 2006; 47(15), 2271-2282.

[54] Kusdiana D, Saka S. Effetcs of water on biodiesel fuel production by supercritical methanol treatment. Bioresource Techonology 2004; 91(3), 289-295.

[55] Wang C, Zhou J, Chen W, Wang W, Wu Y, Zhang J, Chi R, Ying W. Effect of weak acids as a catalyst on the transesterification of soybean oil in supercritical methanol. Energy & Fuels 2008; 22(5), 3479–3483.

[56] Imahara H, Minami E, Hari S, Saka S. Thermal stability of biodiesel in supercritical methanol. Fuel 2007; 87(1), 1-6.

[57] Vieitez I, Silva C, Borges GR, Corazza FC, Oliveira JV, Grompone MA, Jachmanián I. Continuous production of soybean biodiesel in supercritical ethanol–water mixtures. Energy & Fuels 2008; 22(4), 2805-2809.

[58] Vieitez I, Pardo MJ, Silva C, Bertoldi C, Castilhos F, Oliveira JV, Grompone MA, Jachmanián I. Continuous synthesis of castor oil ethyl esters under supercritical ethanol. The Journal of Supercritical Fluids 2011; 56(3), 271–276.

[59] Olivares-Carrillo P, Quesada-Medina J. Synthesis of biodiesel from soybean oil using supercritical ethanol in a one-step catalyst-free process in batch reactor. The Journal of Supercritical Fluids 2011; 58(16), 378-384.

[60] Shin H, Lim S, Bae S, Oh C. Thermal decomposition and stability of fatty acid methyl esters in supercritical methanol. Journal of Analytical and Applied Pyrolysis 2011; 92(2), 332–338.

[61] Anitescu G, Deshpande A, Tavlarides LL. Integrated technology for supercritical biodiesel production and power cogeneration. Energy & Fuel 2008; 22(2), 1391-1399.

[62] Aimaretti N, Manuale DI, Mazzieri VM, Vera CR, Yori C. Batch study of glycerol decomposition in one-stage supercritical production of biodiesel. Energy & Fuel 2009; 23(2), 1076-1080.

[63] Kasim NS, Tsai T, Gunawan S, Ju Y. Biodiesel production from rice bran oil and supercritical methanol. Bioresource Technology 2009; 100(8), 2399-2403.

[64] Marulanda VF, Anitescu G, Tavlarides LL. Investigations on supercritical transesterification of chicken fat for biodiesel production from low cost lipid feedstocks. The Journal of Supercritical Fluids 2010; 54(1), 53–60.

[65] Kusdiana D, Saka S. Two-Step preparation for catalyst-free biodiesel fuel production. Applied Biochemistry and Biotechonology 2004; 113(1), 781-791.

[66] Minami E, Saka S. Kinetics of hydrolysis and methyl esterification for biodiesel production in two-step supercritical methanol process. Fuel 2006; 85(17), 2479-2483.

[67] He H, Tao W, Zhu S. Continuous production of biodiesel from vegetable oil using supercritical methanol process. Fuel 2007; 86(3), 442-447.

[68] Silva C, Weschenfelder TA, Rovani S, Corazza FC, Corazza ML, Dariva C, Oliveira JV. Continuous production of fatty ethyl esters from soybean oil in compressed ethanol. Industrial & Engineering Chemistry Research 2007; 46(16), 5304-5309.

[69] Cao W, Han H, Zhang J. Preparation of biodiesel from soybean oil using supercritical methanol and CO_2 as co-solvent. Process Biochemistry 2005; 40(9), 3148-3156.

[70] Han H, Cao W, Zhang J. Preparation of biodiesel from soybean oil using supercritical methanol and co-solvent. Fuel 2005; 84(4), 347-351.

[71] Hegel P, Mabe G, Pereda S, Brignole EA. Phase transitions in a biodiesel reactor using supercritical methanol. Industrial & Engineering Chemistry Research 2007; 46(19), 6360-6365.

[72] Yin J, Z, Xiao M, Song JB. Biodiesel from soybean oil in supercritical methanol with co-solvent. Energy Conversion and Management 2007; 49(5), 908-912.

[73] Bertoldi C, Silva C, Bernardon JP, Corazza ML, Cardozo Filho L, Oliveira JV, Corazza FC. Continuous production of biodiesel from soybean oil in supercritical ethanol and carbon dioxide as co-solvent. Energy & Fuels 2009; 23(10), 5165-5172.

[74] Trentin CM, Lima AP, Alkimim IP, Silva C, Castilhos F, Mazutti MA, Oliveira JV. Continuous production of soybean biodiesel with compressed ethanol in a microtube reactor using carbon dioxide as co-solvent. Fuel Processing Technology 2011; 92(5), 952-958.

[75] D' Ippolito SA, Yori JC, Iturria ME, Pieck CL, Vera CR. Analysis of a two-step, non-catalytic, supercritical biodiesel production process with heat recovery. Energy & Fuels 2007; 21(1), 339-346.

[76] Silva C, Lima AP, Castilhos F, Cardozo Filho L, Oliveira JV. Non-catalytic production of fatty acid ethyl esters from soybean oil with supercritical ethanol in a two-step process using a microtube reactor. Biomass & Bioenergy 2011; 35(1), 526-532.

[77] Silva C, Borges G, Castilhos F, Oliveira JV, Cardozo Filho L. Continuous production of fatty acid ethyl esters from soybean oil at supercritical conditions. Acta Scientiarum. Technology 2012; 34(2), 185-192.

[78] Chen CH, Chen WH, Chang CMJ, Lai SM, Tu CH. Biodiesel production from supercritical carbon dioxide extracted Jatropha oil using subcritical hydrolysis and supercritical methylation. The Journal of Supercritical Fluids 2010; 52(2), 228-234.

[79] Silva C, Castilhos F, Oliveira JV, Cardozo Filho L. Continuous production of soybean biodiesel with compressed ethanol in a microtube reactor. Fuel Processing Technology 2010; 91(10), 1274-1281.

[80] Andrade SB, Abdala ACA, Silva EA, Cabral VF, Oliveira JV, Cardozo Filho L, Silva C. Non-catalytic production of ethyl esters from soybean oil in a continuous packed bed reactor (Submitted to Fuel Processing Technology).

[81] Glisic S, Skala D. The problems in design and detailed analyses of energy consumption for biodiesel synthesis at supercritical conditions. The Journal of Supercritical Fluids 2009; 49(2), 293-301.

[82] Deshpande A, Anitescu G, Rice PA, Tavlarides LL. Supercritical biodiesel production and power cogeneration: Technical and economic feasibilities. Bioresource Technology 2010; 101(6), 1834-1843.

[83] Johnson DT, Taconi KA. The glycerin glut: Options for the value-added conversion of crude glycerol resulting from biodiesel production. Environmental Progress 2007; 26(4), 338-348.

[84] Zhou C, Beltramini JN, Fan Y, Lu GQ. Chemoselective catalytic conversion of glycerol as a biorenewable source to valuable commodity chemicals. Chemical Society Reviews 2008; 37(1), 527–549.

[85] Marchetti JM, Errazu AF. Possible methods for biodiesel production. Renewable & Sustainable Energy Reviews 2007; 11(6), 1300-1311.

[86] Sasaki T, Omoyubi S, Fumio O. Method for preparing fatty acid esters and fuel comprising fatty acid esters. United States Patent: 2001; 6,187,939.

[87] Crawford JW, Crawford JM, Crafts R. Transesterification of oil to form biodiesels United States Patent: 2007; 059512.

[88] Kiwjaroun C, Tubtimdee C, Piumsomboon P. LCA studies comparing biodiesel synthesized by conventional and supercritical methanol methods. Journal of Cleaner Production 2009; 17(2), 143–153.

[89] Marulanda VF. Biodiesel production by supercritical methanol transesterification: process simulation and potential environmental impact assessment. Journal of Cleaner Production 2012; 33(1), 109-116.

[90] Lee S, Posarac D, Ellis N. An experimental investigation of biodiesel synthesis from waste canola oil using supercritical methanol. Fuel 2012; 91(1), 229-237.

[91] Official Methods and Recommended Practices of the American Oil Chemists' Society (Method AOCS Ce 2-66), 4th ed., edited by R.E. Walker, American Oil Chemists' Society, Champaign; 1990.

[92] Ginosar D, Fox R, Petkovic L M, Christian S. Production of biodiesel using expanded gas solvents. United States Patent: 2006; 121584.

[93] Imahara H, Xin J, Saka S. Effect of CO_2/N_2 addition to supercritical methanol on reactivities and fuel qualities in biodiesel production. Fuel 2009; 88(7), 1329-1332.

[94] Tan KT, Lee KT, Mohamed AR. Effects of free fatty acids, water content and co-solvent on biodiesel production by supercritical methanol reaction. The Journal of Supercritical Fluids 2010; 53(1), 88-91.

[95] Muppaneni T, Reddy HK, Patil PD, Dailey P, Aday C, Deng S. Ethanolysis of camelina oil under supercritical condition with hexane as a co-solvent. Applied Energy 2012; 94(1), 84-88.

[96] Temelli F, King JW, List GR. Conversion of oils to monoglycerides by glycerolysis in supercritical carbon dioxide media. Journal of the American Oil Chemists' Society 1996; 73(6), 699–706.

[97] Ndiaye PM, Franceschi E, Oliveira D, Dariva C, Tavares FW, Oliveira JV. Phase be-
      havior of soybean oil, castor oil and their fatty acid ethyl esters in carbon dioxide at
      high pressures. The Journal of Supercritical Fluids 2006; 37(1), 29-37.

[98] Lanza M, Priamo WL, Oliveira JV, Dariva C, Oliveira D. The effect of temperature,
      pressure, exposure time, and depressurization rate on lipase activity in $SCCO_2$. Ap-
      plied Biochemistry and Biotechnology 2004; 113(1), 181-187.

[99] Pereda S, Bottini SB, Brignole EA. Gas-liquid reactions under supercritical conditions
      – Phase equilibria and thermodynamic modeling. Fluid Phase Equilibria 2002;
      194(30), 493-499.

[100] Busto M, D' Ippolito SA, Yori JC, Iturria ME, Pieck CL, Grau JM, Vera CR. Influence
      of the axial dispersion on the performance of tubular reactors during the noncatalytic
      supercritical transesterification of triglycerides. Energy & Fuels 2006; 20(6),
      2642-2647.

[101] Qiu Z, Zhao L, Weatherley L. Process intensification technologies in continuous bio-
      diesel production. Chemical Engineering and Processing 2010; 49(4), 323-330.

[102] Dewitt S H. Microreactors for chemical synthesis. Current Opinion in Chemical Biol-
      ogy 1999; 3(1), 350-356.

[103] Worz O, Jackel K P, Richter T, Wolf A. Microreactors, a new efficient tool for opti-
      mum reactor design. Chemical Engineering Science 2001; 56(3), 1029-1033.

[104] Fletcher PDI, Haswell SJ, Villar E, Warrington BH, Watts P, Wong SYF, Zhang X. Mi-
      cro reactors: principles and applications in organic synthesis. Tetrahedron 2002;
      58(24), 4735-4757.

[105] Hessel V, Hardt S, Lowe H. Chemical micro process engineering: Fundamentals,
      modelling and reactions. Editora Wiley-VCH Verlag: Germany; 2005.

[106] Ataya F, Dube MA, Ternan M. Transesterification of canola oil to fatty acid methyl
      ester (FAME) in a continuous flow liquid-liquid packed bed reactor. Energy & Fuels
      2008; 22(5), 3551–3556.

[107] Yu X, Wen Z, Lin Y, Tu S, Wang Z, Yan J. Intensification of biodiesel synthesis using
      metal foam reactors. Fuel 2010; 89(11), 3450-3456.

[108] Santacesaria E, Di Serio M, Tesser R, Tortorelli M, Turco R, Russo V. A simple device
      to test biodiesel process intensification. Chemical Engineering and Processing 2011;
      50(10), 1085-1094.

[109] Sun J, Ju J, Zhang L, Xu N. Synthesis of biodiesel in capillary microreactors. Industri-
      al & Engineering Chemistry Research 2008; 47(5), 1398-1403.

[110] Pohar A, Plazl I. Laminar to Turbulent Transition and Heat Transfer in a Microreac-
      tor: Mathematical Modeling and Experiments. Industrial & Engineering Chemistry
      Research 2008; 47(19); 7447–7455.

[111] Ehrfeld W, Lowe H, Hessel V. Microreactors: New Technology for Modern Chemistry; Wiley-VCH: New York; 2000.

[112] Guan G, Kusakabe K, Moriyama K, Sakurai N. Transesterification of sunflower oil with methanol in a microtube reactor. Industrial & Engineering Chemistry Research 2009; 48(3), 1357–1363.

[113] Zhang X, Stefanick S, Villani F J. Application of microreactor technology in process development. Organic Process Research & Developmente 2004; 8(3), 455-460.

[114] Achenbach E. Heat and flow characteristics of packed beds. Experimental Thermal and Fluid Science 1994; 10(1), 17-27.

[115] Su Y. The Intensification of Rapid Reactions for Multiphase Systems in a Microchannel Reactor by Packing Microparticles. AIChE Journal 2011; 57(6), 1409–1418.

[116] European normative EN 14103, Determination of ester and linolenic acid methyl ester contents, issued by Asociación Española de Normalización y Certificación, Madrid; 2003.

[117] Vieitez I, Silva C, Alkimim I, Borges GR, Corazza FC, Oliveira JV, Grompone MA, Jachmanián I. Continuous catalyst-free methanolysis and ethanolysis of soybean oil under supercritical alcohol/water mixtures. Renewable Energy 2010; 35(9), 1976–1981.

[118] Vieitez I, Silva C, Alkimim I, Castilhos F, Oliveira J V, Grompone M A, Jachmanián I. Stability of ethyl esters from soybean oil exposed to high temperatures in supercritical ethanol. The Journal of Supercritical Fluids 2011; 56(3), 265–270.

[119] Vieitez I, Irigaray B, Casullo P, Pardo M, Grompone MA, Jachmanián I. Effect of free fatty acids on the efficiency of the supercritical ethanolysis of vegetable oils from different origins. Energy & Fuels 2012; 26(3), 1946-1951.

[120] Vieitez I, Silva C, Castilhos F, Oliveira JV, Grompone MA, Jachmanián I. Elaboración de biodiesel mediante la transesterificación sin catalizador de aceites en alcoholes supercríticos. Ingeniería Química 2011; 40(1), 10-19.

[121] Kulkarni MG, Dalai AK. Waste cooking oil an economical source for biodiesel: A review. Industrial & Engineering Chemistry Research 2006; 45(9), 2901-2913.

[122] Lam MK, Lee KT, Mohamed AR. Homogeneous, heterogeneous and enzymatic catalysis for transesterification of high free fatty acid oil (waste cooking oil) to biodiesel: A review. Biotechnology Advances 2010; 28(4), 500-518.

[123] Lin L, Cunshan Z, Vittayapadung S, Xiangqian S, Mingdong D. Opportunities and challenges for biodiesel fuel. Applied Energy 2011; 88(4), 1020-1031.

[124] Agência Nacional do Petróleo, Gás Natural e Biocombustíveis. Boletim mensal de biodiesel. Brasília, Brazil, Agência Nacional de Petróleo, Gás Natural e Biocombustíveis, January; 2012. 9p. Retrieved from http://www.anp.gov.br.

[125] Official Methods and Recommended Practices of the American Oil Chemists' Society (Method AOCS 1c-85), 4th ed., edited by R.E. Walker, American Oil Chemists' Society, Champaign; 1990.

[126] Ngamprasertsith S, Sawangkeaw R. Transesterification in Supercritical Conditions. In: Biodiesel - Feedstocks and Processing Technologies. Rijeka: InTech; 2011. p247-268.

[127] Day CY, Chang CJ, Chen CY. Phase equilibrium of ethanol + CO_2 and acetone + CO_2 at elevated pressure. Journal of Chemical & Engineering Data 1996; 41(4), 839-843.

[128] Pohler H, Kiran E. Volumetric properties of carbon dioxide + ethanol at high pressures. Journal of Chemical & Engineering Data 1997; 42(2), 384-388.

[129] Joung SN, Yoo CW, Shin HY, Kim SY, Yoo KP, Lee CS, Huh WS. Measurements and correlation of high-pressure VLE of binary CO_2-alcohol systems (methanol, ethanol, 2-methoxyethanol and 2-ethoxyethanol). Fluid Phase Equilibria 2001; 185(1), 219-230.

[130] Granado ML, Alonso DM, Alba-Rubio AC, Mariscal R, Ojeda M, Brettes P. Transesterification of triglycerides by CaO: Increase of the reaction rate by biodiesel addition. Energy & Fuels 2009; 23(4), 2259–2263.

Chapter 8

Biodiesel Current Technology: Ultrasonic Process a Realistic Industrial Application

Mario Nieves-Soto, Oscar M. Hernández-Calderón,
Carlos Alberto Guerrero-Fajardo,
Marco Antonio Sánchez-Castillo,
Tomás Viveros-García and
Ignacio Contreras-Andrade

Additional information is available at the end of the chapter

1. Introduction

Biodiesel is briefly defined as a renewable fuel derived from vegetable oils or animal fats. Similarly, the American Society of Testing and Materials (ASTM) defines biodiesel as mono-alkyl long-chain fatty acids esters derived from fatty renewable inputs, such as vegetable oils or animal fats. The term "bio" refers to its origin from biomass related resources, in contrast to the traditional fosil-derived diesel, while the term "diesel" refers to its use on engines; as a fuel, biodiesel is typically used as a blend with regular diesel. To date, biodiesel is well recognized as the best fuel substitute in diesel engines because its raw materials are renewable, and it is biodegradable and more environmentally friendly; biodiesel probably has better efficiency than gasoline and exhibits great potential for compression-ignition engines.

Biodiesel was mainly produced from soybean, rapeseed and palm oils, although social and economic considerations have turned attention to second generation biomass raw materials such as *Jatropha curcas* oil [1]. It is well known that biodiesel competitiveness has to be improved, as to compare to curcas oil diesel, to spread out its consumption. Two routes are suggested to overcome this problem; one is related to get cheap raw materials (i.e., triglycerides, nonedible vegetable oils, animal fats and wasted oils), and other one is to reduce processing cost; notoriously both issues are interrelated [1]. The raw material origin is of great relevance because it determines the final biodiesel properties and also the type of process to be used. It is importance to notice that low-cost raw materials usually contain significant amounts of free fatty acids (FFA), which lead to a complex and more expensive final process, e.g. the catalyst depletion is accelerated, the purification costs are increase, and the yield in alkali-catalyzed transesterification is decreased.In the other hand, processing costs could be reduced through simplified operations and eliminating decreased. On the other hand, waste streams. There are several current biodiesel technologies that tried to overcome the issues just indicated. For instance, some plants in Europe produce biodiesel by transesterification using supercritical methanol without any catalyst. In this case, the reaction is very fast (less than 5 min) and the catalyst absence decreases downstream purification costs. However, the reaction requires very high temperature (350–400 ºC) and pressure (100–250 atm) which, in turn, increases the capital and safety costs. Another suggested alternative is the use of heterogeneous catalysts that can be separated more easily from reaction products, and required less harsh reaction conditions than the supercritical methanol process. However, these technologies are still far to produce low cost biodiesel, even if they overcome some problems of the conventional process. In this scenario, new technologies are still required for the transformation of second and third generation biomass raw materials, as well as residual biomass, in sustainable production of biodiesel.

To this respect, recently, an increasing number of applications of ultrasonic processes(US) in chemical transformations have made sonochemistry an attractive area of research and development [13]. The main benefit of US is to enhance chemical reactivity by providing enough energy through out the cavitation phenomenon. The bubble implosions generated in this phenomenon provide sufficient energy to break chemical bonds. Thus, the application of US can completely change the reaction pathways as well as the reaction yield and selectivity. Importantly, the main benefits that can be pointed out from the application of US are the reaction rate increase and the use of less severe operating conditions, as well as shorten induction periods and reduction of reagents amount. An interesting extension of US is the possibility to apply it for the transesterification of vegetable oils to produce biodiesel. Typically, this reaction is kinetically slow and shows mass transfer limitations. Thus, cavitation phenomenon of the US could provide the activation energy required in the reaction as well as the conditions (i.e., mechanical energy) to improve the reaction mixing. In this way, US could provide technical an economic advantage for biodiesel production, as compared to conventional transesterification processes.

In this chapter we report the advantageous application of US for biodiesel lab scale production from *Jatropha curcas* oil (JCO). This proposal is in agreement with the search of optimized, sustainable biodiesel production. The chapter briefly describes the basics of current biodiesel technologies and, in more detail, the fundamentals and benefits provided by sonochemistry to alkaline transesterification process ("sonotransesterification" process). In addition, the experimental setup used for sonotranseterification and the main results to date are also discussed. In general, sonotranseterification shows a significant improve when applied to biodiesel production from JCO; when using a 4.5:1 molar ratio of alcohol/JOC, 25 ºC, atmospheric pressure and 60% of amplitude, yields up to 98% are obtained. Finally, these results are compared to more conventional processes such as supercritical methanol and heterogeneous catalysis for the same raw material. Results are discussed in terms of the advantages/disadvantages of reaction operating conditions, energy demand and process time. Notoriously, sonotransesterification shows significant benefits as compare to conventional technologies, which could be further improved as the process be optimized.

2. Biodiesel

Biodiesel is obtained by transesterification reaction, also known as alcoholysis; in this reaction, vegetable oils (preferably non-edible oils) or animal fats are reacted with a significant excess of alcohol (methanol or ethanol), in the presence of a catalyst (homogeneous, heterogeneous or enzymatic), to form fatty acid alkyl esters (FAME) and glycerol, a valuable by-product for industry [2]. In a conventional biodiesel process (CBP), the alcohol-FAME phase is separated and the alcohol excess is recycled (Figure 1). Next, esters undergo a purification process, consisting of water washing, dry vacuum and subsequent filtering. In this process, importantly, the oil used as raw material must be cleaned and its FFA content must belower than 0.5wt%; otherwise, a pretreatment of the raw material must be carried out. Then, the oil is typically mixed with the alcohol in a 6:1 molar ratio, and 1 to 3% homogeneous catalyst (KOH or NaOH) is added to the reaction mixture. Reactants, including the catalyst, must be anhydrous to avoid soap formation. The reaction is then stirred for 40 to 60 minutes, at temperature between 50 and 60 °C, afterward the reaction is completed [2].

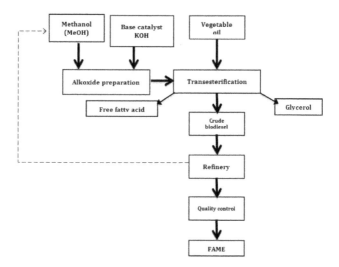

Figure 1. Flow diagram of conventional alkaline homogeneous process for biodiesel production [2].

The overall transesterification chemistry involves an exchange between the alcohol groups (i.e., methanol or ethanol) and glycerol, at given reaction conditions, to produce methyl or ethyl fatty acidsesters (Figure 2). Each fatty acid molecule has the same chemistry configura-

tion [3] and it only differs from other molecules for the carbon chain length or its unsaturation number, which leads to produce FAME with different properties that, in turn, impact
the final biodiesel characteristics such as melting point, oxidation stability, etc. This is the
reason why the raw material quality is suggested to be the key point for the biodiesel process. Figure 2 also shows the well-accepte dreaction pathway. From the thermodynamic point
of view, triglycerides and methanol are well-accepted reaction pathway unable to react at
room temperature and atmospheric pressure (i.e., 25°C and 1 atm, respectively) because of
the extremely low solubility of the alcohol into the oil; for this reason, catalysis plays an important role for the alcoholysis reaction to take place.

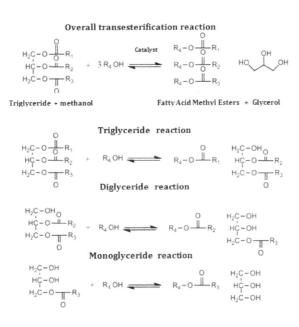

Figure 2. Well-generalized transesterification pathway [3].

Literature [1-4] describes that the first reaction step is the formation of an alkoxide ion (RO⁻)
through proton transfer from the alcohol. Actually, when homogeneous Brönsted basic catalysts (i.e., NaOH, KOH, Na$_2$CO$_3$) are interacts with the alcohol, the following reaction occurs:

$$KOH + CH_3OH \Rightarrow CH_3O^- + [K(OH)H]^+ \tag{1}$$

This alkoxide group then attacks the carbonyl carbon atom of the triglyceride molecule to
form a tetrahedral intermediate ion (step 2); therefore, an alkoxide (NaOCH$_3$, KOCH$_3$) is often directly used as catalyst. This intermediate ion rearranges to generate a diglyceride ion

and alkyl ester molecule (step 3). Next, the diglyceride ion reacts with the protonated base catalyst, which generates a diglyceride molecule and returns the base catalyst to its initial state (step 4). The resulting diglyceride is then ready to react with another alcohol molecule, there by maintaining the catalytic cycle until all the glyceride molecules have been complete converted to biodiesel at 60-80 ºC (Figure 2).

The conventional process (based on homogenous catalysts) has associated several problems, which makes it more expensive when compared to fosil-derived diesel, e.g. raw materials pretreatment and process and cost issues. If raw materials are taken into account, fat and oils cannot directly be used when large amounts of FFA are present. As previously indicated, when alkaline homogeneous catalysts are used, FFA should be less than 0.5 w/w% to avoid high soap formation. Moreover, expensive refinery steps are associated to separate the catalyst and the methanol/biodiesel/glycerol mixture. Generally, water is used to remove alkaline catalysts but this stage makes the overall process less important from environmental point of view. Other relevant issues such as reaction time, mass transfer limitations, optimized set of operating conditions (temperature, pressure, alcohol: oil ratio), determine the economic success of biodiesel production. Regarding the reactor technology, continuous biodiesel process (CBP), especially when equipped with tubular reactors, are always preferred as compared to batch processes. Obviously, this is due to the fact that CBP allow the processing of higher amounts of raw material. However, always that CBP is selected it should be considered the need to incorporate a centrifuge process for glycerol/biodiesel separation, which has a considerable increase in the processing cost. Therefore, an optimum conventional biodiesel process should be conducted at room temperature, atmosphere pressure, avoid water for homogeneous catalysts recovering, and use a low cost glycerol/biodiesel coalescence unit; importantly, the process should reached oil yields over 98%.

Table 1 shows some a comparison of current biodiesel technologies; it is evident from this table that there is a direct connection between complexity and process cost and the quality of final product. As outlined above, conventional process capital cost is low, but processing cost is high because of long reaction time, and separation and purification issues, among others. Regarding the supercritical methanol process, it seems simple and delivers high purity product but, also, capital and operating cost are too high because they are related to severe process conditions. With respect to the use of heterogeneous catalyst, it certainly improves the products separation and purification but, again, this technology is still far to be asuitable economic option because of the high temperature and long reaction time still required for the process. Moreover, another issue to overcome is the design of solid catalysts with appropriate acid sites configuration to improve yield and selectivity, and to decrease catalyst deactivation in hydrous conditions. On the other hand, the US process is less extended and its advantages have not totally documented. However, it could be postulated that the thousands of bubbles formed during the cavitation phenomenon of the US facilitates the formation of a methanol-KOH/oil microemulsion at high temperature, which drastically decreases mass transfer limitations. In this scenario, the transesterification reaction could be carried out within a few seconds, at room temperature (at the "bulk") and atmospheric pressure, thus helping to decrease the process cost. The next section of this chapter describes the basic principles of ultrasound applied to transesterification reaction.

Variable	Homogeneous Catalysis	Heterogeneous Catalysis	Enzymatic Catalysis	Non Catalytic SMP[2]
Reaction time	0.5-4h	0.5-5.5h	1-8h	120-240s
Operation conditions	0.1 MPa, 30-65 ºC	0.1-5 MPa, 30-200 ºC	0.1 Mpa, 35-40 ºC	>25Mpa, >239.4 ºC
Catalyst	Acid/base	Metal oxides o carbonates	Lipase	Non
Free fatty acid	Soap formation	Esters	Esters	Esters
Water	Interfere	No interfere	No interfere	Act as catalyst to the process
Yield	Normal	Low to normal	Low to normal	High
Purification	Difficult	Easy	Easy	Very easy
Downstream	Water	Non	Non	Non
Glycerol purity	Low	Low to normal	Normal	High
Process	Complex	Normal	Simple	Simple
Capital cost	Low	Medium	High	Very high
Operation cost	High	High	Normal	High

Table 1. Comparison of current biodisel technologies for processing biodisel[1] . Source: [9]. [2] SMP: Supercritical methanol process

3. Principle of Ultrasonic Process

Traditionally, sound is a subject studied in physics and it is not a well-met topic in a chemistry course and, so, is somewhat unfamiliar to practicing chemists. However, sono-chemistry, which is defined as the use of sound to promote or enhance chemical reactions, has recently received much attention in several chemical reactions concerning sustainability process [5].

It is known that an acoustic wave is a propagation of pressure oscillation in a given medium (gas, liquid or solid), with the velocity of sound producing both the rarefication and compression phases. Figure 3 shows that sound waves are often disclosed as a series of vertical lines or shaded colors, where line separation or color depth represent the intensity or amplitude of the sine wave; the pitch of the sound depends upon the frequency of the wave. According to the sound spectrum, an ultrasonic wave is an acoustic wave whose frequency is above 20 kHz, which is not audible to human. Hence, when a liquid is irradiated by a strong ultrasonic wave, the pressure at some regions in the liquid becomes negative (expansion) because the acoustic amplitude of the wave is larger than the ambient pressure. Therefore, if the pressure wave propagating through a liquid has enough intensity, formation of vapor bubbles may occur because the gas dissolved in the liquid can no longer be kept dissolved, because the gas solubility is proportional to the pressure; this is known as the *cavitation phenomenon* [11].

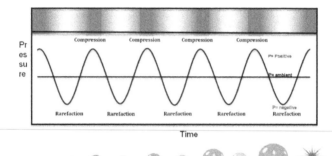

Figure 3. Sound waves interaction with a liquid medium [13]. The bubble growth due to the expansion-compression cycles resulting in the formation of localized "hot spots".

The bubbles formed in the cavitation phenomenon grow from nuclei, over many acoustic cycles, through an elastic process [10]. During the expansion cycle an inflow occurs into the bubble, due to the gradient in gas concentration of the fluid shell surrounding the bubble. As the gas diffusion rate into the bubble is proportional to the concentration gradient of dissolved gas, the net inflow of gas into the bubble is essentially higher during the expansion process. Then, when acoustic bubbles reach a critical size range they undergo a violent collapse. There are three at least theories to explain the chemical effects arising from the collapse of cavitation bubbles:

1. electrical theory,

2. plasma discharge theory and

3. super-critical theory.

Another approach is the "hot spot" theory. This theory suggests that bubbles growth is almost adiabatic up to the collapse. At this point, the gas in the bubble core is rapidly compressed (life time in the order of nanoseconds); hence, temperature of thousands of degrees and pressure of more than hundreds of atmospheres can be locally generated; this is the "hot spot" condition. It is noteworthy that, in addition to the extreme conditions of the "hot spot", a secondary region formed by a thin layer of the liquid surrounding the collapsed bubble, it is also transiently heated, although to a lesser extent; this thin layer is about 200 nm in thickness and may reach a temperature of 1726 °C [11], see a simplified scheme of the "hot spot" model is shown in Figure 4.

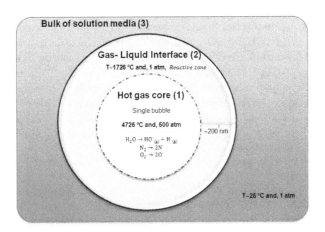

Figure 4. Hot-Spot model in the cavitation process [11].

The physicochemical properties of the solvent and solute, and also the gas in the bubble, have notorious effects on the cavitation phenomenon. Therefore, the sonochemical process is very complicated; it is more frequently influenced by the solvent because cavities are spontaneously formed with solvents having high vapor pressure, low viscosity, and low surface tension. Consequently, as liquid must overcome intermolecular forces to form bubbles, poor cavitation efficiency is obtained when solvents with low vapor pressure, high viscosity, surface tension and density are used. Nevertheless, these kinds of solvents have higher threshold for cavitation but more harsh conditions once cavitation begins; this might help in some chemical reactions [12]. On the other hand, there are several gas phase properties that affect sonochemical cavities, Adewuyi [13] recently reported that heat capacity ratio (also known as polytropic ratio, γ), thermal conductivity and solubility are the most important gas properties. γ is involved with the amount of heat released and, hence, affect the final temperature and pressure produced in the adiabatic compression, according to the following equations [14, 15]:

$$T_{max} = T_0\left[\frac{P_a(\gamma - 1)}{P_v}\right] \tag{2}$$

$$P_{max} = P_v\left[\frac{P_a(\gamma - 1)}{P_v}\right]^{\left[\gamma/\gamma-1\right]} \tag{3}$$

Where T_0= bulk medium temperature, P_v= pressure in the bubble when bubble size is maximum or vapor pressure of the solution, P_a= acoustic pressure in the bubble at the moment of collapse.

Thus, a gas with high thermal conductivity improves the heat transfer from collapsed bubbles to the liquid; this means that it reduces the temperature achieved in an implosion. The solubility of the gas in the liquid is also relevant. The more soluble the gas, the more likely it is to diffuse into the cavitation bubble. Soluble gases should originate the formation of larger number of cavitation nuclei and extensive bubble collapse, because these gases are readily forced back to the liquid phase. Therefore, a decrease of the bulk liquid temperature increases the rate of sonochemical reaction, unlike most chemical reaction systems. This is reasonable because the amount of dissolved gas increases and the vapor pressure of the liquid decreases and, then, less vapor diffuses into the bubble thus cushioning the cavitational collapse; in this condition the implosion more violent.

4. Sonochemical transesterification reaction

There are many aspects that make different the sonochemical and conventional chemical reactions. As already mentioned, the "hot spot" is a suitable concept to explain experimental results in many environmental sonochemistry reactions. This theory considers that reactive species and huge heat are produced from bubble cavitation; each bubble created from the interaction of the ultrasonic wave with the liquid is assumed to be a well-defined microrreactor [13]. Actually, according to the "hot spot" model there are three reactive zones:

1. a huge hot gas core,

2. a gas-liquid interface of approximately 200 nm, and

3. the bulk of the liquid media.

This model is frequently used in aqueous reactions, where solvent or substrate suffer homolytic (symmetrically) bond breakage to produce reactive species, and it assumes that free radicals may be in all the reactive zones. However, this model does not necessarily correspond to the thermodynamic reality of the transesterification reaction, because the system is constituted mainly of triglycerides (TG), small amount of FFA, KOH and methanol. Again, the methanol/oil phase is immiscible creating very large mass diffusional problems. But, in general, the energy generated by the US process produces free radicals, which are very reactive, and a significant amount of heat that improves mass transfer among phases [4]. This combined effect of very reactive species and intimate contact between phases could certainly improve the transesterification reaction rate. In this section, some ideas that further explain experimental results obtained in our laboratories are also discussed.

Figure 5 shows the sono transesterification model, which is an adaptation of Adewuyi's model [13], constrained as follows.

1. Water hydrolysis is not considered as reactive species source because anhydrous conditions should be achieved for biodiesel processing; from our experience with JCO (FFA <1.5-5%), soap formation is not promoted.

2. Relative humidity of air is also dismissed; so, air is dissolved in the methanol and oil phases.

3. The supercritical theory recently proposed by Hua et al. [16] regarding to the transient supercritical water (373 ºC, 22.1 MPa) at the bubble-solution interface is also discarded, because under these conductions the interphase would be considered as a supercritical methanol microrreactor, and then the use of catalyst would become censer.

However, from our lab experience the sonication of methanol/oil mixture without alkaline catalysts does not produce FAME.

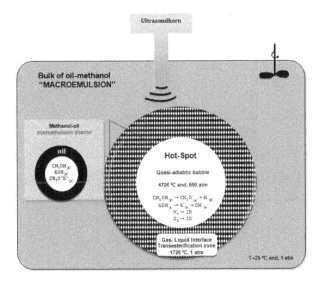

Figure 5. Transesterification cavitation model

The model depicted in Figure 5 assumes that a homogeneous methanol/oil macroemulsion is formed by mechanical mixing. It is very important to note that prior to the sonolysis, the methoxide ion produced, unreacted KOH, and methanol coexist inside the microemulsion. Once that the sound-macroemulsion interaction begins, cavitation is performed with vapor of methanol-KOH and air gas inside the bubbles, carrying out the dissociation reactions of the vapor and gas constituents; then, after several cycles of rarefication and compression, the implosion takes a place involving a significant rate of heat and mass transfer. The surrounding liquid quickly quenches a short-lived, localized entity exposed to high temperature (4226-4726 ºC) and pressure (over 1000 atm). Quenching occurs in few microseconds [17] and very fast cooling rates (about 10^{10}ºC^{-1}). This process has a profound influence on the physical properties of interface (microemulsion), where the transesterification reaction is spontaneously carried out, at local temperatures *ca.* 2000 K, without any diffusional problems.

As already mentioned, TG transesterification by basic catalysis consists of three consecutive, reversible reactions (Figure 2). In the reaction sequence, TG is converted stepwise to diglyceride, to monoglyceride and, finally, to glycerol, accompanied with the liberation of an ester at each step. The reaction mechanism of TG transesterification shown in Figure 6 indicates that in the catalyst-TG interaction the key step is the nucleophilic attack of the alkoxide ion, originating a different reaction chemical. The conventional transesterification process has been associated to a mass-transfer controlled regime occurring at the beginning of reaction. In addition, as the reaction proceeds and ester products act as emulsifiers, two rate-limiting steps change over time. One step is kinetically controlled and it is characterized by a sudden surge in product formation; the second step is reached once equilibrium is found near the reaction completion [19]. Importantly, in the sono transesterification model showed in Figure 5, neither mass transfer nor kinetic reaction are rate-limiting steps, but rather the chemical equilibrium.

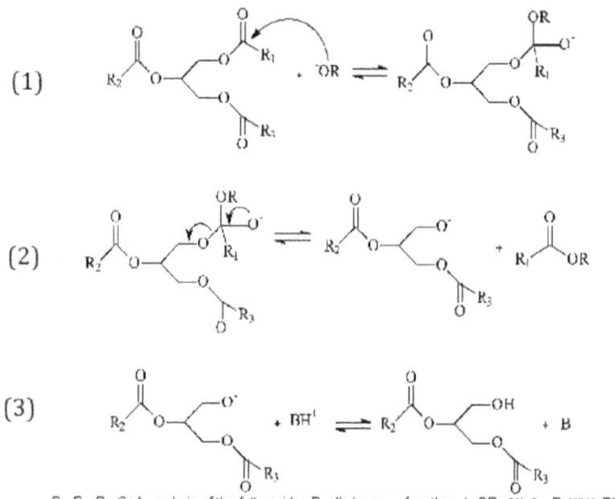

R_1, R_2, R_3: Carbon chain of the fatty acids; R: alkyl group of methanol; ^-OR: CH_3O^-; B: KOH; BH^+: $[K(OH)H]^+$

Figure 6. Homogeneous based-catalyzed reaction mechanism for triglyceride (TG) transesterification: methoxide ion form by dissociation of potassium hydroxide into methanol and it is encapsulated into TG-methoxidemicroemulsion, then: (1) CH_3O^-attacks nucleophilically to carbonyl group on TG, which leads to the tetrahedral intermediate formation; (2) intermediate breakdown; (3) regeneration of CH_3O^- active species. These steps are repeated twice to complete TG transesterification, according to Loreto et al. [18].

5. Comparison of experimental biodiesel processing technologies from *Jatropha curcas*

5.1. Why *Jatropha curcas*?

Current feedstock for biodiesel production plants derive from a great biomass variety, including first generation biomass raw materials such as vegetable oils (e.g., soybean, cottonseed, palm, peanut, rapeseed/canola, sunflower, safflower and coconut oils), animal fats (usually tallow) as well as spent or waste oils (e.g., used frying oils). But, given the fact that the use of vegetable oils has been strongly questioned, the use of second- and third-generation biomass feedstock is continuously growing. Among the raw materials coming from nonedible crops for humans, a key issue is their availability near to the biodiesel production plant. In this scenario, our research group is interested to use *Jatropha curcas* oil as feedstock. *Jatropha curcas L. (JC)* is a stress-tolerant ruderal, drought-resistant, oil-bearing small tree, which is well adapted to tropical, semi-arid regions and marginal sites. *JC* propagates easily and can be established quickly in a wide variety of soils with different agroclimatic conditions and does not put pressure on fertile agricultural land or natural ecosystems. In addition, *JC* is characterized for a short gestation period, low seed cost and, importantly, for the multiple uses that may have different parts of the plant [20, 21]. *JC* has received a lot of attention as a source of renewable energy, because its seeds contain 27–40% nonedible oil with a high quality of fatty acid profile (Table 2), which can be easily converted into biodiesel that meets American and European Standards (Table 3).

Fatty acid	Systematic name	Structure	wt %
Lauric acid	Dodecanoic acid	C12	-
Mysteric acid	Tetradcanoic acid	C14	0-0.1
Palmitic acid	Hexadecanoic acid	C16	14.1- 15.3
Palmitolileic acid	Cis-9-hexadecanoic acid	C16:1	0-1.3
Stearic acid	Octadecanoic acid	C18	3.7-9.8
Oleic acid	Cis-9-Octadecanic acid	C18:1	34.3-45.8
Linoleic acid	Cis-9-cis-12-Octadecanoic acid	C18:2	29-44.2
Linolenic acid	Cis-6-cis-9-cis-12-Octadecanoic acid	C18:3	0-0.3
Arachidice acid	Ecosanoic acid	C20	0-0.3
Behenic acid	Docosanoic acid	C22	0-0.22
Gadoleic acid		C24	14
Saturated	-	-	**21.1**
Unsaturated	-	-	**78.9**

Table 2. Fatty acids profile of *Jatropha curcas* oil [22]

In terms of availability, JC easily grows in Norwest Mexico, where our lab is located. For this reason, we set a research project to evaluate the potential of the local JC variety as source of renewable energy (i.e., biodiesel production). Notoriously, we also seek the utilization of valuable byproducts or residues of the conversion of *JC* oil to biodiesel; for instance, fruit husk and seed shell may lead to production of energetic pellets, part of the harvested seed shell may be used to produce humic acid, a biofertilizer; from the seed kernel, not only oil and subsequently biodiesel may be produced, but also protein flour for poultry, sheep, shrimp and tilapia (Figure 7). Once that biodiesel is produced a significant amount of glycerol become available, and we look for the production of high-added value chemical derive from glycerol catalytic conversion. Results presented hereby concern to biodiesel production, in particular the development of alternative strategies to improve the efficiency of the transesterification reaction and to decrease the overall processing cost. In this way, the proposed integrated approach clearly contributes for the development of sustainable biomass conversion processes.

Parameter	JC Biodiesel	Diesel	USA ASTM	Europe EN
Density at 15°C (g mL⁻¹)	0.91	0.85	0.88	0.8-0.9
Kinematicviscosityat 40°C (mm²s⁻¹)	3.43	2-8	1.9 - 6	3.5 - 5
Cetane number	52	47.5	"/47	>51
FAME content (%)	>99	0	-	> 96.5
Sulfur (ppm)	0	<5	<15	< 10
Flash point (°C)	186	>61.5	>93	>101
Acid number (mgKOHg⁻¹)	Depend of process		≤ 0.5	≤ 0.5

Table 3. USA and Europe international standards for biodiesel [23].

Figure 7. Productive chain for non-toxic *JC research* project in Norwest Mexico [23].

5.2. Materials and methods

5.2.1. Physical chemical JC oil characterization

The study used JC from selected elite germoplsms and cultivated in three zones of Sinaloa, Mexico. The approach used to obtain *Jatropha curcas* oil (JCO) was the well-established cold pressing followed by solvent oil extraction. The JCO physicochemical properties studied in this work included: fatty acid profile, acid index (AI), saponification index (SI), peroxide index (PI), and iodine index (II), which were obtained following the methodologies suggested by the Association of Analytical Communities, AOAC.

The quality criteria for the production of biodiesel are specified in EN 14214. In particular, method EN 14103 specifies the FAME content, which is used to profile the vegetable or animal oil feedstock used in biodiesel production. EN 14103 requires calibration of all FAME components by relative response to a single compound, methyl heptadecanoate. This requires the measurement of accurate weights for each sample and the addition of an internal standard. The FAME range for which the method is intended lies between C14:0 and C24:1. A modified EN 14103 chromatographic method was used. In this method, FAME analysis was carried out in a 6890N Agilent Gas Chromatograph (GC), equipped with a capillary split/splitless injector and a selective 5973 Agilent mass spectrometer detector. A 1 μL split injection (split ratio 50:1) was made to a Supelco omega wax column (bonded polyethylene glycol), using 1 mlmin⁻¹of helium into the column as carrier. Samples were injected via an auto sampler series 7683 also from Agilent technologies. A good resolution and peak shape was obtained when using the following oven temperature program: The initial temperature, 100 °C was kept for 2 min; then a heating rate of 4 °C min⁻¹ was used to increase the temperature to 240 °C and, finally, this temperature was kept for 10 min. For identification and calibration of the individual FAME, the Supelco standard "37 Component FAME Mix" was used. The response and retention time of each component was experimentally determined. Then, the calibration was verified by both, the analysis of a calibration-check standard and the database of mass spectrum reported by the National Institute of Standards and Technology (NIST). Results of analyses were then compared with the certificate of analysis, verifying the quality of the calibration. The standard preparation for this technique consisted of the dilution of the FAME standard into 4 mL of n-heptane. The sample preparation was also quite simple with 100 μL of biodiesel feedstock into 4 mL of n-heptane. Finally, concentration reports were based on the area percentage rather than a mass percentage, to simplify the calculations.

On the other hand, quantitative determination of free and total glycerin in biodiesel (B100) was also carried out by gas chromatography, followed by a modified methodology proposed by the ASTM D6584-10aε[1].The same Agilent GC system was also used for this analysis, the only difference being the use of a MS detector. ADB-5 ms column from Agilent Technologies was used for free and total glycerin analysis, which is equivalent in chromatographic efficiency and selectivity to that of the MET-Biodiesel capillary column of Sigma Aldrich.

5.2.2. Transesterification procedure

5.2.2.1. Conventional process

Conventional alkaline transesterification was conducted in a 2-necked glass reactor (100 mL, Aldrich). A homogeneous reaction mixture was obtained by using plate stirrers, and a constant reaction temperature was kept by using isolated bath vessels equipped with a stainless steel coils. The reaction temperature was fixed by using of a heater/cooler recirculation isothermal bath (Fisher Scientific 3016). Figure 8 shows that each reactor was connected to-cooled straight glass condenser to avoid alcohol leaks; water at 5°C from another isothermal bath (Fisher Scientific 3028) was used as cooling fluid.

Figure 8. Transesterification reaction system for the conventional process

Anhydrous methanol (Sigma-Aldrich, 99.8 %,) and KOH reagent grade (Sigma-Aldrich, 90%) were used for all experiment of this study. The stirrer was fixed at 600 rpm, and the temperature at 40, 60, 70 or 90 C), a methanol: JCO molar ratio was 3:1 or 6:1. Previously to each reaction, methanol and KOH solutions were prepared according to the proposed molar ratio. Then, the reaction volume was fixed to 50 mL of JCO. After the desired reaction temperature was reached, a preheated methanol-catalyst solution was added to start the reaction. Reaction mixture was sampled after 15, 45, 60, 90, and 120 min. These samples were quenched by a sudden immersion of the sample to a plastic container at 0 C, for 15 min. Then, reaction products were purified according to the methodology suggested by Cervantes [24](Figure 9), and the biodiesel yield was determined by means of the following equation:

$$Yield = \frac{wt\ B_{100}}{wt\ oil} x\ 100 \tag{4}$$

5.2.2.2. Heterogeneous process

As reported elsewhere [25], JCO transesterification was also conducted by using ZnO, Al_2O_3 and $ZnO-Al_2O_3$ mixed oxide powders as catalysts. The objective was to compare the heterogeneous catalytic conversion of the same JCO. In this case, the catalytic activity was measured in a Parr 4560 stirred tank reactor, operated at 1000 rpm, 250°C, P= 14.7 atm. A methanol: JCO molar ratio of 6:1 and a 3 wt% of catalysts (based on JCO weight) were used. Previously to the reaction, the reactor was uploaded with the 50 ml of JCO, the required methanol to achieve the 6:1 molar ratio, and 1.36 g of catalyst. Then, the reactor was purged with nitrogen (Praxair, reagent grade) for 3 min to avoid JCO burning. The reaction time was 1 h and then the reactor mixture was suddenly cooled to room temperature. The product separation included the following steps.

1. Catalysts removal by means of vacuum filtration.

2. Methanol recovery by using a rotary evaporator at the same condition indicated in Figure 9.

3. Glycerol and biodiesel separation by centrifugation, using the same condition indicated in Figure 9.

At the end, the biodiesel yield was calculated by using equation 4

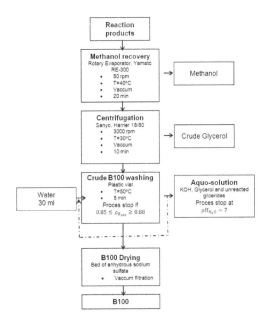

Figure 9. Biodiesel purification process for a conventional alkaline transesterification process.

5.2.2.3. Supercritical methanol process

Non-catalytic transesterification process was evaluated by means of the supercritical methanol reaction. This process was also carried out using the 4560 Parr stirred tank reactor. The effect of both, methanol: JOC molar ratio (40:1 and 60:1) and temperature (250, 300, and 350°C) was evaluated using nitrogen as co-solvent. Once the required reagents amounts were charged to the reactor, the air was vented with nitrogen and the stirrer was fixed at 1000 rpm. Next, the temperature was increased until the desired set point; in this process the pressure increased but not enough to reach the methanol supercritical point. Therefore, additional nitrogen was loaded to ensure 14 MPa. As an alternative to decrease the drastic operation conditions N_2 was used as co-solvent. The reaction took place over 30 min, sampling the mixture every 5 min through the liquid reactor valve. After the reaction was finished, the reactor was suddenly cooled to room temperature. The product separation included the following steps.

1. Methanol recovery by using a rotary evaporator at the same condition indicated in Figure 9.

2. Glycerol and biodiesel separation by decantation.

At the end, the biodiesel yield was calculated by using equation 4.

5.2.2.4. Ultrasonic process

The sonotransesterification of JCO was conducted by using a highly efficient Hielscher Ultrasonic processor, model UP200 HS. This equipment was used to generate mechanical vibrations by means of the reversed piezoelectric effect (electric excitation), with frequency of 24 kHz, and a control range of 1 kHz. The vibrations were amplified by the S14 sonotrode fitted to the horn and formed as a $\lambda/2$ vibrators, and transferred via its end face to the JCO.

To optimize the sonotransesterification reaction, the effect acoustic power density (N), sonication time (or reaction time), and methanol: JCO molar ratio (MR) were evaluated at room temperature (25°C) and ambient pressure (1 atm). Reaction temperature was controlled by using an isothermal bath (Fisher Scientific 3016). The continuous sonication of the reaction mixture was conducted using N=105 Wcm^{-2} and a molar ratio of 6:1, following the approach described in Figure 9. Reaction time was fixed at 1, 2, 4, 6, 8, 10, 15, 20, 25 or 30 min. Next, the methanol: JCO molar ratio was evaluated varied to 3:1, 4:1 and 6:1. For the smaller reaction time and molar ratio, the acoustic power density effect was evaluated at 42, 63, 73.5, 84, 94.5, and 105 Wcm^{-2}. When the best set of parameters was found, an experiment was conducted again to determine the biodiesel quality.

5.3. Results and discussions

5.3.1. Physical chemical JC oil characterization

The *Jatropha curcas* oil obtained from non-toxic, harvested seed in Northwest Mexico, seems to be an excellent candidate for biodiesel production due to its high quality. Table 4 includes the basic JCO physicochemical characteristics that back up this quality. The iodine index is a measurement of the oils unsaturation degree; a higher iodine index corresponds to higher degree of unsaturation [26], and probably leads to oxidation and viscosity problems. The JCO iodine index was 28.75 cg I_2g^{-1}, which is well below the maximum specified value (120 cg $I_2 g^{-1}$) for biodiesel as indicated in the EN14214 specification. The limitation of unsaturated fatty acids is convenient because heating higher unsaturated fatty acids results in polymerization of glycerides, leading to the formation of deposits or to deterioration of the lubricant [27]. Fuels with this characteristic (e.g Sunflower, soybean and safflower oil) are also likely candidates to produce thick sludge's in the sump of the engine, when fuel seeps down the sides of the cylinder into crankcase [26]. The JCO iodine index could was caused by the high content of unsaturation fatty acid such as oleic and linoleic acid (Table 5).

Test	Parameter[1]
Appearance	Yellowish transparent
Free fatty acid (%)	1.51 ± 0.10
Density at 15ºC (gml⁻¹)	0.92 ± 0.01
Acid index (mg KOH g⁻¹)	3.07 ± 0.12
Saponification index (mg KOH g⁻¹)	180.92 ± 2
Iodine index (cg $I_2 g^{-1}$)	28.75 ± 0.1
Peroxide index (meq $O_2 Kg^{-1}$)	18.5 ± 0.7

Table 4. Physical chemical properties of *Jatropha curcas* Oil. [1] Standard desviation measured from triplicate determinations.

In on another hand, JCO peroxide index was 18.5 meqg⁻¹, that is higher than the index recently reported in the literature for crude seed Jatropha oil, 1.93 meqg⁻¹ [26] and 2.5 meqg⁻¹[28]. Despite this high peroxides index, JCO upholds the good quality of biodiesel purposes. The JCO saponification index was 181 mg KOH g⁻¹, which suggested that JCO was mostly normal triglycerides, and very useful in biodiesel production due to its low FFA content (1.15wt%). The content of FFA was assessed from the acid index (AI) measurement, taking into account the composition showed in Table 5. The acid index of 3.07 mg KOH g⁻¹reported in Table 3 was lower than the values reported by other authors (10 - 14 mg KOH g⁻¹) for crude JCO [29, 30]; this could be attributed to the change of local environmental conditions where by the *Jatropha curcas* plant was grown. Therefore, acid index becomes a very important parameter to determine the most convenient processing route of a given FAME; this means that oils can undergo a pretreatment or direct transesterification as a function of FFA amount.

Compound	Estructure	wt %
Palmitic	16:00	23.992
Stearic	18:00	7.224
Oleic	18:01	41.368
Linoleic	18:02	27.186

Table 5. Faty acid composition de *Jatropha curcas* Oil determine by MS-CG.

The properties of triglyceride and biodiesel are determined by the amounts of each fatty acid present in the molecules. Chain length and number of double bonds determine the physical characteristics of both fatty acids and triglycerides [3]. Nevertheless, transesterification does not alter the fatty acid composition of the feedstocks, and this composition plays an important role in some critical parameters of the biodiesel, as cetane number and cold flow properties. Therefore, measuring fatty acid profile of JCO was another important target of this study. These results are shown in Table 5.

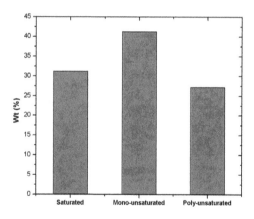

Figure 10. Type of fatty acids in *Jatropha curcas* oil from the Norwest of México

There are three main types of fatty acids that can be present in a triglyceride which is satu-rated (Cn:0), monounsaturated (Cn:1) and polyunsaturated with two or three double bonds (Cn:2,3). Ideally, the vegetable oil should have low saturation and low polyunsaturation, that is, be high in monounsaturated fatty acids, as shown in Figure 10. Vegetable oils rich in polyunsaturated (linoleic and linolenic) acids, such as soybean and sunflower oils [26], usually produce methyl ester fuels with poor oxidation stability. In the other hand, vegetable oils with high degree of unsaturation (Cn:2,3) lead to a product with high freezing point, poor flow characteristics and may become solid (e.g palm oil) at low temperatures, although they may perform satisfactorily in hot climates. The main fatty acids in the JCO used in this study were the oleic, linoleic, palmitic and the stearic fatty acids. The predominant acids were monounsaturated (41.36%), polyunsaturated (27.18%) and saturated fatty acid (31.21%) (Figure 10). This result was in agreement with the reported by Akbar [26], although it was slightly different in terms of saturated and polyunsaturated compounds for the JCO from Malaysia. Thus, JCO can be classified as oleic–linoleic oil. Compared to others vegetable oil JCO had highest oleic acid contain than palm oil, palm kernel, sunflower, coconut, and soybean oil.

5.3.1. Jatropha curcas oil transesterification

Three current biodiesel technologies were evaluated and compared with the conventional homogeneous transesterification, using the JCO characterized above. The main objective was to evaluate the potential advantages of sonotransesterification in terms of operating conditions, transesterification rate and processing steps and costs.

Conventional alkaline transesterification

According to the overall transesterification pathway shown in Figure 1, stoichiometrically, JCO methanolysis requires three moles of methanol for each mole of oil. Since the transesterification of triglycerides is a reversible reaction, excess methanol shifts the equilibrium towards the direction of ester formation. As it is evident from Figure 11, the maximum yield for the conventional alkaline transesterification process (84%) was reached after 15 min reaction time; afterwards no signification variations were observed. In addition, when the methanol: JCO molar ratio was increased from 3:1 to 6:1, no major differences were found within the first 15 min; however, a higher biodiesel yield was observed in the experiment with a 6:1 molar ratio toward the end of the reaction. On the other hand, results shown in table 5 indicate that temperature effect is not important. These results correspond to the biodiesel yield evaluated after 15 min. Thus, the higher biodiesel yield was found at 40 °C, and then it decreased to around 73 – 75 % for temperatures between 60 and 90 °C.

Current results of the conventional process disclosed in Figure 11 and Table 6 suggested a significant improve to the conventional alkaline transesterification process, because the reaction yield was enhanced at a shorter reaction time (40 min as compared 60 min) and temperature (40 °C as compared to 60ºC) for industrial application [3]. A shorter reaction time can be translated to a continuous process with a shorter resident time and then, the possibility to reduce costs at the reaction stage. However, a higher JCO conversion is needed to ensure a sustainable process. Moreover, the biodiesel purification process is still a problem because it implies long times and it is energy demanding.

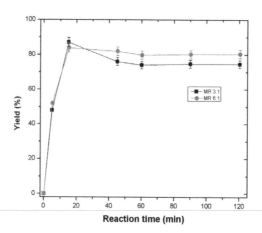

Figure 11. Progress of transterification reaction as function of methanol:JCO ratio at 40°C

Temperature, °C	Yield, % (At 15 min of reaction time)
40	84.0
60	73.06
70	73.60
90	75.30

Table 6. The effect of temperature on the performance of alkaline transesterification of JCO by conventional process.

Supercritical methanol process

Thus, as an alternative of the problems indicated above, the supercritical methanol process (SMP), using nitrogen as co-solvent, was conducted. Figure 12 shows that the best set of operating conditions for this non-catalytic process were: methanol: JCO mol ratio of 40:1 and 350ºC. Under these conditions, a biodiesel yield *ca.* 60% was obtained. From Figure 13, it can be observed that after 20 min the equilibrium was reached for the transesterification reaction for both molar ratios studied: 40:1 and 60:1. This is a very promising result if it is compared with reported for palm [31] and soybean oils [32], where biodiesel yields up to 84% were obtained under very high pressure, 40 and 35Mpa, respectively.

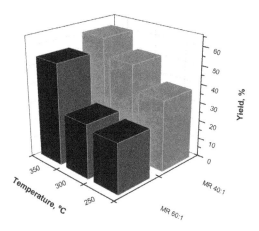

Figure 12. Effect of temperature and methanol: JCO molar ratio on the yield of Biodiesel obtained by supercritical methanol process at 14Mpa and 30 min.

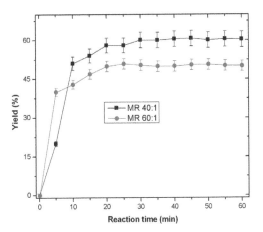

Figure 13. Progress of transterification reaction as function of methanol: JCO ratio at T= 350ºC for supercritical methanol process

Importantly, supercritical methanolysis did not require any kind of catalyst, and no pretreatment to remove water or FFA was used in this work. A very simple separation processes – evaporation and layer separation – were used for biodiesel purification. Our findings agree with the literature that supercritical process is simpler and faster than conventional alkaline transesterification for biodiesel production. In addition, since wastewater was not introduced by pretreatment or washing processes, the supercritical process is environmental friendly. However, to date, high investment and energy cost are still required due to high temperature and pressure of the supercritical state. Another issue with economic implications is the large methanol needed to enhance the forward reaction without catalyst. It could expected that these costs are comparable to those of the pretreatment and separation process of the conventional alkaline transesterification process. Clearly, as the methanol demand be decreased, and the operating conditions be more moderate, the economic feasibility of supercritical methanol process would be possible.

Heterogeneous process

As indicated in the previous section, three heterogeneous powder catalysts, ZnO, Al_2O_3 and ZnO- Al_2O_3 mixed oxides supported on SBA-15 were evaluated for transesterification reaction. Figure 14 shows our best results to date, when experiments were conducted with a methanol: JCO molar ratio of 6:1, 250ºC, and 3 wt % of catalyst. Results were collected after 1 h of reaction time. Under these conditions, the equilibrium biodiesel yield (83%) was reached for the supported Al_2O_3 catalysts. Importantly, no catalysts deactivation was observed for at least 10 runs (without regeneration treatment). It is noteworthy that Al_2O_3 is traditionally used as support instead of active phase due to its poor catalytic activity for transesterification [33]. In fact, in our experiments Al_2O_3 itself showed no more than 5% of FAME yield, but the it showed a totally different catalytic performance when it was well dispersed on SBA-15. On the other hand, several supported basic catalysts have also been reported in the literature -sodium [33] or potassium [34] loaded on a support (normally alumina), using several precursors and treated at high calcination temperatures (500–600ºC). The catalysts showed good activities (80-90 % biodiesel yield) at low temperatures (70-90ºC), but no data were reported about their stability. K_2CO_3 supported on both MgO and Al_2O_3 provided good results for rapeseed oil transesterification with methanol at 60–63 ºC, but K_2CO_3 leached into the solution.

Meantime, pure ZnO and ZnO supported on Al_2O_3 have also been reported as good transesterification catalyst. In experiments performed in a packed-bed reactor at 225-230 ºC, 91.4% and 94.3% of FAME yields were obtained, for 1 and 7 h, respectively [35]; in this case, no Zn leaching was practically observed (5 ppm). In addition, no data about catalysts has been reported. In our case, experiments conducted with ZnO and ZnO/Al_2O_3 showed biodiesel yields below 75 %. The most promising results found for the $Al_2O_3/SBA15$ have to be studied in detail to optimize the catalyst formulation.

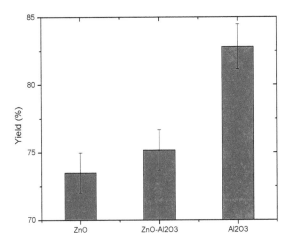

Figure 14. Yield of biodiesel of transesterification of JCO with MR=6, and 250ºC, 1 h and 3 wt.% of each heterogeneous catalyst.

Sonotransesterification

Experimental results of JCO sonotransesterification are shown in Figure 15. The first issue that became evident was that sonotransesterification was much faster than the conventional alkaline transesterification. Thus, in just 1 minute of reaction time the maximum FAME yield (*ca.* 65%) was reached for the experiment conducted with a methanol: JCO molar ratio of 6:1, an acoustic power density (N) of 105 Wcm^{-2} and temperature of 25ºC. Moreover, Figure 16 shows that for 1 min of reaction time, a reduction of the methanol: JCO molar ratio from 6:1 to 4:1 increased the biodiesel yield. Under these conditions, a 71 % biodiesel yield was obtained. Notoriously, the later molar ratio is closer to the stoichiometric one, thus helping to decrease the excess of alcohol required by the other biodiesel technologies under comparison in this study. These results clearly showed the following advantages for the sonotransesterification process: a shorter processing time is required, a lower amount of alcohol is required (almost the stoichiometric amount), and the experiment is conducted at room temperature and atmospheric pressure.

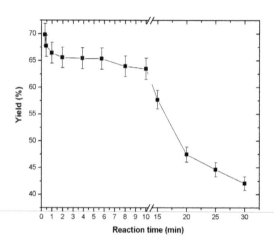

Figure 15. Effect of the sonication time on the yield of biodiesel by sonotransterification reaction with MR of 6:1, room temperature, and acoustic power density of 105 Wcm⁻².

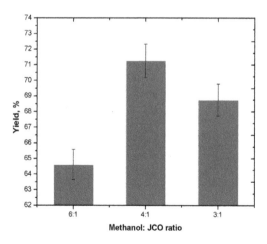

Figure 16. Effect of the methanol:JCO molar ratio on the yield of biodiesel by sonotransterification reaction at room temperature,1 min of reaction time, and acoustic power density of 105 Wcm⁻².

Despite of the important advantages initially found for the sono transesterification process in this work, the biodiesel yield had to be increased to make it atractive from the industrial point of view. To this respect, a more detail study of the acoustic power effect was conducted. Figure 17 shows that acoustic sonocation power had a significant effect on yield. For an N of 64 Wcm^{-2}, coupled with the best set of parameters used in previous experiments, a FAME yield up to 96% was reached at room temperature. The reason why a higher transesterification rate was obtained with the ultrasonic process was already outlined in the previous sections. Briefly, the huge local temperature generated in the "hot spot" formed during the cavitation phenomenon favors the formation of highly reactive species and promotes mass transfer. These issues are the key to improve the transesterification reaction rate because under the experimental ultrasonic conditions the process is not affected by mass transfer or by kinetic limitations, but rather by the equilibrium condition.

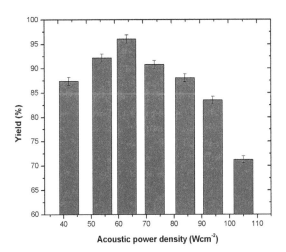

Figure 17. Effect of the acoustic power density on the yield of biodiesel by sonotransterification reaction with MR of 4:1, room temperature, and one minute of reaction time.

In this scenario, sono transesterification becomes a very attractive process to be implemented in a continuous industrial process. Thus, results found in Figure 17 were used to configure a continuous US process with a tubular sonorreactor, using a resident time of 1 min. In this case, a constant yield of 96% was reached. Importantly, the quality of the biodiesel obtained in this experiment, overcame the quality of biodiesel with international standards (Table 7).

Parameter	Value
FAME content[a] (%)	98
Density at 15°C (gml⁻¹)	0.84
Acid index (mgKOHg⁻¹)	0.5
Total glycerin (wt. %)	<0.2
Free glycerin (wt. %)	<0.02

Table 7. Physical chemical properties of biodiesel obtained by sonotransesterification under continuous process with optimized conditions operated at room temperature. [a] after purification process

6. Conclusions

Nowadays, the conversion of non-edible and residual biomass feedstock into biofuels is already considered a suitable alternative for the generation of alternative energy sources. In particular, transesterification of oils and fats is a well-known technology, and the production of biodiesel is continuously growing using second- and third-generation biomass raw materials. The main technology used in the industrial production of biodiesel is based on the alkaline transesterification of vegetable oils with methanol. However, the problems related with this technology (mainly in operating conditions and product purification) are the driving force for research in the field of heterogeneous catalysis for biodiesel production and for the development of non-catalytic process under supercritical fluids. The use of heterogeneous catalysts and supercritical methanol process for transesterification reaction seem to be attractive for industrial application because theses simpler processes have a beneficial impact in the process economy. In particular, industry is making great research efforts to find the optimum catalyst formulation and to decrease the drastic operating conditions of supercritical methanol process. However, these technologies are still far to be economically attractive. A more recent alternative is the new ultrasound-assisted method for biodiesel production, which has to be tested and optimization for this particular application.

In this work, we evaluated and compared the performance of four technologies for the transesterification of JCO obtained from JC grown in Northwest Mexico: conventional alkaline catalyst (KOH), heterogeneous powder catalysts (ZnO, ZnO/Al₂O₃ and Al₂O₃/SBA 15), supercritical methanol and sonotransesterification. Results showed that the ultrasonic method has significant advantages as compared to the other three methods. Notoriously, ultrasonication reduced the transesterification reaction time to 1 min at room temperature and atmospheric pressure, as compared to 1-6 h in conventional processing under more drastic operating conditions. We suggest that this result could be explained with the proposed sonotransesterification cavitational model where by diffusional problems are eliminated. Our results demonstrated that acoustic power density and methanol: JOCmolratio are the most sensitive parameters to increase FAME yield for JCO; at the best set of experimental conditions, the biodiesel yield is higher than that obtained by conventional methods. Importantly, the ultrasound-assisted method was also effectively used for continuous production of biodiesel by using a plug flow reactor; the physicochemical properties of the biodiesel produced, such as acid value, density, FAME content, total and free glycerin were within the limits of ASTM and EN standards.

In summary, sonotransesterification is faster and easier to handle than conventional transesterification processes. The sonoreactor is significantly cheaper, and the process works under safer and less energy demanding conditions (e.g room temperature and ambient pressure). The major advantages of the current ultrasonic system include operational simplicity, short reaction time, high conversion and reusability. In summary, ultrasonic irradiation is a faster alternative that leads to higher product yield, and with the real possibility to benefit the process economy. Thus, the ultrasonic process discussed in this work established the basis for the development a sustainable process for biodiesel production, although some issues are still to be solved; for instance, if water is avoided in the purification process,the overall process would be even more environmentally friendly.

Acknowledgements

The authors want to thank to FORDECyT for financial support with the project 146409, and Universidad Autónoma de Sinaloa throughout the PROFAPI project No. 2011/47.

Author details

Mario Nieves-Soto[1], Oscar M. Hernández-Calderón[2], Carlos Alberto Guerrero-Fajardo[3], Marco Antonio Sánchez-Castillo[4], Tomás Viveros-García[5] and Ignacio Contreras-Andrade[6*]

*Address all correspondence to: ica@uas.edu.mx

1 Facultad de Ciencias del Mar, Universidad Autónoma de Sinaloa, Mazatlán, México

2 Facultad de Ciencias Químico Biológicas, Universidad Autónoma de Sinaloa, Culiacán, México

3 Facultad de Química, Universidad Nacional de Colombia, Bogotá, México

4 Universidad Autónoma de San Luis Potosí, San Luis Potosí, México

5 Departamento de Ingeniería de Procesos e Hidráulica, Universidad Autónoma Metropolitana-Iztapalapa, México

6 Facultad de Ciencias Químico Biológicas, Universidad Autónoma de Sinaloa, Culiacán, México

References

[1] Di Serio, M., Tesser, R., Pengmei, L., & Santacesaria, E. (2008). Heterogeneous Catalysts for Biodiesel Production. *Energy & Fuels*, 22(1), 207-217.

[2] Demirbas, A. (2008). Biodiesel: A Realistic Fuel Alternative for Diesel Engines. *London: Springer-Verlang.*

[3] Drapcho, C. M., Nhuan, N. P., & Walker, T. H. (2008). Biofuel Engineering Process Technology. *New York: McGraw-Hill.*

[4] Lee, D. W., Park, Y. M., & Lee, K. Y. (2009). Heterogeneous Base Catalysts for Transesterification in Biodiesel Synthesis. *Cata. Surv. Asia.*, 13(1), 63-77.

[5] Cravotto, G., & Cintas, P. (2012). Introduction to Sonochemistry. *In: Chen D, Sharma SK, Mundhoo A. (ed) Handbook on Applications of Ultrasound: Sonochemistry for Sustainability. New York: Taylor & Francis*, 23-40.

[6] Kumar, D., Kumar, G., & Singh, P. C. (2010). Ultrasonic-Assisted Transesterification of Jatropha curcas Oil Using Solid Catalyst, Na/SiO2. *Ultrasonic Sonochemistry.*, 17(1), 839-844.

[7] Lee, S. B., Lee, J. D., & Hong, I. K. (2011). Ultrasonic Energy Effect on Vegetable Oil Based Biodiesel Synthetic Process. *Ind. Eng. Chem*, 17(1), 138-143.

[8] Kumar, D., Kumar, G., Johari, R., & Kumar, P. (2012). Fast, Easy Ethanomethanolysis of Jatropha curcas Oil for Biodiesel Production Due to the Better Solubility of Oil with Ethanol in Reaction Mixture Assisted by Ultrasonication. *Ultrasonic Sonochemistry.*, 19(1), 816-822.

[9] Soto, S., Fajardo, C. A., Sierra, F. E., Valdez-Ortiz, A., & Contreras-Andrade, I. (2012). Biodiesel Production by Supercritical Alcohols. *Summited to Revista Mexicana de IngenieríaQuímica.*

[10] Teo, B. M., Grieser, F., & Ashokkumar, M. (2012). Application of Ultrasound to Polymer Synthesis. *In Chen D, Sharma SK Mudhoo A. (ed) Handbook on Applications of Ultrasound: Sononchemistry for Sustainability. New York: Taylor & Francis*, 475-500.

[11] Yasui, K., Tuziuti, T., Sivakumar, M., & Iida, Y. (2004). Sonoluminiscence. *Applied Spectroscopy Reviews*, 39(3), 399-436.

[12] Chowdhury, P., & Viraraghavan, T. (2009). Sonochemical Degradation of Chlorinated Organic Compounds, Phenolic Compounds and Organic Dyes- A Review. *Science of the Total Environment*, 407(1), 2474-2492.

[13] Adewuyi, Y. G. (2011). Sonochemistry: Environmental Science and Engineering Application. *Ind. Eng. Chem. Res*, 40(1), 4681-4715.

[14] Noltingk, B. E., & Neppiras, E. A. (1950). Cavitation Produced by Ultrasonic. *Proc. Phys. Soc. London, Ser. B.*, 63(1), 674.

[15] Neppiras, E. A. (1980). Acustic Cavitation. *Phys. Rep.*, 61(1), 159.

[16] Hua, I., Hochemer, R. H., & Hoffmann, M. R. (1995). Sonolytic Hydrolysis of p-Nitro-phenyl Acetate: The Role of Supercritical Water. *J. Phys. Chem.*, 99(1), 2335-2342.

[17] Thangavadivel, K., Mengharaj, M., Mudhoo, A., & Naidu, R. (2012). Degradation of Organic Pollutants Using Ultrasound. *In Chen D, Sharma SK Mudhoo A. (ed) Handbook on Applications of Ultrasound: Sononchemistry for Sustainability. New York: Taylor & Francis*, 447-470.

[18] Loreto, E., Liu, Y., Lopez, D. E., Suwannakarn, K., Bruce, D. A., & Goodwin, J. G. (2005). Synthesis of Biodiesel via Acid Catalysis. *Ind. Eng. Chem. Res.*, 44(1), 5353-5363.

[19] Freeman, B., Butterfield, R. O., & Pryde, E. H. (1986). Transesterification Reaction Kinetics of Soybean Oil. *J. Am. Oil Chem. Soc.*, 63(1), 1375-1380.

[20] Wouter, M. J., Mathijs, E., Verchot, L., Singh, V. P., Aerts, R., & Muys, B. (2007). Jatropha Biodiesel Fueling Sustainability? *Biofuels, Bioprod. Bioref.*, 1(1), 283-291.

[21] Kumar, A., & Sharma, S. (2008). An Evaluation of Multipurpose Oil Seed Crop for Industrial Use (Jatropha curcas L.). A Review. *Ind. Crop. Prod.*, 28(1), 1-10.

[22] Jain, S., & Sharma, M. P. (2010). Prospects of Biodiesel from Jatropha curcas in India: A Review. *Renewable and Sustainable Energy Reviews.*, 14(1), 763-771.

[23] Council for Economic Development of Sinaloa (CODESIN) (2012). http://www.codesin.org.mx/node/169accessed 29 july).

[24] Cervantes, E. (2012). Síntesis de Biodiesel a Partir de Aceite de Jatropha curcas L.: Cinética de Reacción y Evaluación Fisicoquímica. *Thesis, Universidad Autónoma de Sinaloa, Culiacán, Sinaloa.*

[25] Carrillo, L. G. (2012). Síntesis y Caracterización de Catalizadores de ZnO-Al2O3/ SBA-15 Aplicados a la Producción de Biodiésel a Partir de Jatropha curcas. *Thesis, Universidad Autónoma de Sinaloa, Culiacán, Sinaloa.*

[26] Akbar, E., Yaakob, Z., Kamarudin, S. K., Ismail, M., & Salimon, J. (2009). Characteristic and Composition of Jatropha curcas Oil Seed from Malaysia and its Potential as Biodiesel Feedstock. *European Journal of Scientific Research*, 29(3), 396-403.

[27] Mittelbach, M., & Remschmidt, C. (2005). Biodiesel: The Comprehensive Handbook. *2nd edition). Graz, Austria.*

[28] Huerga, I. R. (2010). Producción de Biodiesel a Partir de Cultivos Alternativos: Experiencia con Jatropha curcas. *Master Thesis. Instituto en Catálsis y Petroquímica (INCAPE), Santa Fe, Argentina.*

[29] Lu, H., Liu, Y., Zhou, H., Yang, Y., Chen, M., & Liang, B. (2009). Production of Biodiesel from Jatropha curcas L. Oil. *Computers and Chemical Engineering*, 33(1), 1091-1096.

[30] Vyas, A., Subrahmanyam, N., & Patel, P. A. (2009). Production of Biodisel Through Transestrification of Jatropha curcas Oil Using KNO3/Al2O3 Solid Catalyst. *Fuel*, 88(1), 625-628.

[31] Song, E. S., Lim, J. W., Lee, H. S., & Lee, Y. W. (2008). Transesterificaction of RBD Palm Oil Using Supercritical Methanol. *J. of Supercritical Fluids*, 44(1), 356-363.

[32] Olivares-Carrillo, P., & Quesada-Medina, J. (2011). Synthesis of Biodiesel from Soybean Oil Using Supercritical Methanol in a One-Step Catalyst-Free Process in Batch Reactor. *J. of Supercritical Fluids*, 58(1), 378-384.

[33] Kim, H. J., Kang, B. S., Kim, M. J., Park, Y. M., Kim, D. K., & Lee, J. S. (2004). Transesterification of Vegetable Oil to Biodiesel Using Heterogeneous Base Catalyst. *Cata. Today*, 93-95(1), 315-320.

[34] Xiea, W., Peng, H., & Chen, L. (2006). Transesterification of Soybean Oil Catalyzed by Potassium Loaded on Alumina as a Solid-Base Catalyst. *Appl. Catal. A*, 300(1), 67.

[35] Stern, R., Hillion, G., Rouxel, J., & Leporq, J. (1999). Patent US(5,908,946)

Chapter 9

Lipase Applications in Biodiesel Production

Sevil Yücel, Pınar Terzioğlu and Didem Özçimen

Additional information is available at the end of the chapter

1. Introduction

Because of the global warming and depletion of fossil fuels, in recent years, intensive inves‐ tigations are carried on for providing the greater use of sustainable biofuels instead of fossil fuels. Biomass, which various biofuels are produced from, has an important role among oth‐ er alternative energy sources including wind energy, solar energy, geothermal energy, etc.

Biodiesel is one of the important biofuels and a clean energy source as an alternative to pe‐ troleum-based diesel fuels. Biodiesel has some advantages and disadvantages. Transporta‐ bility, high combustion efficiency, low sulphur and aromatic content, high cetane number and biodegradability are advantages of the biodiesel [1]. Disadvantages of biodiesel are high viscosity, lower energy content, high cloud and pour point, high nitrogen oxide emission, lower engine speed and power, injector cooking, high price and engine erosion [2].

The flash point of biodiesel is higher than diesel fuel. This feature is important for fuel stor‐ age and transportation in the way of safety. Cetane number of biodiesel (~50) is higher than diesel fuel [3]. Biodiesel does not include aromatic and sulphur content and contains oxygen at the rate of 10-11% by mass [4]. Cetane number is an important factor to determine the quality of diesel fuel, especially ignition quality of diesel fuel. In other words, it determines the ignition tendency of fuel when being injected into engine. Ignition quality of biodiesel is determined by the structure of methyl ester [5].Viscosity is also an important factor for bio‐ diesel. Viscosity affects mostly fuel injection equipment and the increase of fuel viscosity changes the viscosity at low temperatures. High viscosity has an negative effect on fuel spray atomization [6]. Amounts of elements and compounds in biodiesel and diesel fuel are present in Table 1 [7]. Biodiesel has more polar structure than diesel fuel because of the oxy‐ gen, which is an electronegative element present in its structure, and therefore biodiesel has higher viscosity comparing with diesel fuel. In addition, elemental oxygen content is respon‐ sible for lower heating value of biodiesel when compared with diesel fuel. [7-9]. Biodiesel can be used in its pure form or when mixed with diesel fuel in certain proportions. Most common biodiesel blends are B2 (2 % biodiesel, 98 % diesel), B5 (5 % biodiesel, 95% diesel), B20 (20 % biodiesel, 80 % diesel) [10].

	Biodiesel Content (%)	Diesel Content (%)
Carbon	79.6	86.4
Hydrogen	10.5	13.6
Oxygen	8.6	-
Nitrogen	1.3	-
C/H	7.6	6.5
n-Aliphatics	15.2	67.4
Olephenics	84.7	3.4
Aromatics	-	20.1
Naphtens	-	9.1

Table 1. The comparison of elemental and chemical content of diesel and biodiesel [7]

The transesterification reaction can be influenced by several factors including molar ratio of alcohol, catalyst, presence of water, free fatty acid in oil samples, temperature, time and agitation speed. In this context, an understanding of the factors affecting the process is very important to make economically and environmentally biodiesel production [11].

To accelerate reaction rate, transesterification process is carried out in the presence of catalysts. So, biodiesel production is made by using chemical or enzymatic catalysts. Compared to chemical, enzymatic reaction is more attractive because of ability of make a high quality product, simplify the separation of products, mild reaction conditions, the reuse of the catalyst and especially environmental impact, although high conversion and reaction rate are obtained with chemical catalysts [11-14]. Lipase is important enzyme catalyst that catalyzes esterification and transesterification reaction to produce methyl esters (biodiesel). Figure 1 presents the enzymatic transesterification reaction [15].

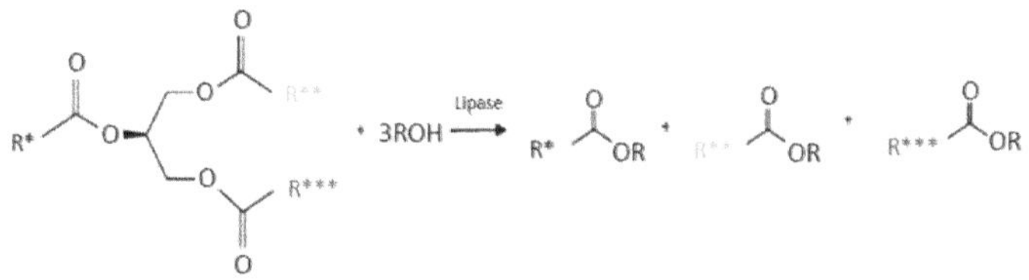

Figure 1. Enzymatic transesterification reaction [15].

In this study, enzymatic approach for biodiesel production was reviewed, and especially the usage of lipases in biodiesel production and factors affecting the effectiveness of lipase in reaction were explained in detail.

2. Lipases in biodiesel production

Biocatalyst based biotechnological applications are receiving increasing attention. Lipases (triacylglycerol acylhydrolases, EC 3.1.1.3) are the important biocatalysts because of their excellent biochemical and physiological properties. Lipases are the hydrolytic enzymes that can be used in various industrial applications for alcoholysis, acidolysis, amynolysis and hydrolysis reactions. Biodiesel production is one of the stunning applications of lipase. Lipase catalyzed biodiesel production was reported first by Mittelbach [16]. Lipase-catalyzed transesterification takes place in two steps, which involves hydrolysis of the ester bond and esterification with the second substrate [15]. A ping-ping bi bi mechanism generally used for kinetic studies of enzyme catalyzed transesterification.

Lipases can be isolated from many species of plants (papaya latex, oat seed lipase, and castor seed lipase), animals (pig's and human pancreatic lipases), bacteria, filamentous fungi and yeast [17-19]. For industrial enzyme production generally microorganisms are preferred because of their shortest generation time [20]. The other advantages of microorganisms can be listed as high yield of conversion of substrate into product, great versatility to environmental conditions and, simplicity in genetic manipulation and in cultivation conditions [20]. Although lipases from different sources are able to catalyze the same reaction, bacterial and fungal lipases are mostly used in biodiesel production such as *Aspergillus niger,Candida antarctica, Candida rugosa, Chromobacterium Viscosum, Mucor miehei, Pseudomonas cepacia, Pseudomonas fluorescens, Photobacterium lipolyticum, Rhizopus oryzae, Streptomyces sp., and Thermomyces lanuginose* [21]. *Candida rugosa*, obtained from yeast, is the most used microorganism for lipase production [22]. Recently, *Streptomyces sp.* was investigated as a potent lipase producing microbe for biodiesel production and found applicable in the field of biodiesel [23].

Specificity of lipases has a great importance in the selection of the usage area of lipases. Lipases can be divided into three groups due to their specificity as 1,3-specific lipases, fatty acid-specific lipases and nonspecific lipases. Especially, 1,3-specific lipases which release fatty acids from positions 1 and 3 of a glyceride and hydrolyze ester bonds in these positions such as *Aspergillus niger, Rhizopus oryzae and Mucor miehei* catalyze transesterification reactions efficiently [20,24]. The study of Du et al. [25], showed that higher yield (90%) was achieved for biodiesel production by using a sn-1,3-specific lipase, *Thermomyces lanuginosa* immobilized on silica gel (Lipozyme TL IM). Thus, the use of sn-1,3- specific lipases can give rise to biodiesel yield of above 90% under appropriate conditions [24]. Substrate specificity of lipases is also a crucial factor towards the biodiesel production which acts on the choice of the proper enzyme based on the composition of raw materials by consisting in the capability of distinguishing structural features of acyl chains [20,24]. Lipases from *Pseudomonas fluorescens, Pseudomonas cepacia, Candida rugosa, Candida antarctica and Candida cylindracea* are suitable for transesterification reaction by displaying both wide substrate specificity and regiospecificity [24].

2.1. Immobilization of lipases

The immobilization of enzymes, which is attracting worldwide attention, was firstly report-ed in 1971 at Enzyme Engineering Conference [26]. During the past decade, chemical modi-fication, physical modification, and gene expression techniques have been developed to obtain more economic, active, selective, or stable lipases. Immobilization is a modification method that can be defined as attaching the enzyme onto an insoluble solid support materi-al [18]. By immobilization more operational and temperature stable lipases can be obtained and also lipases can be reused in the reactions. In addition, reusability of lipases will be a possible solution to the high cost of the enzymes and make them suitable for applications in industrial scale. The comparison of free enzymes and immobilized enzymes is given in Ta-ble 2. Methods for enzyme immobilization can be classified as adsorption, covalent bonding, entrapment, and cross-linking. The selection of method and support material is a prominent factor for obtaining an efficient lipase. The results of comparative studies revealed that the same lipase molecule can show very different catalytic activities after immobilization onto different supports [27].

Characteristics	Free Enzyme	Immobilized Enzyme
Price	High	Low
Efficiency	Low	High
Activity	Unstable	Stable
Reusability and recovery	Not possible	Possible
Tolerance to temperature, pH, etc.	Low	High
To separate from the substrate	Difficult	Easy
To separate from the product	Difficult	Easy

Table 2. The comparison of free enzyme and immobilized enzyme [19]

2.1.1. Adsorption technique

Adsorption is the adhesion of lipase on the surface of the adsorbent by weak forces, such as van der Walls, ionic and hydrophobic interactions, or dispersion forces [28]. Immobilization via adsorption method is the simply mixing of an aqueous solution of enzyme with the car-rier material for a period and washing away the excess enzyme from the immobilized en-zyme on the carrier after a time [29]. The level of adsorption is strictly related to the pH, temperature and ionic strength. Adsorption is the most widely employed method besides other methods because of its special commercial advantages and simplicity. Adsorption is the only reversible enzyme immobilization method. The advantages of adsorption is mild and easy preparing conditions, low cost, no need for chemical additives, the carrier can be recovered for repeated use, and high activity [30].

Various types of carriers used in immobilization of lipases. Acrylic resin, celite, polypropylene and textile membrane are broadly used carriers. Some of the reported results of adsorption technique based immobilized enzymes used in biodiesel production are summarized in Table 3. As can be seen from table generally the biodiesel yields using the enzymes obtained by adsorption method are higher than 85%. Novozym 435 is a commercial lipase, which is obtained by immobilization of *Candida antartica* lipase on acrylic resin and is a good catalyst that provides biodiesel yield higher than 90% with vegetable oil or waste cooking oil as feedstock [31]. The other commercialized lipase is known as *Candida sp.* 99–125 lipase immobilized on textile membrane, which can catalyze lard, waste oil and vegetable oils with higher yields that is more than 87% [31]. Besides many advantages of immobilization by adsorption method, the main disadvantage is that desorption of the lipase from the carrier occurs because of the weak interactions between the enzyme and support.

2.1.2. Covalent binding technique

Another approach is covalent binding technique, which is the formation of covalent bonds between the aldehyde groups of support surface and active amino acid residues on the surface of the enzyme [29]. A variety of supports have been used such inorganic materials, natural polymers (agarose, chitin and chitosan), synthetic polymers (hydrophobic polypeptides,nylon fibers) and Eupergit® (made by copolymerization of N,N'-methylene-bis-(methacrylamide), glycidyl methacrylate, allyl glycidyl ether and methacrylamide) for immobilization of lipases by covalent binding [56].The main advantage of covalent binding method is obtaining thermal and operational stable enzymes because of strong interactions between the lipase and the carrier [31]. The comparison of biodiesel production performance using immobilized lipase via covalent binding method is summarized in Table 4. Chitosan is a promising carrier as a natural polymer due to its membrane forming and adhesion ability, high mechanical strength and facility of forming insoluble in water thermally and chemically inert films [57]. Xie and Wang [58], reported a technique for immobilization of *Candida rugosa* lipase on magnetic chitosan microspheres for transesterification of soybean oil.The immobilized enzyme was determined as an effective biocatalyst for the transesterification reaction due to giving a good conversion of soybean oil and retaining its activity during the four cycles [58].

Using two immobilized lipases with complementary position specificity instead of one lipase is a new approach to produce a cost effective biodiesel [19]. Lipase from *Rhizopus orizae* and *Candida rugosa* was covalently bound to the silica, which was used to produce biodiesel from crude canola oil. Under optimum conditions, the conversion rate of degummed crude canola oil to fatty acid methyl esters was 88.9%, which is higher than the conversion obtained by free enzyme mixture (84.25%) [59].

Lipase Source	Carrier	Acid/Oil Source	Alcohol	Maximum Performance (%)	Reference
Burkholderia sp. C20	Alkyl-functionalized Fe₃O₄–SiO₂	Olive	Methanol	92 (conversion)	[32]
Candida antartica	Acrylic resin	Soybean	Methanol	92(yield)	[33]
Candida antartica	Acrylic resin	Soybean and rapeseed	Methanol	98.4 (conversion)	[34]
Candida antartica	Acrylic resin	Soybean and rapeseed	Methanol	"/>95 (conversion)	[35]
Candida antarctica B	Granular activated carbon	Palm	Isobutanol	100 (conversion)	[36]
Candida sp. 99–125	Textile membrane	Lard	Methanol	87.4 (yield)	[37]
Candida sp. 99–125	Textile (cotton) membrane	Salad	Methanol	96 (conversion)	[38]
Candida sp. 99–125	Textile membrane	Crude rice bran	Methanol	87.4 (yield)	[39]
Candida rugosa and Pseudomonas fluorescens	Acurel	Palm	Ethanol	89 (yield)	[40]
Chromoacterlum viscosum	Celite-545	Jatropha	Ethanol	92 (yield)	[41]
Geobacillus thermocatenulatus	Poly-hydroxybutyrate beads	Babassu	Ethanol	100 (yield)	[42]
Pseudomonas aeroginosa	Celite	Soybean	Methanol	80(yield)	[43]
Pseudomonas cepacia	Celite	Jatropha	Ethanol	98 (yield)	[44]
Pseudomonas cepacia	Electrospun polyacrylonitrile fibers	Rapeseed	n-butanol	94 (conversion)	[45]
Pseudomonas cepacia	Polystyrene	Sapium sebiferum	Methanol	96.22 (yield)	[46]
Pseudomonas cepacia	Ceramic beads	Waste cooking	Methanol	40 (yield)	[47]
Pseudomonas fluorescens	Porous kaolinite particle	Triglyceride triolein	1-propanol	"/>90 (conversion)	[48]
Pseudomonas fluorescens and Pseudomonas cepacia	Polypropylene powder	Soybean	Methanol	58 37 (yield)	[49]
Penicillium expansum	Resin D4020	Waste	Methanol	92.8 (yield)	[50]
Rhizomucor miehei	Hydrophilic resins	Olive husk	Ethanol	-	[51]
Rhizomucor miehei	Silica	Waste cooking	Methanol	91.08 (yield)	[52]
Rhizopus oryzae	Macroporous resin HPD-400	Pistacia chinensis bge seed	Methanol	94 (yield)	[53]
Saccharomyces cerevisiae	Mg–Al hydrotalcite	Rape	Methanol	96 (conversion)	[5454]
Thermomyces lanuginosus (Lipozyme TL IM)	Hydrotalcite	Waste cooking	Methanol	95 (yield)	[55]

Table 3. Comparison of biodiesel production performance using immobilized lipase via adsorption method

Lipase Source	Carrier	Acid/Oil Source	Alcohol	Maximum Performance (%)	Reference
Burkholderia cepacia	Niobium Oxide (Nb₂O₅)	Babassu	Ethanol	74.13 (yield)	[60]
Burkholderia cepacia	Polysiloxane–Polyvinyl Alcohol (SiO₂–Pva)	Babassu Beef Tallow	Ethanol	100 89.7 (yield)	[60]
Candida rugosa	Chitosan Microspheres	Soybean	Methanol	87 (conversion)	[58]
Candida rugosa	Chitosan Powder	Rapeseed Soapstock	Methanol	95 (conversion)	[61]
Enterobacter aerogenes	Silica	Jatropha	Methanol	94 (yield)	[62]
Porcine pancreatic	Chitosan Beads	Salicornia	Methanol	55 (conversion)	[63]
Pseudomonas fluorescens	Toyopearl Af-Amino-650m Resin	Babassu	Ethanol	94.9 (yield)	[64]
Rhizopus oryzae	Resin Amberlite Ira-93	Pistacia Chinensis Bge Seed	Methanol	92 (yield)	[63]
Rhizopus oryzae	Polystyrene Polymer(Amberlite Ira-93)	Soybean	Methanol	90.05 (yield)	[65]
Rhizopus Orizae +Candida rugosa	Silica	-	Methanol	"/>98 (conversion)	[66]
Rhizopus orizae +Candida rugosa	Silica	Crude Canola	Methanol	88.9 (conversion)	[59]
Thermomyces lanuginosus	Olive Pomace	Pomace	Methanol	93 (yield)	[67]
Thermomyces lanuginosus	Polyglutaraldehyde Activated Styrene-Divinylbenzene Copolymer	Canola	Methanol	97 (yield)	[68]
Thermomyces lanuginosus	Toyopearl Af-Amino-650m Resin	Palm	Ethanol	100 (yield)	[64]
Thermomyces lanuginosus	Polyurethane Foam	Canola	Methanol	90 (yield)	[69]
Thermomyces lanuginosus	Aldehyde-Lewatit	Soybean	Ethanol	100 (conversion)	[70]
Thermomyces lanuginosus	Magnetic Fe₃O₄ Nano-Particles	Soybean	Methanol	90 (conversion)	[71]

Table 4. Comparison of biodiesel production performance using immobilized lipase via covalent binding method

2.1.3. Entrapment technique

Entrapment method is based on capturing of the lipase within a polymer network that retains the enzyme but allows the substrate and products to pass through [72]. This method can be simply defined as mixing an enzyme with a polymer solution and then crosslinking the polymer to form a lattice structure that captures the enzyme [29]. Entrapment is often used for industrial applications because the method is fast, cheap and can be carried out under mild conditions [73]. Entrapment can be divided into three categories such as gel or fiber entrapping and microencapsulation [74]. A number of supports have been investigated such as alginate, celite, carrageenan, resins, acrylic polymers etc. Some carriers used for entrapment and the biodiesel production yields obtained by these enzymes are displayed in Table 5. A disadvantage of entrapment method is the mass transfer problem due to the act of support as a barrier, so the lipase became effective only for low molecular weight substrates [19,75].

Lipase Source	Carrier	Acid/Oil Source	Alcohol	Maximum Performance (%)	Reference
Burkholderia cepacia	K-Carrageenan	Palm	Methanol	100 (conversion)	[76]
Burkholderia cepacia	Phyllosilicate Sol–Gel	Tallow and Grease	Ethanol	94 (yield)	[77]
Burkholderia cepacia	Mtms-Based Silica Monolith Coated With Butyl-Substituted Silicates	Jatropha	Methanol	95 (yield)	[78]
Candida antarctica	Celite®	Triolein	Methanol	60 (conversion)	[79]
Candida rugosa	Calcium Alginate Matrix	Palm	Ethanol	83 (yield)	[80]
Candida rugosa	Activated Carbon	Palm	Ethanol	85 (conversion)	[81]
Pseudomonas cepacia	Hydrophobic Sol–Gel	Soybean	Methanol	67 (conversion)	[82]
Pseudomonas fluorescens Mtcc 103	Alginate	Jatropha	Methanol	72 (yield)	[83]
Via Encapsulation Method					
Burkholderia cepacia	Silica Aerogels	Sunflower Seed	-	56 (conversion)	[84]
Burkholderia cepacia	K-Carrageenan	Palm	Methanol	100 (conversion)	[85]
Candida antartica	Silica Aerogels	Sunflower Seed	Methanol	90 (conversion)	[86]

Table 5. Comparison of biodiesel production performance using immobilized lipase via entrapment method

2.1.4. Cross linking technique

Cross-linking is another method for immobilization that can be defined as the interaction of a three dimensional network within enzyme, coupling reagent, and carrier [19]. The advantage of cross-linking is obtaining stable lipases due to the strong interaction between the lipase and the carrier. On the other hand, the cross-linking conditions are intense and the immobilized lipase shows lower activity [31].

The high free fatty acid content of waste cooking oil form water by esterification with alcohol which cause agglomeration of lipase and lowering biocatalysis efficiency [87]. Hence, free *Geotrichum sp.* lipase was not a suitable enzyme catalyst for transesterification of waste cooking oil. Yan et al. [87], report a modification procedure for preparation of cross-linked *Geotrichum sp.* The obtained lipase exhibited improved pH and thermostable stability compared to free lipase. The relative biodiesel yield was 85% for transesterification of waste cooking oil with methanol.

Kumari et al. [88] studied the preparation of *Pseudomonas cepacia* lipase cross-linked enzyme aggregates. It was shown that cross linked lipases has a greater stability than free enzymes to the denaturing conditions. The enzyme also used to catalyze madhuca indica oil, which's transesterification is difficult by chemical routes due to its high free fatty acid content. As a result, 92% conversion was obtained after 2.5 h.

Immobilization of *Candida rugosa* lipase on fine powder of Scirpus grossus L.f. by glutaraldehyde by cross linked technique for biodiesel production from palm oil, as already investigated by Kensingh et al. [89]. It was concluded that immobilized lipase yielded higher conversion of biodiesel than that of free lipase.

Lorena et al. [90] investigated the immobilization of the *Alcaligenes spp.* lipase on polyethylenimine agarose, glutaraldehyde agarose, octyl agarose, glyoxyl agarose, Sepabeads® by the aggregation and crosslinking method. The transesterification of canola oil was achieved with a yield 80% using a six-step addition of methanol and lipase immobilized on Sepabeads® by the aggregation method.

All these methods are shown schematically in Figure 2.

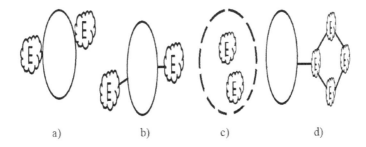

a)   b)   c)   d)

Figure 2. Schematic diagram of enzyme immobilization methods: a)Adsorption method b)Covalent binding method c)Entrapment method d) Crosslinking method

2.1.5. Whole cell immobilization

The applicability of lipases for the bulk production of fuels was limited significantly by the high cost of lipases [91]. Utilizing microbial cells such as fungi, bacteria, and yeasts cells containing intracellular lipase instead of extracellular lipases (free and immobilized lipase) is an easier and a cost effective way of enzymatic biodiesel production. Compared to conventional enzymatic processes, the use of whole cells provides excellent operational stability and avoids the complex procedures of isolation, purification and immobilization [91,92]. The general preparation steps for immobilized extracellular enzymes and whole cell enzymes showed in Figure 3. Biomass support particles have been used for immobilization of whole cells.

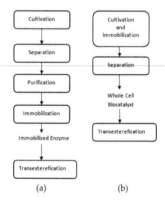

Figure 3. The preparation steps of a) immobilized extracellular lipase and b) whole cell biocatalyst

Aspergillus and Rhizopus have been most widely used as whole cell biocatalyst. Ban et al. [93], used first a whole cell biocatalyst, immobilized Rhizopus oryzae IFO4697 (a 1,3-positional specificity lipase) cells within biomass support particles, for the production of biodiesel and 91.1% methyl ester content was attained which was a similar result as that using the extracellular lipase. Many researchers have experimented on the use of whole cells to catalyze transesterification reaction summarized in Table 6.

A technique using glutaraldehyde cross-linking treatment on whole cell catalyst for methanolysis of soybean oil was developed by Sun et al. [94]. The glutaraldehyde cross linking treatment resulted in higher methanol tolerance and high catalytic activity (with the ratio of methanol to oil reaching 3). Also, a novel methanol addition strategy was proposed as stepwise addition of different amounts of methanol (1.0, 1.2, 1.5, and 2.0M equivalent of oil) every 24 h. It was found that the highest methyl ester yield could reach 94.1% after 24 h reaction by 1.2 mol, 1.5 M and 1.2 mol methanol additions at 0, 8, and 14 h. In general, the whole cell catalyzed process is slower than extracellular lipase catalyzed process. Sun et al. [94], also reported that the reaction time could be shortened by this way. It is clear that significant reduction in the cost of biodiesel production can be achieved by combining the whole cell biocatalyst process with stepwise addition of methanol.

Lipase Source	Carrier	Acid/Oil Source	Alcohol	Maximum Performance (%)	Reference
Aspergillus niger	BSPs[a]	Waste Cooking	Methanol	86.4 (yield)	[96]
Aspergillus niger	Polyurethane BSPs[a]	Palm	Methanol	>90 (yield)	[97]
Aspergillus niger	BSPs[a]	Palm	Methanol	87 (yield)	[98]
Aspergillus oryzae NS4	BSPs[a]	Soybean	Methanol	98 (conversion)	[99]
A. oryzae carrying r-CALB[b]	BSPs[a]	Palm Soybean	Methanol	85 90 (conversion)	[100]
Aspergillus oryzae expressing r-FHL[c]	BSPs[a]	Rapeseed	Methanol Ethanol 1-Propanol 1-Butanol	96 (yield) 94 (yield) 96 (yield) 97 (yield)	[101]
Escherichia coli BL21	-	Rapeseed	Methanol	97.7 (conversion)	[102]
Rhizopus chinensis CCTCC M201021	-	Soybean	Methanol	"/>86 (yield)	[103]
Rhizomucor miehei displaying Pichia pastoris	-	Soybean	Methanol	83.14 (yield)	[104]
Rhizopus oryzae IFO 4697	BSPs[a]	Refined Rapeseed Crude Rapeseed, Acidified Rapeseed	Methanol	~60(yield) ~60(yield) ~70(yield)	[105]
Rhizopus oryzae IFO 4697	BSPs[a]	Soybean	Methanol	~90(yield)	[106]
Rhizopus oryzae IFO 4697	BSPs[a]	Soybean	Methanol	~85(yield)	[107]
Rhizopus oryzae IFO4697	-	Soybean	Methanol	71 (conversion)	[108]
Rhizopus oryzae IFO 4697 and Aspergillus oryzae niaD300 (combined use)	BSPs[a]	Soybean	Methanol	~100 (conversion)	[109]

Lipase Source	Carrier	Acid/Oil Source	Alcohol	Maximum Performance (%)	Reference
Rhizopus oryzae ATCC 24563	-	Soybean (Free Fatty Acid Content 5.5%)	Methanol	97 (conversion)	[110]
Rhizopus oryzae IFO 4697	BSPs[a]	Soybean	Methanol	72 (yield)	[111]
Rhizopus oryzae	Polyurethane foam BSPs[a]	Soybean	Methanol	90 (conversion)	[112]
Rhizopus oryzae	BSPs[a]	Jatropha Curcas	Methanol	80 (conversion)	[113]
Rhizopus oryzae IFO 4697	-	Soybean	Methanol	86 (yield)	[114]
Rhizopus oryzae	BSPs[a]	Rapeseed	Methanol, Ethanol, 1-Propanol, 1-Butanol	83, 79, 93, 69 (yield)	[101]
Serratia marcescens YXJ-1002	-	Grease	Methanol	97 (yield)	[115]

[a]BSPs: Biomass support particles

[b]r-CALB: Candida antarctica lipase B

[c]r-FHL: Fusarium heterosporum lipase

Table 6. Comparison of biodiesel production performance using whole cell biocatalysts

Whole cell biocatalysts will be a way to industrialization of biodiesel production but the limited mass transfer efficiency of product and substrate is a hurdle to further investigations [95].

3. Feedstocks

The main aim of researches is to obtain a biodiesel, which will have a competitive price compared to other conventional sources of energy [116]. At this point, selecting the feedstock, represents more than 75-80% of the overall biodiesel production cost, is a vital step to ensure a cost effective biodiesel production. Different kinds of feedstock with varied range of edible and inedible vegetable oil, animal fats, waste oil, microbial oil and microalgae oil can be used for enzyme catalyzed transesterification [117].

3.1. Vegetable oils

Vegetable oils are candidates as alternative fuels for diesel engines with their high heat content [118]. But, direct use of vegetable oils is not possible because of the high kinematics viscosity of them which are varies in the range of 30–40 cSt at 38 °C and are about 10 times higher than of diesel fuel (Grade No. 2D) leads to many problems [118,119]. Therefore, modification of vegetable oil is necessary and the valuable product of this modification is named "biodiesel". The edible vegetable oils such as soybean [120,121], sunflower [122-124], palm [81,125], corn [126], cottonseed [127], canola [68,69,128] and olive [129,130] oils have been widely used in enzymatic transesterification. In developed countries, edible oils constitute more than 95% of biodiesel production feedstock because the produced biodiesel from these oils have properties very similar to petroleum-based diesel [131]. Also, the country and its climate, the oil percentage and the yield per hectare are effective parameters in selecting the potential renewable feedstock of fuel [118,132]. For example, while rapeseed oil prevailing the EU production, soybean oil prevailing the US and Latin American production, and palm oil mainly being used in Asia [133].

Inedible oils do not find a place in human consumption due to including toxic components. Therefore, inedible oils do not compete with food crops. Thus, inedible vegetable oils are an alternative feedstock for biodiesel production. Babassu (Orbinya martiana), Jatropha curcas (Linnaeus), neem (Azadiracta indica), polanga (Calophyllum inophyllum),karanja (Pongamia pinnata), rubber seed tree (Hevea brasiliensis), mahua (Madhuca indica and Madhuca longifolia), tobacco (Nicotina tabacum), silk cotton tree, etc. are promising inedible vegetable oil sources. Jatropha curcas is an attractive feedstock between various oil bearing seeds as it has been developed scientifically and found to give better biodiesel yield and productivity [134]. Crude Jatropha oil contains about 14% of free fatty acid that is too high for alkaline catalyzed biodiesel production [118]. However, high free acid content is not a problem in the production process of biodiesel via using enzyme catalysts. Besides Jatropha curcas, 26 species of fatty acid methyl ester of oils of including Azadirachta indica, Calophyllum inophyllum, and Pongamia pinnata were found most suitable for use as biodiesel, which adjust to the major specification of biodiesel standards of European Standard Organization, Germany, and USA [135]. Modi et al. reported conversion of crude oils of Pongamia pinnata (karanj), Jatropha curcas (jatropha) via immobilized Novozym 435 to biodiesel fuel with yield 90, and 92.7%, respectively [136].

3.2. Animal oils/fats

Animal fats are another group of feedstock for biodiesel production. Animal fats used to produce biodiesel via enzymatic route include lard [137], lamb meet [138] and beef tallow [139]. Animal fats are economically feasible feedstocks compared to vegetable oils. Animal fat methyl ester also has many favorably properties such as non-corrosive, high cetane number, and renewable [140,141]. However, animal fats saturated compounds lead to a tendency to oxidation and crystallization unacceptably at high temperatures [142].

3.3. Waste oils/fats

In general, around the world only half of the discharged edible oils recycled as animal feed or as raw material for lubricant and paint and the remainder is discharged into the environment [143]. Hence, the use of waste oils/fats for biodiesel production is very important to reduce and recycle the waste oil [143], to eliminate the environment and human health risk caused by waste oils [144] and to lower the biodiesel production cost. Waste cooking oil, animal fats, yellow grease, brown grease obtained from highly oxidized yellow grease or recovered waste grease from plumbing trap and waste sludge or soap-stock from the vegetable oil refining process were the major sources of waste oil have been used for biodiesel production [145]. The selection of a catalyst to be used for the production of biodiesel fuel is mostly influenced by the amount of free fatty acid content in various feedstocks [146]. The lipase-catalyzed reaction is a promising method for converting waste oils which contains high percentage of free fatty acids and high water content, into biodiesel with high yield [145]. It has been reported that Novozym 435 is capable of converting the used olive oils [129].

3.4. Algae oils

There is a considerable interest in the use of algae (micro and macro) oils for synthesis of biodiesel. Because these oils are cheap raw materials besides animal fats and have rapid growth rate and productivity when compared to conventional forestry, agricultural crops, high lipid content, tolerance for poor quality water, smaller land usage up to 49 or 132 times less when compared to rapeseed or soybean crops [142,147]. The smaller land usage brings the advantage of reducing the competition for arable soil with other crops, in particular for human consumption [147]. However, there are still some drawbacks for utilization of algae for biodiesel production. A considerable investment in technological development and technical expertise is needed to optimize the microalgae harvesting and oil extraction processes, to use cheap sources of CO_2 for culture enrichment [147]. Algae oils contain about 20-40% oil [148]. Several researchers have been experimented on microalgal oils as raw material for biodiesel production. Tran et al. [130], investigated the conversion of microalgal oil from Chlorella vulgaris ESP-31 to biodiesel by using immobilized Burkholderia lipase and a high fatty acid methyl esters conversion efficiency of 97.25 wt% oil (or 58.35 wt % biomass) was obtained for 48 h reaction. It is proposed that microalgal oil has good potential for application in the commercial production of biodiesel. The enzymatic conversion of microalgal oils to biodiesel in ionic liquids was firstly studied by Lai et al. [149]. Four microalgae two strains of Botryococcus braunii (BB763 and BB764), Chlorella vulgaris, and Chlorella pyrenoidosa have been catalyzed by two immobilized lipases, Penicillium expansum lipase and Candida antarctica lipase B (Novozym 435), in two solvent systems: an ionic liquid (1-butyl-3-methylimidazolium hexafluorophosphate, [BMIm][PF6]) and an organic solvent (tert-butanol). Penicillium expansum lipase was found more efficient for this application and the ionic liquid [BMIm] [PF6] showed a greater conversion yield (90.7% and 86.2%) obtained relative to the one obtained in the commonly used organic solvent tert-butanol (48.6% and 44.4%).

4. The effect of reaction parameterson enzymatic transesterification

4.1. The effect of temperature on enzymatic transesterification

Enzymatic transesterification takes place at low temperatures varying from 25 to 60°C. In general, initially the rate of reaction increases with rise in reaction temperature, because of an increase in rate constants with temperature and less mass transfer limitations [150,151]. Nevertheless, increased temperature after the optimum temperature promotes to denaturation and higher thermal deactivation of the enzyme, since it decreased the catalytic activity [152].

Various researches have been carried out to find out the effect of temperature on biodiesel production with immobilized enzymes. It is clear that immobilization provide more temperature resistance compared to free enzymes due to supplying a more rigid external backbone for lipase molecule [150,151]. However, optimum temperature is specific for each production. The studies about the effect of temperature for enzymatic transesterification are shown in Table 7.

Lipase	Oil Source	Alcohol	Performed Temperatures In The Range (°C)	Optimum Temperature (°C)	Reference
Immobilized Aspergillus niger	Palm	Methanol	25-50	40	[153]
Immobilized Aspergillus niger	Waste Cooking	Methanol	25-50	30	[154]
Immobilized Burkholderia cepacia	Babassu	Ethanol	39-56	39	[155]
Candida antarctica	Cotton Seed	T-Butanol	30-50	50	[156]
Candida antarctica	Acid	Methanol	30-50	30	[157]
Immobilized Candida Sp. 99–125	Salad	Methanol	27-50	40	[158]
Candida Sp. 99–125	Waste Cooking	Methanol	35-50	40-50	[159]
Immobilized Enterobacter aerogenes	Jatropha	T-Butanol	30-55	55	[160]
Immobilized Enterobacter aerogenes	Crude Rapeseed	Ethanol	25-50	35	[161]
Lipozyme RM IM	Soybean	Butanol	20-50	30	[162]
Lipozyme RM IM	Soybean	Methanol and Ethanol	40–60	50	[163]
Lipozyme RM IM	Soybean Oil Deodorizer Distillate	Ethanol	45-78	50	[164]
Lipozyme TL IM	Rapeseed	N-Butanol	30-60	40	[165]
Lipozyme TL IM	Soybean	Ethanol	20-50	35	[162]

Lipase	Oil Source	Alcohol	Performed Temperatures In The Range (°C)	Optimum Temperature (°C)	Reference
Lipozyme TL IM	Palm	Ethanol	30-78	50	[166]
Novozyme 435	Rapeseed	Methanol	25-55	40	[167]
Novozyme 435	Tung and Palm	Methanol and Ethanol	45-55	55	[168]
Novozym 435	Cottonseed	-(Dimethyl Carbonate As Organic Solvent)	30-55	50	[169]
Novozym 435	Canalo	Methanol	25-65	38	[170]
Novozym 435	Olive	Methanol	30-70	40	[129]
Novozym 435	Soybean	T-Amyl	30-60	40	[171]
Novozym 435	Sunflower	Methanol	25-65	45	[172]
Novozym 435	Stillingia	Methanol	30-60	40	[173]
Novozym 435	Cotton Seed	Methanol	30-70	50	[174]
Novozym 435, Lipozyme TL IM and Lipozyme RM IM	Soybean	Ethanol	25-60	25	[175]
Immobilized Penicillium expansum	Waste	T-Amyl	25-55	35	[176]
Immobilized Pseudomonas cepacia	Soybean	Methanol and Ethanol	25–60	35	[177]
Pseudomonas cepacia	Soybean	Methanol	20–60	30	[178]
Immobilized Pseudomonas fluorescens	Triolein	1-Propanol	40-70	60	[48]
Pseudomonas fluorescens	Soybean	Methanol	30-60	40	[49]
Rhizopus chinensis CCTCC M201021	Soybean	Methanol	30-40	30	[179]
Thermomyces lanuginosus	Canola	Methanol	30-70	40	[69]

Table 7. Data on optimum temperature for enzymatic biodiesel production

4.2. The effect of water content on enzymatic transesterification

Water content is one of the key factors for enzymatic transesterification reaction that have a strong effect on lipase's active three-dimensional conformational state [21,180]. Biocatalysts, needs a small amount of water to retain their activities [181]. Lipase has an unique feature on the water-oil interface, and the lipase activity depends on this interface. The presence of an oil–water interface required because it provides a suitable environment for enzyme activation which occurs due to the unmasking and restructuring the active site through conformational changes of the lipase molecule [182,183]. When the addition of water increased, the amount of water available for oil to form oil–water droplets also increases, hence increasing the available interfacial area [182]. Thus, enzymatic activity can not be possible in a water free media. However, excess water cause reverse reaction of hydrolysis. The amount of required water, to provide an optimum enzyme activity, differs according to the type of enzyme and reaction medium composition. Enzymes, substrates, organic solvent and also immobilized support have a crucial role on optimal water activity for lipase [184]. Optimum water content not only provides keeping the hydrolysis of ester linkages at the minimum level, but also ensures the highest degree of transesterification [24]. Thus, a better control of water content is very important for enzymatic process.

Water activity (a_w) is defined as free (boundness) water in the system, which is a ratio of vapor pressure over the given system versus that over pure water [24]. Thermodynamic water activity is the best predictor of reaction rate that can be determined in any phase by different kinds of sensors such as holographic sensor, Weiss LiCl humidity sensor [180,185]. Also, several methods have been developed for control of water activity, for example, equilibration with saturated salt solutions [186], addition of salt hydrate pairs [187,188] and introduction of air or nitrogen into the reactor [189]. Recently, Peterson et. al. developed a practical way for control of water activity in large-scale enzymatic reactions by using a programmable logic controller. On the other hand, percentage water content is another expression which is used widely in transesterification, generally assayed by Karl-Fischer coulometer.

In general, lipases show higher activity with higher water activities in solvent free systems instead of Candida antarctica lipase (Novozym 435) [184]. For Candida sp. 99–125 lipase, the optimum water content is 10–20% based on the oil weight to maintain the highest transesterification activity [31].

Salis et al., investigated production of oleic acid alkyl esters by using Pseudomonas cepacia and determined that a_w in the range 0.4–0.6, 1-butanol:triolein 3:1 – were the best conditions to reach maximum enzymatic activity. It was also found that at the higher values of water activity, no hydrolysis reaction was occurred [190].

Noureddini and Philkana [82] tested immobilized Pseudomonas cepacia for the transesterification of soybean oil with methanol and ethanol and observed that increased addition of water provide a considerable increase in the ester yield. The optimal conditions were determined for processing 10 g of soybean oil by 475 mg lipase in 1 h as 1:7.5 oil/methanol molar ratio, 0.5 g water in the presence of methanol that resulted in 67 % yield and 1:15.2 oil/ethanol molar ratio, 0.3 g water in the presence of ethanol that resulted in 65% yield.

Al-Zuhair et al. studied the esterification of n-butyric acid with methanol in the presence of Mucor miehei lipase, and found similar results with literature [191] that higher water content, makes lipase more efficient [182].

Shah and Gupta used immobilized Pseudomonas cepacia lipase for ethanolysis of Jatropha oil and noted that the best yield 98% gained by in the presence of 4–5% (w/w) water in 8 h. The yield was only 70% in absence of water [44].

Kawakami et al. determined the effect of water content for transesterification of Jatropha oil and methanol to characterize Burkholderia cepacia lipase immobilized in an n-butyl-substituted hydrophobic silica monolith. The authors reported that biodiesel yield reached 90% with water content of 0.6% (w/w) after 12 h using a stoichiometric mixture of methanol and oil (3:1) [78].

Chen et al. investigated the effect of water content for production of biodiesel with oleic acid with methanol catalyzed by soluble lipase NS81020, produced by modification of Aspergillus oryzae microorganism, in the biphasic aqueous-oil systems and found that the esterification yield is low if the water was scant. The higher reaction rate and fatty acid methyl ester yield was obtained with 10 wt % water by oleic acid weight [192].

It is clear that during the past decade numerous investigations have been made to determine the optimal water content for transesterification. As a result, the necessary amount of water content is an important factor to create an interfacial surface between oil and water and to ensure optimal enzymatic activity. Also, water has a strong influence on structural integrity, active site polarity, and protein stability of lipase [21,193]. However, it differs from enzyme to reaction conditions.

4.3. The effect of acyl acceptors on enzymatic transesterification

Methanol, short chain alcohol, usually used as an acyl acceptor due to its low price and availability. Insoluble and a relatively high amount of methanol with respect to oil, have a negative influence on the stability of lipases and could be solved by a stepwise addition of the alcohols [15, 194]. To eliminate inhibitory effects of methanol some co-solvents are added to the reaction mixture. Tert-butanol is one of the important co-solvents which is added to enzymatic reaction. Usage of tert-butanol, a polar solvent, is also a possible solution for eliminating the inhibitory effects of methanol and glycerol (both of them soluble in tert-butanol) and suggested instead of using butanol [195]. Liu et al. [196], transesterified waste baked duck oil by three different commercial immobilized lipases (Novozym 435, Lipozyme TLIM and Lipozyme RMIM) with different monohydric alcohols (methanol, ethanol, propanol, isopropanol, isobutanol, isoamyl alcohol) and fusel oil-like alcohol mixture (containing 15% isobutanol, 80% isoamyl alcohol, 5% methanol) in solvent-free and tert-butanol systems. It was reported that each lipase presented a different kinetic pattern depending on the monohydric alcohols. The results showed that Lipozyme TL IM and Novozym 435 gave high conversion rate with isobutanol and isoamyl alcohol either in solvent-free or in tert-butanol system. Thus, the combined use of lipases, Novozym 435 and Lipozyme TLIM, as catalyst and fusel oil-like mixture as raw material for biodiesel synthesis was found effective in view of cost saving of biodiesel production [195].

Recently, novel acyl acceptors were investigated such as ethyl acetate, methyl acetate, butyl acetate, vinyl acetate [197], dimethyl carbonate [198]. Du and coworkers demonstrated the positive effect of methyl acetate, on enzymatic activity of Novozym 435 and found that lipase could be reused directly without any additional treatment [199]. The advantage of using methyl acetate is that the cost of the catalyst can be reduced dramatically due to the longer operational life and reusability of lipase. The byproduct of the system is triacetylglycerol, which does not have any negative effect on the fuel property, and also no glycerol produced [200]. Hence, these advantages will provide industrial implementation of enzymatic biodiesel production. Dimethyl carbonate is another promising alternative acyl acceptor, which is eco-friendly, odorless, cheap, non-corrosive, and non-toxic [200]. The transesterification reaction is irreversible, because carbonic acid monoacyl ester, the intermediate compound, immediately decomposes to carbon dioxide and alcohol [200]. The fatty acid methyl ester yield is higher for lipase-catalyzed transesterification of vegetable oils with dimethyl carbonate besides conventional acyl acceptors (methanol and methyl acetate) [200]. Only, the higher price of acyl acceptor besides alcohols is a disadvantage [194].

4.4. Effects of the solvent on enzymatic transesterification reaction

In enzymatic transesterification reaction, excess of alcohol increases reaction efficiency, but if alcohol doesn't dissolve in reaction medium it can disrupt the enzyme activity. Methanol and vegetable oil in the values close to 1:1 molar ratio forms a solution in 40°C. Solvent is added into the reaction medium to increase the solubility of alcohol and thus it allows first step enzymatic transesterification by blocking degradation lipase catalytic activity [24]. To overcome deactivation of lipase activity and improve the lipase activity, various organic solvents have been used for enzymatic biodiesel synthesis. These solvents have been listed in Table 8. Cyclohexane, n-hexane, tert-butanol, petroleum ether, isooctane and 1,4-dioxane are mainly studied hydrophilic and hydrophobic organic solvents in enzymatic biodiesel production. In organic solvent medium, overall alcohol is added at the beginning of the reaction. In solvent free reaction medium, alcohol is added in several portions to prevent enzyme activity with high alcohol concentration [24].

Hexane is generally preferred because of its low cost and easily availability in the market. Some studies were performed in hexane solvent systems with soybean and tallow oil using monohydric alcohols [70,201, 202]. Nelson et al. performed transesterification of tallow with monohydric alcohols by Lipozyme IM 60 (M. miehei) and Novozyme SP435 (C. antarctica) in hexane and a solvent-free system. They compared the transesterification yields of two different systems. The yields with higher than 95% were obtained with methanol, ethanol and butanol with Lipozyme IM 60 lipase under hexane system (Table 8) while reaction yields under solvent-free system were 19% for methanol, 65.5% for ethanol, and 97.4% for isobutanol [201]. Similar results were found by Rodrigues et al. [70]. They compared the yields of transesterification of soybean with ethanol by Lipozyme TL IM. In the presence of n-hexane with 7.5:1 molar ratio of ethanol:soybean oil, the transesterification conversion was found to be as 100% while in solvent-free system the yield was 75%. At stoichiometric molar ratio, the yield was 70% conversion after 10 h of reaction in both systems. Transesterification conversion was obtained as 80% by three stepwise addition of ethanol, while a two step ethanolysis produced 100% conversion after 10 h of reaction in both solvent and solvent-free systems.

In enzyme catalyzed reaction, both alcohol amount and low glycerol solubility in biodiesel have negative effects on enzyme activity. Deposit of glycerol coating the immobilized catalyst is formed during the process, which reduces the enzymes activity [203]. The solubility of methanol and glycerol in hydrophobic solvents is low. For this reason, this problem may occur in hydrophobic solvent system.

The enzymatic alcoholysis of triglyceride also was studied with petroleum ether, isooctane, cyclo hexane,1,4-dioxane (Table 4-2) [16,48,204]. Iso et al. [48], reported that when methanol and ethanol were used as alcohol in enzymatic transesterification, the reactions need an appropriate organic solvent [48]. On the other hand, the reaction could be performed without solvent when 1-propanol and 1-butanol was used. They also used, benzene, chloroform and tetrahydrofuran as solvent and immobilized P. fluorescens lipase as catalyst at 50°C to compare the results that of the 1,4 dioxane. The highest enzymatic activity was observed with 1,4-dioxane. The enzymatic activity increased with the high amount of 1,4-dioxane. But high conversion of oil (app.90%) to biodiesel was obtained with high proportion of 1,4 dioxane(90%). Although usage of high amount of solvents is not preferable in industry solvents can be recovered together with methanol after transesterifiation reaction.

Hydrophilic organic solvents can interact with water molecule in enzyme and this may affect the catalytic activity of enzyme. However, as shown in Table 8 high performance was ensured with hydrophilic solvents such as 1,4-dioxane and tert-butanol [48,156, 205-208]. Some studies were performed in the presence of t-butanol solvent because of positive effects on enzymatic catalyzed reaction. T-butanol has moderate polarity so methanol and glycerol are easily soluble in tertiary butanol. Solubility of methanol prevents enzyme inhibition and solubility of glycerol prevents accumulation on the enzyme carrier material. Another advantage of this solvent is sinteric hindrance. Due to this property, tert-butanol is not accepted by the lipase. High yield and conversions were obtained in the presence of t-butanol with various vegetable oils and immobilized lipases shown in Table 4-2. For example, Liu et al., [196] studied biodiesel synthesis by immobilized lipases in solvent-free and tert-butanol media. Each lipase showed a different conversion depending on the monohydric alcohols and immobilized lipase in solvent-free medium and tert-butanol system. For methanolysis, regardless of the lipase type, the conversion rate is higher in tert-butanol than that in solvent-free medium. Novozym 435 showed higher conversion rate with straight monoalcohols in tert-butanol medium. Lipozym RM IM and Lipozyme TL IM showed lower conversion with straight and branched monoalcohols (except methanol) in solvent free system. Similar results were obtained by Halim and Kamaruddin [208], in transesterification of waste cooking palm oil using various commercial lipases (Lipozyme RM IM, Lipozyme TL IM and Novozyme 435) in tert-butanol as reaction medium. Novozyme 435 was found to be more effective in catalyzing the transesterification with methanol in in-tert-butanol medium. It was also been demonstrated that even 3:1 methanol to oil molar ratio didn't inhibit the Novozyme 435 in tert-butanol system. Du et al. [209], showed that Lipozyme TL IM could be used without loss of lipase activity for 200 batches in tert-butanol system. Li et al. [210], used acetonitrileand tert-butanol mixture as co-solvent in transesterfication of stillingia oil with methanol. The highest biodiesel yield (90.57%) was obtained in co-solvent with 40% tert-butanol and 60% acetonitrile (v/v) with co-solvent. They also reported that co-solvent (as a mixture)enhance the tolerance of lipase to the methanol than the pure tert-butanol.

Solvent	Oil	Alcohol	Lipase	Temp/ Time	Reaction mixture	Performance (%)	Ref.
Tert-butanol	Cotton seed	Methanol	Novozyme 435 (Candida antarctica)	50 °C / 24h	13.5% meth., 54% oil 32.5% tert-butanol, Lipase:1.7% (wt of oil)	97 (yield)	[156]
Tert-butanol	Cotton seed	Methanol	Pancreatic lipase	37 °C / 4 h	Methanol :oil mol ratio:1:15 Lipase:0.5% enzyme (wt of oil) water conc.5% (wt of oil)	75–80 (conversion)	[205]
Tert-butanol	Rapeseed	Methanol	Novozyme435 & Lipozyme TL IM	35 °C / 12 h	Methanol: oil mol. ratio 4:1 tert-butanol/oil vol. 1:1 Lipase: 3% Lipozyme TL IM 1% Novozym 435 (wt of oil)	95 (conversion)	[206]
Tert-butanol	Soybean and deodorizer distillate	Methanol	Lipozyme TL IM Novozym 435	40°C/ 12 h	Methanol:oil molar ratio 3.6:1 Lipase :3% Lipozyme TL IM 2% Novozym 435 tert-butanol: 80% (wt of oil)	84 (yield)	[207]
Tert-butanol	Waste cooking palm	Methanol	Novozyme 435	40°C / 12 h	Methanol:oil mol. ratio 4:1, Lipase:4% (wt of oil)	88(yield)	[208]
Tert-butanol	Waste baked duck	Methanol	Novozym 435 Lipozyme TL IM	45 °C / 20 h	Methanol:oil mol. ratio 4:1, Lipase: 5 wt%(wt of oil)	85.4, 78.5, (conversion)	[196]
Hexane	Tallow	Methanol Ethanol Propanol	Lipozyme IM 60	45 C/ 5 h	0.34 M tallow in hexane (8 mL), Lipase: 10 (wt of oil) 200rpm	94.8, 98.0, 98.5 (conversion)	[201]

Solvent	Oil	Alcohol	Lipase	Temp/ Time	Reaction mixture	Performance (%)	Ref.
Hexane	Soybean	Methanol	Lipozyme IM 77	36.5°C/ 3h	Methanol :oil mol ratio:3.4:1 Lipase:0.9BAUN*of lipase; water 5.8% (wt% of oil)	92.2 (yield)	[202]
Hexane	Soybean	Ethanol	Lipozyme TL IM	30 °C/ 10 h	Ethanol:oil mol.ratio:7.5:1 Lipase: 15 %(wt of oil). 4% water	100 (conversion)	[70]
Cyclo hexane	Sunflower	Methanol	Lipase AK Lipozyme TL IM Lipozyme RM IM	40°C/ 24 h	Volume of organic solvent/ oil: 2 ml/0.2 mmol Lipase: 10% (wt of oil)	65, 75, 35 (conversion)	[204]
Acetonitrile 60%and 40% t-butanol (v/v)	Stillingia	Methanol	Novozym 435 and Lipozyme TL IM	40°C/ 24h	Methanol:oil mol ratio: 6.4:1 Lipase: 4% (w/w) of multiple-lipase (1.96% Novozym 435+2.04% Lipozyme TL IM)	90.57 (yield)	[210]
Petroleum ether	Sunflower	Ethanol	Lipozyme IM Lipase AK	45°C / 5h	Ethanol:oil mol. ratio:11:1 Lipase:20% (wt of oil)	82, 99, (yield)	[16]
I-octane	Sunflower	Methanol	Lipase AK Lipozyme TL, IM Lipozyme RM,IM	40 °C	Methanol: oil mol ratio::3:1 Vol. of organic solvent/oil: 2 ml/0.2 mmol	80, 65, 60, (yield)	[204]
1,4-dioxane	Triolein	Methanol	Lipase AK	50°C / 80h	Methanol:oil mol. ratio: 3:1 90% solvent	~70 (conversion)	[48]

*BAUN:Batch Acidolysis Units Novo

Table 8. Effect of the solvent on the performance of enzymatic transesterification reaction

Although positive effects of the usage of the solvents on the transesterification reaction, some drawbacks has also been known) such as; extra reactor volume, solvent toxicity and emissions, solvent recovery and loss cost [133].

4.5. The effect of molar ratio of alcohol to oil on enzymatic transesterification

Biodiesel yield always increased due to the molar excess of alcohol over fatty acids in triglycerides in traditional transesterification system [15]. The transesterification reaction is reversible and so, an increase in the amount of one of the reactants will result in higher ester yield and minimally 3 molar equivalents of methanol are required for the complete conversion of methyl ester [174]. Conversely, for enzyme catalyzed transesterification, insoluble excess methanol which exists as fine droplets demonstrates negative effects on enzyme activity and also decrease the production yield [211]. The reaction medium is an important factor during the determination of the optimum molar of alcohol to oil. The inactivation of lipases occurs by contact with insoluble alcohol because the highly hydrophilic alcohol eliminates the layer of essential water from the enzymes [212]. Thus, stepwise addition of alcohol is a potential approach for ratio optimizing the molar ratio in solvent free systems [15]. Whilst, higher reaction rates could be obtained with a slight excess of alcohol in organic solvent systems [15].

The two-step reaction system was reported to avoid the inactivation of the lipase by addition of excess amounts of methanol in the first-step reaction, and by addition of vegetable oil and glycerol in the second-step reaction [213]. Watanabe et al. [213], used a two-step reaction system for methyl esterification of free fatty acids and methanolysis of triacylglycerols using immobilized Candida antarctica lipase. The first step reaction was methyl esterification of free fatty acids that was performed by treating a mixture of 66 wt % acid oil and 34 wt % methanol with 1 wt % immobilized lipase. The second step reaction was conducted to convert triacylglycerols to fatty acid methyl esters. In this step, a mixture of 52.3 wt % dehydrated first-step product, 42.2 wt% rapeseed oil, and 5.5 wt% methanol using 6 wt% immobilized lipase in the presence of additional 10 wt % glycerol was treated. The contents of fatty acid methyl esters was 91.1wt.% after the second step reaction was repeated by the use of immobilized lipase for 50 cycles using recovered glycerol.

Moreno-Pirajan and Giraldo [81], added different amounts of alcohol varied from 2.7 to 13.7 molar equivalents for methanol and from 5.7 to 26.7 molar equivalents for ethanol, based on the moles of triglycerides toward the transesterification of palm oil catalyzed by Candida rugosa lipase and 10.4 molar ratio for all alcohols to palm oil was determined as optimal alcohol requirement resulted in 85 mol% of methyl esters yield with n-butanol.

Lipase catalyzed esterification of palmitic acid with ethanol in the presence of Lipozyme IM 20 in a solvent free medium was investigated by Vieira et al. [212]. Different acid/alcohol molar ratios were tried as 0.16, 0.5, 1.0, 1.5, and 1.84. The best result was obtained with 0.5 acid/alcohol molar ratio.

Zaidi et al. [214], explained the correlation existing between the kinetic parameters and the chain-length of the substrates in esterification of oleic acid using nylon-immobilized lipase in n-hexane. It is observed that the inhibition coefficient of the alcohol increased from 0.034

to 0.42 mol l^{-1}, when the number of carbon atoms increased from 1(methanol) to 18 (oleyl alcohol), respectively.

Dizge and Keskinler [69], used immobilized Thermomyces lanuginosus lipase to produce biodiesel with canola oil with methanol and investigated the role of substrate molar ratio. The biodiesel production was conducted at 1:1, 1:2,1:3,1:4,1:5;1:6 and 1:10 oil/alcohol molar ratios at 40°C. The highest methyl ester yield (85.8%) was obtained at the oil/methanol molar ratio of 1:6. Two important result from this study can be concluded as (i) an increase in the number of moles of methanol resulted in an increase in the ester production, (ii) when the formation of esters reached a maximum level the further increases in the methanol concentrations cause a decrease in the formation of esters due to enzyme inactivation.

Thus, the actual amount of alcohol needed varies significantly depending on the origin of the lipase and fat.

5. Reactors for enzymatic transesterification

Through the industrialization of enzymatic biodiesel production, it is necessary to show the applicability of enzymes in reactor systems. Various reactors, including batch reactors, packed bed reactors and supercritical reactors have been investigated by researchers. Most of the investigations on enzymatic synthesis of biodiesel have been performed in batch reactors and packed bed reactors.

Batch reactors are simple designs used in the laboratory. In batch reactors, methanol shows a good dispersion in the oil phase. But the physical agitation caused by shear stress from the stirring would disrupt the enzyme carrier which shortens the enzymes life [31]. On the other hand, batch operation is labor intensive, and not suitable for automation [215]. Packed bed reactors are alternative of batch reactors which are substantially faster and more economical continuous reactors [216]. A packed-bed reactor system is most widely used in biotechnology, as it is easy to operate and scale up these systems. In addition, these systems have high bed volume. The most important advantage of these systems is that it is lowering shear stress on immobilized enzymes which leads to long-term enzyme stability [217]. Furthermore, stepwise addition of alcohol can be performed to reduce the inactivation of the enzyme caused by excess alcohol. One of the encountered problems with an immobilized lipase is the inhibition of the enzyme due to the cloggage of the catalyst by accumulation of the glycerol by-product inside the reactor [218]. Also, the separation of glycerol which remains in the bottom of the reactor can be achieved in a simple way by using more than one column. Recently, a packed-bed reactor system, in which a reactant solution is pumped through a column containing biomass support particles immobilized recombinant Aspergillus oryzae and the effluent from the column is recycled into the same column with a stepwise addition of methanol was developed by Yoshida et al. [219]. In this system, lipase retains its activity for five batch cycles and 96.1% methyl ester content was obtained with a residence time of 140 min per pass and stepwise addition of 4.25 molar equivalents of methanol to oil for 6 passes. The methanolysis of soybean oil in packed bed reactor system using

Rhizopus oryzae whole cell was studied by Hama et al. [112]. The final methyl ester content was over 90% at a flow rate of 25 l/h in the first cycle and also, after 10 cycles approximately 80% conversion was achieved. Wang et al. [216], developed Pseudomonas cepacia lipase – Fe_3O_4 nanoparticle biocomposite based packed bed reactors. A single-packed-bed reactor and the four-packed-bed reactor were used to produce biodiesel by using refined soybean oil. A high conversion rate (over 88%, 192 h) and great stability was achieved with the four-packed-bed reactor compared to single-packed-bed reactor. It is considered that the four-packed-bed reactor supplied a longer residence time of the reaction mixture in the reactor and lowered the inhibition of the lipase by products [216]. By this way, the reaction efficiency was improved. Additionally, the cost of biodiesel production can be reduced by the effective recycling of the enzyme catalysts [184].

Supercritical reactors also have been investigated by researchers for enzymatic biodiesel production. D. Oliveira and J. V. Oliveira [220], produced biodiesel from palm kernel oil in the presence of Novozym 435 and Lipozyme IM in supercritical carbon dioxide in the temperature range of 40–70 °C and from 60 to 200 bar using a water concentration of 0–10 wt % and oil/ethanol molar ratios from 1:3 to 1:10. Lipozyme IM showed better results and the highest reaction conversion was obtained as 77.5 %. It was observed that lipase structure changed at pressures beyond 200 bar. Madras et al. [221], synthesized biodiesel from sunflower oil in supercritical carbon dioxide catalyzed by Novozym. However, the obtained conversions, when the reaction was conducted in supercritical methanol and ethanol at the optimum conditions, were 23 and 27%, respectively [221]. Enzymatic transesterification of lamb meat fat in supercritical carbon dioxide was investigated by Taher et al. [222].The maximum conversion (49.2%) was obtained at 50°C, with 50% Novozym 435 loading, 4:1 molar ratio, within 25 h reaction. Supercritical reactors could not commercialized according to the low conversion rate and cost of the system.

Consequently, packed bed reactor systems seem to be a practical transesterification reactor system with high transesterification efficiency. These systems will bring industrial scale up enzymatic biodiesel production in an economic way.

6. Conclusion

Today, the growing energy necessity and environmental pollution problem requires the use of renewable alternative energy sources to become less dependent on fossil resources. As known, biodiesel is an important alternative energy resource and seems to be the fuel of future because it is an environmentally friendly, nontoxic, renewable, and biodegradable fuel.

Conventionally, biodiesel production is achieved by mainly alkaline or acid catalysts. The interest in the use of biocatalyst for biodiesel production has been an increasing trend due to its many advantages.

Biodiesel have been shown to be effectively produced by enzymatic catalyst and also, numerous researches have been performed to obtain highly active lipases and to optimize process conditions for biodiesel production. Besides many advantages, to produce biodiesel by enzyme catalysts on an industrial scale, it is necessary to reduce the high cost of enzymes and obtain lipases with better features. The immobilization of lipases and genetic engineering methods seems to be an attractive way to obtain more active, stable, and reusable lipases in organic solvents and alcohols. Also, selection of alternative acyl-acceptors is an option for eliminating the negative effects of methanol on lipase activity.

It can be concluded that in enzyme catalyzed biodiesel production significant progresses have been made but further improvements such as novel reactor design should be addressed and emphasized in the future research in order to ensure industrial enzymatic biodiesel production. By making novel improvements, much attention will be focused on enzyme usage in biodiesel production, and especially lipase reactions will be applied much more in this area.

Author details

Sevil Yücel[1*], Pınar Terzioğlu[1,2] and Didem Özçimen[1]

*Address all correspondence to: yuce.sevil@gmail.com syucel@yildiz.edu.tr

1 Yıldız Technical University, Faculty of Chemical and Metallurgical Engineering, Bioengineering Department, Istanbul, Turkey

2 Mugla Sıtkı Koçman University, Faculty of Sciences, Chemistry Department, Mugla, Turkey

References

[1] Demirbas A.Biodiesel: a realistic fuel alternative for diesel engine. London:Springer Publishing Co.;2008.

[2] Demirbas A. Progress and recent trends in biodiesel fuels. Energy Conversion and Management 2009; 50: 14–34.

[3] Demirbas M F, Balat M.Recent advances on the production and utilization trends of bio-fuels: a global perspective. Energy Convers Manage. 2006; 47: 2371–81.

[4] Chetri A B, Watts K C, Islam M R. Waste cooking oil as an alternate feedstock for biodiesel production. Energies 2008; 1: 3–18.

[5] Bamgboye A I, Hansen A C. Prediction of cetane number of biodiesel fuel from the fatty acid methyl ester (FAME) composition. Int Agrophys 2008; 22: 21–9.

[6] Balat M, Balat H. A critical review of biodiesel as a vehicular fuel. Energy Convers Manage, 2008; 49: 2727–41.

[7] Kulkarni BM, Bujar BG.Shanmukhappa S. Investigation of acid oil as a source of bio-diesel. Indian J Chem Tech 2008; 15: 467–71.

[8] Gerpen JV, Shanks B, Pruszko R, Clements D, Knothe G.Biodiesel Production Tech-nology. National Renewable Energy Laboratory. CO, USA; 2004.

[9] Capehart B L. Encyclopedia of energy engineering and technology. CRC Press, Tay-lor & Francis LLC; 2007.

[10] Pandey A. Handbook of plant-based biofuels.CRC Press, Taylor & FrancisLLC; 2008.

[11] Ghaly A E, Dave D, Brooks M S, Budge S.Production of Biodiesel by Enzymatic Transesterification: Review. American Journal of Biochemistry and Biotechnology 2010; 6 (2): 54-76.

[12] Mittelbach M, Remschmidt C. Biodiesel-the comprehensive handbook. Boersedruck Ges.m.b.H.69–80; 2004.

[13] Jegannathan KR, Abang S, Poncelet D, Chan E S, Ravindra P.Production of Biodiesel Using Immobilized Lipase—A Critical Review. Critical Reviews in Biotechnology 2008; 28: 253–264.

[14] Akoh C C, Chang S W, Lee G C, Shaw J F. Enzymatic approach to biodiesel produc-tion. J. Agric. Food. Chem. 2007; 55: 8995–9005.

[15] Fan X. Enzymatic biodiesel production – the way of the Future, Lipid Technology 2012; 24(2).

[16] Mittelbach M. Lipase catalyzed alcoholysis of sunflower oil. J. Am. Oil Chem. Soc. 1990;67:168-170.

[17] Hasan F, Shah A A, Javed S, Hameed A. Enzymes used in detergents: lipases. African Journal of Biotechnology 2010; 9(31) : 4836-4844.

[18] Villeneuve P, Muderhwa J M, Graille J, Haas MJ. Customizing lipases for biocataly-sis: a survey of chemical, physical and molecular biological approaches. Journal of Molecular Catalysis B: Enzymatic 2000; 9 :113–148.

[19] Zhang B, Weng Y, Xu H, Mao Z. Enzyme immobilization for biodiesel production. Applied Microbiol Biotechnol 2012; 93:61–70. DOI 10.1007/s00253-011-3672-x

[20] Ribeiro B D, De Castro A M, Coelho M A Z, Freire D M G. Production and Use of Lipases in Bioenergy: A Review from the Feedstocks to Biodiesel Production. En-zyme Research 2011;1-16.

[21] Yahya A R M, Anderson W A, Moo-Young M. Ester synthesis in lipase catalyzed re-actions. Enzyme and Microbial Technology 1998; 23:438–450.

[22] Freire D M G, Castilho L R. Lipases em Biocat′alise. Enzimas em Biotecnologia. Pro-ducao, Aplicacoes e Mercado 2008;1: 369–385.

[23] Cho S S, Park D J, Simkhada J R, Hong J H, Sohng J K, Lee O H, Yoo J C. A neutral lipase applicable in biodiesel production from a newly isolated Streptomyces sp. CS326. Bioprocess Biosyst Eng 2012; 35:227–234.

[24] Antczak M S, Kubiak A, Antczak T, Bielecki S. Enzymatic biodiesel synthesis – Key factors affecting efficiency of the process. Renewable Energy 2009; 34:1185–1194.

[25] Du W, Xu Y, Liu D, Li Z. Study on acyl migration in immobilized lipozyme TL catalyzed transesterification of soybean oil for biodiesel production. J Mol Catal B: Enzym 2005;37:68–71.

[26] Hartmeier W. Immobilized Biocatalysts: An Introduction. Springer-Verlag, Berlin; 1988.

[27] Gomes FM, De Paula A V, Silva SG, De Castro HF. Determinação das propriedades catalíticas em meio aquoso e orgânico da lipase de Candida rugosa imobilizada em celulignina quimicamente modificada por carbonildiimidazol. Quim. Nova 2006; 29, 710–718.

[28] Jegannathan KR, Abang S, Poncelet D, Chan ES, Ravindra P. Production of biodiesel using immobilized lipase-a critical review. Crit Rev Biotechnol 2008; 28(4):253–264.

[29] Costa S A, Azevedo H S, Reis R L. Biodegradable Systems in Tissue Engineering and Regenerative Medicine .Enzyme Immobilization in Biodegradable Polymers for Biomedical Applications, CRC Press LLC; 2005.

[30] Fukuda H, Kondo A, Noda H. Biodiesel fuel production by transesterification of oils. J Biosci Bioeng 2001; 92(5):405–416. doi:10.1016/S1389-1723(01)80288-7

[31] Tan T, Lu J, Nie K, Deng L, Wang F, Biodiesel production with immobilized lipase: A review. Biotechnology Advances 2010; 28: 628–634.

[32] Tran D T, Chen C L, Chang J S. Immobilization of Burkholderia sp. lipase on a ferric silica nanocomposite for biodiesel production. Journal of Biotechnology 2012; 158: 112– 119.

[33] Du W, Xu Y, Liu D, Zeng J. Comparative study on lipase-catalyzed transformation of soybean oil for biodiesel production with different acyl acceptors. Journal of Molecular Catalysis B: Enzymatic 2004; 30: 125–129.

[34] Shimada Y, Watanabe Y, Samukawa T, Sugihara A, Noda H, Fukuda H, Tominaga Y. Conversion of Vegetable Oil to Biodiesel Using Immobilized Candida antarctica Lipase. J9045, JAOCS 1999; 76: 789–793.

[35] Watanabe Y, Shimada Y, Sugihara A, Noda H, Fukuda H, Tominaga Y. Continuous Production of Biodiesel Fuel from Vegetable Oil Using Immobilized Candida antarctica Lipase. J9352, JAOCS 2000;77:355–360.

[36] Naranjo J, Córdoba A, Giraldo L, García V, Moreno-Parajàn J C. Lipase supported on granular activated carbon and activated carbon cloth as a catalyst in the synthesis of biodiesel fuel. J. Mol. Catal. B: Enzymatic2010; 66: 166–171.

[37] Lu J, Nie K, Xie F, Wang F, Tan T. Enzymatic synthesis of fatty acid methyl esters from lard with immobilized Candida sp. 99–125. Process Biochem 2007;42:1367–70.

[38] Nie K, Xie F, Wang F, Tan T. Lipase catalyzed methanolysis to produce biodiesel: optimization of the biodiesel production. J Mol Catal B Enzym 2006;43:142–7.

[39] Li Z, Deng L, Lu J, Guo X, Yang Z, Tan T. Enzymatic synthesis of fatty acid methyl esters from crude rice bran oil with immobilized Candida sp. 99-125. ChineseJournal of Chemical Engineering 2010; 18:870-875.

[40] Tongboriboon K, Cheirsilp P, H-Kittikun A. Mixed lipases for efficient enzymatic synthesis of biodiesel from used palm oil and ethanol in a solvent-free system. J. Mol. Catal. B: Enzymatic 2010; 67: 52–59.

[41] Shah S, Sharma S, Gupta M N.Biodiesel preparation by lipase-catalyzed transesterification ofjatropha oil. Energy & Fuels2004;18: 154-159.

[42] MendesA A, Oliveirac P C, Velezb A M, Giordanob R C, Giordanob R L C, De Castro H F.International Journal of Biological Macromolecules 2012; 50 :503– 511.

[43] Ji Q, Xiao S, He B, Liu X. Purification and characterization of an organic solvent-tolerant lipase from Pseudomonas aeruginosa LX1 and its application for biodiesel production. J. Mol. Catal. B: Enzymatic 2010; 66: 264–269.

[44] Shah S, Gupta MN. Lipase catalyzed preparation of biodiesel from Jatropha oil in a solvent free system. Process Biochem 2007;42:409–14.

[45] Sakai S, Liu Y, Yamaguchi T, Watanabe R, Kawabe M, Kawakami K. Production of butyl-biodiesel using lipase physically-adsorbed onto electrospun polyacrylonitrile fibers. Bioresource Technology 2010; 101 : 7344–7349.

[46] Li Q, Yan Y. Production of biodiesel catalyzed by immobilized Pseudomonas cepacia lipase from Sapium sebiferum oil in micro-aqueous phase. Applied Energy2010;87: 3148–3154.

[47] Al-Zuhair S. Enzymatic Production of Bio-Diesel from Waste Cooking Oil Using Lipase. The Open Chemical Engineering Journal 2008; 2: 84-88.

[48] Iso J M, Chen B, Eguchi M, Kudo T, Shrestha S. Production of biodiesel fuel from triglycerides and alcohol using immobilized lipase. Journal of Molecular Catalysis B: Enzymatic 2001; 16: 53–58.

[49] Salis A, Pinna M, Monduzzi M, Solinas V, Comparison among immobilized lipases on macroporous polypropylene toward biodiesel synthesis. Journal of Molecular Catalysis B: Enzymatic 2008; 54: 19–26.

[50] Li N, Zong M, Wu H. Highly efficient transformation of waste oil to biodiesel by immobilized lipase from Penicillium expansum. Process Biochemistry 2009; 44: 685–688.

[51] De Paola M G, Ricca E, Calabrò V, Curcio S, Iorio G. Factor analysis of transesterification reaction of waste oil for biodiesel production. BioresourceTechnology 2009; 100:5126–513.

[52] Chen Y, Xiao B,Chang J, Fu Y, Lv P, Wang X. Synthesis of biodiesel from waste cooking oil using immobilized lipase in fixed bed reactor. Energy Conversionand Management 2009; 50:668–673.

[53] Li X, He X Y, Li Z L, Wang Y D, Wang C Y, Shi H, Wang F. Enzymatic production of biodiesel from Pistacia chinensis bge seed oil using immobilized lipase. Fuel 2012; 92: 89–93.

[54] Zeng H, Liao K, Deng X, Jiang H, Zhang F. Characterization of the lipase immobilized on Mg–Al hydrotalcite for biodiesel. Process Biochemistry 2009; 44:791–798.

[55] Yagiz F, Kazan D, Akin N A. Biodiesel production from waste oils by using lipase immobilized on hydrotalcite and zeolites. Chem. Eng. J. 2007; 134:262–267

[56] Knezevic Z, Milosavic N, Bezbradica D, Jakovljevic Z, Prodanovic R. Immobilization of lipase from Candida rugosa on Eupergit® C supports by covalent attachment. Biochemical Engineering Journal 2006; 30: 269–278.

[57] Stoytcheva M, Montero G, Toscano L, Gochev V , Valdez B. The immobilized lipases inbiodiesel production. Biodiesel – Feedstocks and Processing Technologies,InTech.

[58] Xie W, Wang J. Immobilized lipase on magnetic chitosan microspheres for transesterification of soybean oil. Biomass and Bioenergy 2012; 3 6: 373-380.

[59] Jang M G, Kim D K, Park S C, Lee J S, Kim S W. Biodiesel production from crude canola oil by two-step enzymatic processes. Renewable Energy 2012; 42: 99-104.

[60] Da Ros P, Silva G, Mendes A, Santos J, Castro H. Evaluation of the catalytic properties of Burkholderia cepacia lipase immobilized on non-commercial matrices to be used in biodiesel synthesis from different feedstocks. Bioresource Technology 2010; 101:5508–5516.

[61] Shao P, Meng X, He J, Sun P. Analysis of immobilized Candida rugosa lipase catalyzed preparation of biodiesel from rapeseed soapstock. Food and BioproductsProcessing 2008; 86: 283–289.

[62] Kumari A,Mahapatra P, Garlapati V, Banerjee R. Enzymatic transesterification of Jatropha oil. Biotechnology for Biofuels2009; 2:1, doi:10.1186/1754-6834-2-1

[63] Desai P D, Dave A M, Devi S. Alcoholysis of salicornia oil using free and covalently bound lipase onto chitosan beads. Food Chemistry 2006; 95: 193–199.

[64] Mendes A, Giordano R C, Giordano R L C, Castro H. Immobilization and stabilization of microbial lipases by multipoint covalent attachment on aldehyderesin affinity: Application of the biocatalysts in biodiesel synthesis. J. Mol. Catal. B:Enzymatic 2011; 68: 109–115.

[65] Wang Y, Shen X, Li Z, Li X, Wang F, Nie X, Jiang J. Immobilized recombinant Rhizopus oryzae lipase for the production of biodiesel in solvent free system. Journal of Molecular Catalysis B: Enzymatic 2010; 67: 45–51.

[66] Ho L J, Lee D H, Lim J S, Um B H, Park C, Kang S W, Kim S W. Optimization of the process for biodiesel production using a mixture of immobilized Rhizopus oryzae and Candida rugosa lipases. J. Microbiol. Biotechnol. 2008; 18:1927–1931.

[67] Yücel Y. Biodiesel production from pomace oil by using lipase immobilized onto olive pomace. Bioresource Technology 2011; 102: 3977–3980.

[68] Dizge N, Keskinler B, Tanriseven A. Biodiesel production from canola oil by using lipase immobilized onto hydrophobic microporous styrene-divinylbenzene copolymer. Biochem Eng J 2009;44:220–5.

[69] Dizge N, Keskinler B. Enzymatic production of biodiesel from canola oil using immobilized lipase. Biomass and Bioenergy 2008; 32:1274–1278.

[70] Rodrigues R, Pessela B,Volpato G, Fernandez-Lafuente R, Guisan J, Ayub M. Two step ethanolysis: A simple and efficient way to improve the enzymatic biodiesel synthesis catalyzed by an immobilized–stabilized lipase from Thermomyces lanuginosus. Process Biochemistry 2010; 45: 1268–1273.

[71] Xie W, Ma N. Enzymatic transesterification of soybean oil by using immobilized lipase on magnetic nano-particles. Biomass and Bioenergy2010; 34: 890-896.

[72] O'Driscoll, K. F. (1976) Techniques of enzyme entrapment in gels. In: Methods in Enzymology, volume XLIV, (Mosbach K., ed.), Academic Press, New York, NY, 169–183.

[73] Ikeda Y, Kurokawa Y, Nakane K, Ogata N.Entrap-immobilization of biocatalysts on cellulose acetate-inorganic composite gel fiber using a gel formation of cellulose acetate–metal (Ti, Zr) alkoxide.Cellulose2002; 9: 369–379.

[74] Brena B M, Batista-Viera F. Immobilization of Enzymes,Methods in Biotechnology: Immobilization of Enzymes and Cells, Second Edition, Totowa, NJ: Humana Press Inc.

[75] Murty V R, Bhat J, Muniswaran P K A. Hydrolysis of oils by using immobilized lipase enzyme: a review.Biotechnol. Bioprocess Eng. 2002; 7: 57-66.

[76] Jegannathan K, Jun-Yee L, Chan E, Ravindra P. Design an immobilized lipase enzyme for biodiesel production. J. Renewable and Sustainable Energy 2009; 1: 063101-1 -063101-8.

[77] Hsu A, Jones K, Marmer WN, Foglia TA. Production of alkyl esters from tallow and grease using lipase immobilized in a phyllosilicate sol–gel. J Am Oil Chem 2001;78:585–8.

[78] Kawakami K, Oda Y, Takahashi R. Application of a Burkholderia cepacia lipase immobilized silica monolith to batch and continuous biodiesel production with a stoi-

chiometric mixture of methanol and crude Jatropha oil. Biotechnology for Biofuels 2011; 4:42.

[79] Meunier S, Legge R. Evaluation of diatomaceous earth as a support for sol–gel immobilized lipase for transesterification. J. Mol. Catal. B: Enzymatic 2010; 62, 54–58.

[80] Sawangpanya N, Muangchim C, Phisalaphong M. Immobilization of lipase on Ca-CO_3 and entrapment in calcium alginate bead for biodiesel production,Sci. J. UBU 2010; 1(2): 46-51.

[81] Moreno-Pirajan J C, Giraldo L. Study of immobilized candida rugosa lipase for biodiesel fuel production from palm oil by flow microcalorimetry. Arabian Journal of Chemistry 2011; 4: 55–62.

[82] Noureddini H, Gao X, Philkana R S. Immobilized Pseudomonas cepacia lipase for biodiesel fuel production from soybean oil. Bioresource Technology 2005; 96: 769–777.

[83] Devanesan M G, Viruthagiri T, Sugumar N, Transesterification of jatropha oil using immobilized Pseudomonas fluorescens. African Journal of Biotechnology 2007; 6 (21): 2497-2501.

[84] Orçaire O, Buisson P, Pierre A. Application of silica aerogel encapsulated lipases in the synthesis of biodiesel by transesterification reactions. J. Mol. Catal. B:Enzymatic 2006; 42: 106–113.

[85] Jegannathan K R, Jun-Yee L, Chan E S, Ravindra P. Production of biodiesel from palm oil using liquid core lipase encapsulated in K-carrageenan.Fuel 2010; 89: 2272–2277.

[86] Nassreddine S, Karout A, Christ M, Pierre A. Transesterification of a vegetal oil with methanol catalyzed by a silica fibre reinforced aerogel encapsulated lipase. Applied Catalysis A: General 2008; 344: 70–77.

[87] Yan J, Yan Y, Liu S, Hu J, Wang G, Preparation of cross-linked lipase-coated microcrystals for biodiesel production from waste cooking oil. Bioresource Technology 2011; 102 : 4755–4758.

[88] Kumari V, Shah S, Gupta M N. Preparation of biodiesel by lipase-catalyzed transesterification of high free fatty acid containing oil from Madhuca indica. Energy & Fuels2007; 21:368-372.

[89] Kensingh P, Chulalaksananukul W, Charuchinda S. Lipase immobilization on Scirpus grossus L.f. fiber support byglutaraldehyde-crosslinked technique for biodiesel synthesis. Special Abstracts / Journal of Biotechnology 2010; 150S: 1–576.

[90] LorenS, Wilson L, Andrés I. Immobilization of the Alcaligenes spp. lipase to catalyze the transesterification of fatty acids to produce biodiesel. New Biotechnology 2009; 25S.

[91] Fukuda H, Hama S, Tamalampudi S, Noda H. Whole-cell biocatalysts for biodiesel fuel production. Trends in Biotechnology 2008;26:12.

[92] Li W, Du W, Liu D, Yao Y. Study on factors influencing stability of whole cell during biodiesel production in solvent-free and tert-butanol system. Biochemical Engineering Journal 2008; 41:111–115.

[93] Ban K, Kaieda M, Matsumoto T,Kondo A, Fukuda H. Whole-cell biocatalyst for biodiesel fuel production utilizing Rhizopus oryzae cells immobilized within biomass support particles. Biochem. Eng. J. 2001; 8:39–43.

[94] Sun T, Du W, Liu D, Dai L. Improved catalytic performance of GA cross-linking treated Rhizopus oryzae IFO 4697 whole cell for biodiesel production. Process Biochemistry 2010; 45: 1192–1195.

[95] Gog A, Roman M, Tos M, Paizs C, Irimie F D. Biodiesel production using enzymatic transesterification- Current state and perspectives. Renewable Energy 2012; 39: 10-16.

[96] Xıao M, Obbard J P. Whole cell-catalyzed transesterification of waste vegetable oil. GCB Bioenergy 2010; 2: 346–352.

[97] Xıao M, Qi C, Obbard J P. Biodiesel production using Aspergillus niger as a whole-cell biocatalyst in a packed-bed reactor.GCB Bioenergy 2011; 3: 293–298.

[98] Xıao M, Mathew S, Obbard J P. A newly isolated fungal strain used as whole-cell biocatalyst for biodiesel production from palm oil.GCB Bioenergy 2010; 2:45–51.

[99] Adachi D, Hama S, Numata T, Nakashima K, Ogino C, Fukuda H, Kondo A. Development of an Aspergillus oryzae whole-cell biocatalyst coexpressing triglyceride and partial glyceride lipases for biodiesel production. Bioresource Technology 2011; 102:6723–6729.

[100] Adachi D, Hama S, Nakashima K, Bogaki T, Ogino C, Kondo A. Production of biodiesel from plant oil hydrolysates using an Aspergillus oryzae whole-cell biocatalyst highly expressing Candida antarctica lipase B. Bioresource Technology 2012;in press: http://dx.doi.org/10.1016/j.biortech.2012.06.092

[101] Koda R, Numata T, Hama S, Tamalampudi S, Nakashima K, Tanaka T, Ogino C, Fukuda H, Kondo A. Ethanolysis of rapeseed oil to produce biodiesel fuel catalyzed by Fusarium heterosporum lipase-expressing fungus immobilized whole-cell biocatalysts. Journal of Molecular Catalysis B: Enzymatic 2010; 66: 101–104.

[102] Gao B, Su E, Lin J, Jiang Z, Ma Y, Wei D. Development of recombinant Escherichia coli whole-cell biocatalyst expressing a novel alkaline lipase-coding gene from Proteus sp. for biodiesel production. Journal of Biotechnology 2009; 139:169–175.

[103] He Q, Xu Y, Teng Y, Wang D. Biodiesel production catalyzed by whole-cell lipase from Rhizopus chinensis. Chin J Catal 2008; 29(1): 41–46.

[104] Huang D, Han S, Han Z, Lin Y. Biodiesel production catalyzed by Rhizomucor miehei lipase-displaying Pichia pastoris whole cells in an isooctane system. Biochemical Engineering Journal 2012; 63: 10– 14.

[105] Li W, Du W, Liu D. Rhizopus oryzae IFO 4697 whole cell catalyzed methanolysis of crude and acidified rapeseed oils for biodiesel production in tert-butanol system. Process Biochemistry 2007; 42: 1481–1485.

[106] Sun T, Du W, Zeng J, Dai L, Liu D. Exploring the effects of oil inducer on whole cell-mediated methanolysis for biodiesel production. Process Biochemistry 2010; 45: 514–518.

[107] Sun T, Du W, Zeng J, Liu D. Comparative study on stability of whole cells during biodiesel production in solvent-free system. Process Biochemistry 2011; 46: 661–664.

[108] Matsumoto T, Takahashi S, Kaieda M, Tanaka M U A, Fukuda H, Kondo A. Yeast whole-cell biocatalyst constructed by intracellular overproduction of Rhizopus oryzae lipase is applicable to biodiesel fuel production. Appl Microbiol Biotechnol 2001;57:515–520.

[109] Arai S, Nakashima K, Tanino T, Ogino C, Kondo A, Fukuda H. Production of biodiesel fuel from soybean oil catalyzed by fungus whole-cell biocatalysts in ionic liquids. Enzyme and Microbial Technology 2010; 46: 51–55.

[110] Lin Y H, Luo J J, Hwang S C J, Liau P R, Lu W J, Lee H T. The influence of free fatty acid intermediate on biodiesel production from soybean oil by whole cell biocatalyst. Biomass and Bioenergy 2011;35 : 2217-2223.

[111] Li W, Du W, Liu D. Optimization of whole cell-catalyzed methanolysis of soybean oil for biodiesel production using response surface methodology, Journal of Molecular Catalysis B: Enzymatic 2007;45 :122–127.

[112] Hama S, Yamaji H, Fukumizu T, Numata T, Tamalampudi S, Kondo A, Noda H, Fukuda H. Biodiesel-fuel production in a packed-bed reactor using lipase-producing Rhizopus oryzae cells immobilized within biomass support particles. Biochemical Engineering Journal 2007; 34: 273–278.

[113] Tamalampudi S, Talukder M R, Hama S, Numata T, Kondo A, Fukuda H. Enzymatic production of biodiesel from jatropha oil: a comparative study of immobilized-whole cell and commercial lipases as a biocatalyst. Biochemical Engineering Journal 2008; 39: 185–189.

[114] Zeng J, Du W, Liu X, Liu D, Dai L, Study on the effect of cultivation parameters and pretreatment on Rhizopus oryzae cell-catalyzed transesterification of vegetable oils for biodiesel production. Journal of Molecular Catalysis B: Enzymatic 2006; 43:15–18.

[115] Li A, Ngo T P N, Yan J, Tian K, Li Z. Whole-cell based solvent-free system for one-pot production of biodiesel from waste grease. Bioresource Technology 2012; 114: 725–729.

[116] Atabani A E, Silitonga A S, Badruddin I A, Mahlia TMI, Masjuki HH, Mekhilef S. A comprehensive review on biodiesel as an alternative energy resource and its characteristics. Renewable and Sustainable Energy Reviews 2012; 16: 2070– 2093.

[117] Demirbas A. Progress and recent trends in biodiesel fuels. Energy Conversion and Management 2009; 50:14–34.

[118] Koh M Y, Ghazi T I M. A review of biodiesel production from Jatropha curcas L. Oil. Renewable and Sustainable Energy Reviews 2011;15: 2240–2251.

[119] Singh S P, Singh D. Biodiesel production through the use of different sources and characterization of oils and their esters as the substitute of biodiesel: a review. Renewable and Sustainable Energy Reviews 2010;12:200–16.

[120] Mander P, Cho S S, Simkhada J R, Choi Y H, Park D J, Yoo J C. An organic solvent–tolerant lipase from Streptomyces sp. CS133 for enzymatic transesterification of vegetable oils in organic media. Process Biochemistry 2012; 47:635–642.

[121] Huang D, Han S, Han Z, Lin Y. Biodiesel production catalyzed by Rhizomucor miehei lipase-displaying Pichia pastoris whole cells in an isooctane system. Biochemical Engineering Journal 2012; 63: 10– 14.

[122] Karout A, Pierre A C. Partial transesterification of sunflower oil with ethanol by a silica fiber reinforced aerogel encapsulated lipase. J Sol-Gel Sci Technol 2009; 52:276–286.

[123] De los Ríos A P, Hernández Fernández FJ, Gómez D, Rubio M, Víllora G. Biocatalytic transesterification of sunflower and waste cooking oils in ionic liquid Media. Process Biochemistry 2011; 46: 1475–1480.

[124] Ognjanovic N, Bezbradica D, Knezevic-Jugovic Z. Enzymatic conversion of sunflower oil to biodiesel in a solvent-free system: Process optimization and the immobilized system stability. Bioresource Technology 2009; 100: 5146–5154.

[125] Talukder M M R, Das P, Fang T S, Wu J C. Enhanced enzymatic transesterification of palm oil to biodiesel. Biochemical Engineering Journal 2011; 55: 119–122.

[126] Mata T M, Sousa I R B G, Vieira S S, Caetano N S. Biodiesel Production from Corn Oil via Enzymatic Catalysis with Ethanol. Energy Fuels 2012;26: 3034–3041.

[127] Chattopadhyay S, Karemore A, Das S, Deysarkar A, Sen R. Biocatalytic production of biodiesel from cottonseed oil: Standardization of process parameters and comparison of fuel characteristics. Applied Energy 2011; 88: 1251–1256.

[128] Jang M G, Kim D K, Park S C, Lee J S, Kim S W. Biodiesel production from crude canola oil by two-step enzymatic processes. Renewable Energy 2012; 42: 99-104.

[129] Sanchez F, Vasudevan PT. Enzyme catalyzed production of biodiesel from olive oil. Appl Biochem Biotechnol2006; 135:1–14.

[130] Tran D T, Yeh K L, Chen C L, Chang J S. Enzymatic transesterification of microalgal oil from Chlorella vulgaris ESP-31 for biodiesel synthesis using immobilized Burkholderia lipase. Bioresource Technology 2012; 108:119–127.

[131] Leung D Y C, Wu X, Leung M K H. A review on biodiesel production using catalyzed transesterification. Applied Energy 2010;87:1083–95.

[132] Knothe G, Dunn R O. Biodiesel: an alternative diesel fuel from vegetable oils or animal fats, In: Industrial uses of vegetable oils. USA, Champaign. AOCS Press, ISBN 1-893997-84-7,2005.

[133] Nielsen P M, Brask J, Fjerbaek L. Enzymatic biodiesel production: Technical and economical considerations. Eur. J. Lipid Sci. Technol. 2008;110: 692–700.

[134] Jain S, Sharma MP. Prospects of biodiesel from Jatropha in India: a review. Renewable and Sustainable Energy Reviews 2010;14:763–71.

[135] Mohibbe A M, Amtul W, Nahar N M. Prospects and potential of fatty acid methyl esters of some non-traditional seed oils for use as biodiesel in India. Biomass Bioenergy 2005;29:293–302.

[136] Modi M K, Reddy J R C, Rao B V S K, Prasad R B N. Lipase-catalyzed mediated conversion of vegetable oils into biodiesel using ethyl acetate as acyl acceptor. Bioresour Technol 2007;98:1260–4.

[137] Lu J, Nie K, Xie F, Wang F, Tan T. Enzymatic synthesis of fatty acid methyl esters from lard with immobilized Candida sp. 99–125. Process Biochem 2007;42:1367–70.

[138] Al-Zuhair S, Hussein A, Al-Marzouqi A H, Hashim I. Continuous production of biodiesel from fat extracted from lamb meat in supercritical CO_2 media. Biochemical Engineering Journal 2012; 60: 106– 110.

[139] Da Ros P C M, De Castro H F, Carvalho A K F,. Soares C M F, De Moraes F F, Zanin G M. Microwave-assisted enzymatic synthesis of beef tallow biodiesel. Ind Microbiol Biotechnol 2012; 39:529–536.

[140] Guru M, Koca A, Can O, Cinar C, Sahin F. Biodiesel production from waste chicken fat based sources and evaluation with Mg based additive in a diesel engine. Renew Energy 2010;35:637–43.

[141] Balat M. Potential alternatives to edible oils for biodiesel production – A review of current work. Energy Conversion and Management 2011; 52: 1479–1492.

[142] Karatay S E, Donmez G. Microbial oil production from thermophile cyanobacteria for biodiesel production. Applied Energy 2011; 88: 3632–3635.

[143] Watanabe Y, Shimada Y, Sugihara A, Tominaga Y. Enzymatic conversion of waste edible oil to biodiesel fuel in a fixed-bed bioreactor. J9830, JAOCS 2001; 78: 703–707.

[144] Chen G, Ying M, Li W. Enzymatic conversion of waste cooking oils Intoalternative fuel-biodiesel. Applied Biochemistry and Biotechnology. 129–132: Humana press; 2006.

[145] Huynh L H, Kasim N S, Ju Y H. Biodiesel production from waste oils. Biofuels: Alternative Feedstocks and Conversion Processes. 375 -392, Elsevier Inc; 2011.

[146] Atadashi I M, Aroua M K, Aziz A R A, Sulaiman N M N, Production of biodiesel us-
       ing high free fatty acid feedstocks. Renewable and Sustainable Energy Reviews 2012;
       16: 3275– 3285.

[147] Mata T M, Martins A A, Caetano N S. Microalgae for biodiesel production and other
       applications: A review. Renewable and Sustainable Energy Reviews 2010;14: 217–
       232.

[148] Borugadda V B, Goud V V. Biodiesel production from renewable feedstocks: Status
       and opportunities. Renewable and Sustainable Energy Reviews 2012; 16: 4763–4784.

[149] Lai J Q, Hu Z L, Wang P W, Yang Z, Enzymatic production of microalgal biodiesel in
       ionic liquid [BMIm][PF6].Fuel 2012; 95: 329–333.

[150] Leca M, Tcacenco L, Micutz M, Staicu T.Optimization of biodiesel production by
       transesterification of vegetable oils using lipases.Romanian Biotechnological Letters
       2010; 15(5): 5618-5630.

[151] Al-Zuhair S. Production of biodiesel: possibilities and challenges.Biofuels, Bioprod.
       Bioref.2007;1: 57–66.

[152] Luković N, Knežević-Jugović Z, Bezbradica D. Biodiesel fuel production by enzymat-
       ic transesterification of oils: recent trends, Challenges and Future Perspectives, Alter-
       native Fuel,Intech

[153] Xıao M, Mathew S, Obbard JP. A newly isolated fungal strain used as whole-cell bio-
       catalyst for biodiesel production from palm oil. GCB Bioenergy 2010; 2: 45–51.

[154] Xıao M, Obbard JP. Whole cell-catalyzed transesterification of waste vegetable oil.
       GCB Bioenergy 2010; 2 : 346–352.

[155] Freitas L, Da Ros PCM, Santos JC, Castro HF. An integrated approach to produce bi-
       odiesel and monoglycerides by enzymatic interestification of babassu oil (Orbinya
       sp). Process Biochemistry 2009; 44: 1068–1074.

[156] Royon D, Daz M, Ellenrieder G, Locatelli S. Enzymatic production of biodiesel from
       cotton seed oil using t-butanol as a solvent. Bioresour Technol2007; 98:648-653.

[157] Watanabe Y, Nagao T, Nishida Y, Takagi Y, Shimada Y. Enzymatic production of fat-
       ty acid methyl esters by hydrolysis of acid oil followed by esterification. J Am Oil
       Chem Soc 2007; 84:1015–1021. DOI 10.1007/s11746-007-1143-4

[158] Nie K, Xie F, Wang F, Tan T. Lipase catalyzed methanolysis to produce biodiesel:
       Optimization of the biodiesel production. Journal of Molecular Catalysis B: Enzymat-
       ic 2006; 43: 142–147.

[159] Chen Y, Xiao B, Chang J, Fu Y, Lv P, Wang X. Synthesis of biodiesel from waste cook-
       ing oil using immobilized lipase in fixed bed reactor. Energy Conversion and Man-
       agement 2009; 50: 668–673.

[160] Kumari A, Mahapatra P, Garlapati V K, Banerjee R. Enzymatic transesterification of Jatropha oil.Biotechnology for Biofuels2009; 2:1.

[161] Zhang Z, Guan C. A promising alternate lipase for biodiesel fuel production. Front. Agric. China 2010; 4(2): 129–136, DOI 10.1007/s11703-010-0005-2.

[162] Rodrigues R C, Volpato G, Wada K, Ayub M A Z. Enzymatic synthesis of biodiesel from transesterification reactions of vegetable oils and short chain alcohols. J Am Oil Chem Soc 2008; 85:925–930 DOI 10.1007/s11746-008-1284-0

[163] Bernardes O L, Bevilaqua J V, Leal M C M R, Freire D M G, Langone M A P. Biodiesel fuel production by the transesterification reaction of soybean oil using immobilized lipase.Applied Biochemistry and Biotechnology 2007;136–140.

[164] Souza M S, Aguieiras E C G, Silva M A P, Langone M A P. Biodiesel synthesis via esterification of feedstock with high content of free fatty acids. Appl Biochem Biotechnol 2009;154:253–267.

[165] Kazanceva I, Makarevičienė V, Kazancev K. Application of biotechnological method to biodiesel fuel production using n-butanol.Environmental Research, Engineering and Management 2011;2(56): 35 – 42.

[166] Matassoli A L F, Corrêa I N S, Portilho M F, Veloso C O, Langone M A P. Enzymatic synthesis of biodiesel via alcoholysis of palm oil. Appl Biochem Biotechnol 2009; 155:347–355.

[167] Jeong G T, Park D H. Lipase-catalyzed transesterification of rapeseed oil for biodiesel production with tert-butanol. Appl Biochem Biotechnol 2008; 148:131–139, DOI 10.1007/s12010-007-8050-x

[168] Wang Y N, Chen M H, Ko C H, Lu P J, Chern J M, Wu C H, Chang F C. Lipase catalyzed transesterification of tung and palm oil for biodiesel. World Renewable Energy Congress 2011;Sweden,8-13 May.

[169] Su E Z, Zhang M J, Zhang J G, Gao J F, Wei D Z. Lipase-catalyzed irreversible transesterification of vegetable oils for fatty acid methyl esters production with dimethyl carbonate as the acyl acceptor. Biochemical Engineering Journal 2007; 36:167–173.

[170] Chang H M, Liao H F, Lee C C, Shieh C J. Optimised synthesis of lipase-catalyzed biodiesel by Novozym 435. J Chem Technol Biotechnol 2005; 80:307–312.

[171] Zheng Y, Quan J, Ning X, Zhu L M, Jiang B, He Z Y. Lipase-catalyzed transesterification of soybean oil for biodiesel production in tert-amyl alcohol. World J Microbiol Biotechnol 2009; 25:41–46.

[172] Ognjanovic N, Bezbradica D, Knezevic-Jugovic Z. Enzymatic conversion of sunflower oil to biodiesel in a solvent-free system: Process optimization and the immobilized system stability. Bioresource Technology 2009; 100: 5146–5154.

[173] Liu Y, Xin H l, Yan Y J. Physicochemical properties of stillingia oil: Feasibility for biodiesel production by enzyme transesterification. Industrial Crops and Products 2009; 30: 431–436.

[174] Köse O, Tüter M, Aksoy H A. Immobilized Candida antarctica lipase-catalyzed alcoholysis of cotton seed oil in a solvent-free medium. Bioresource Technology 2002; 83: 125–129.

[175] Hernandez-Martın E, Otero C. Different enzyme requirements for the synthesis of biodiesel: Novozym 435 and Lipozyme TL IM. Bioresource Technology 2008; 99: 277–286.

[176] Li N W, Zong M H, Wu H. Highly efficient transformation of waste oil to biodiesel by immobilized lipase from Penicillium expansum.Process Biochemistry 2009; 44: 685–688.

[177] Noureddini H, Gao X, Philkana RS.Immobilized Pseudomonas cepacia lipase for biodiesel fuel production from soybean oil. Bioresource Technology 2005; 96: 769–777.

[178] Gamba M, Lapis A A M, Dupont J. Supported ionic liquid enzymatic catalysis for the production of biodiesel. Adv. Synth. Catal. 2008; 350: 160 – 164.

[179] Qin H, Xu Y, Teng Y, Wang D. Biodiesel Production Catalyzed by whole-cell lipase from Rhizopus chinensis.Chinese Journal of Catalysis 2008; 29: 1.

[180] Petersson A E V, Adlercreutz P, Mattiasson B. A water activity control system for enzymatic reactions in organic media. Biotechnology and Bioengineering. 2007;97 (2): 1, ,DOI 10.1002/bit.21229.

[181] Bommarius A S, Riebel-Bommarius B R. Biocatalysts:Fundamentals and Applications. John Wiley & Sons, 266; 2000.

[182] Al-Zuhair S, Jayaraman K V, Krishnan S, Chan W H. The effect of fatty acid concentration and water content on the production of biodiesel by lipase. Biochemical Engineering Journal 2006; 30: 212–217.

[183] Al-Zuhair S, Hasan M, Ramachandran K B. Kinetic hydrolysis of palm oil using lipase. Proc Biochem2003; 38:1155–1163.

[184] Fjerbaek L, Christensen K V, Norddahl B. A review of the current state of biodiesel production using enzymatic transesterification. Biotechnology and Bioengineering 2009; 102 (5):1. DOI 10.1002/bit.22256

[185] Vermue M H, Tramper J. Bıocatalysıs ın non-conventıonal medıa: medıum engıneerıng aspects. Pure &Appl. Chem. 1995; 67(2): 345-373.

[186] Ujang Z, Vaidya A M. Stepped water activity control for efficient enzymatic interesterification. Appl Microbiol Biotechnol 1998; 50(3):318–322.

[187] Bell G, Halling P J, May L, Moore BD, Robb DA, Ulijn R, Valivety RH. 2001. Methods for measurement and control of water in non-aqueous biocatalysis. In: Vulfson EN,

Halling PJ, Holland HL, editors. Methods in Biotechnology. Totowa, NJ: Humana Press. p 105–126.

[188] Rosell C M, Vaidya A M. Twin-core packed-bed reactors for organic phase enzymatic esterification with water activity control. Appl Microbiol Biotechnol 1995; 44(3–4): 283–286.

[189] Won K, Lee S B. On-line conversion estimation for solvent-free enzymatic esterification systems with water activity control. Biotechnol Bioprocess Eng 2002; 7(2):76–84.

[190] Salis A, Pinna M, Monduzzi M, Solinas V. Biodiesel production from triolein and short chain alcohols through biocatalysis. Journal of Biotechnology 2005; 119: 291–299.

[191] Tweddell R J, Kermasha S, Combes D, Marty A. Esterification, interesterification activities of lipase from Rhizopus niveus and Mucor miehei in three different types of organic media: a comparitive study. Ezyme Microb. Technol. 1998; 22: 439–445.

[192] Chen X, Du W, Liu D. Effect of several factors on soluble lipase-mediated biodiesel preparation in the biphasic aqueous-oil systems. World J Microbiol Biotechnol 2008; 24:2097–2102,DOI 10.1007/s11274-008-9714-6

[193] Atadashi I M, Aroua M K, Abdul Aziz A R, Sulaiman N M N. The effects of water on biodiesel production and refining technologies: a review. Renewable and Sustainable Energy Reviews 2012; 16: 3456– 3470.

[194] Robles-Medina A, González-Moreno P A, Esteban-Cerdán L, Molina-Grima E. Biocatalysis: Towards ever greener biodiesel production. Biotechnology Advances 2009; 27: 398–408.

[195] Ranganathan S V, Narasimhan S L, Muthukumar K. An overview of enzymatic production of biodiesel.Bioresource Technology 2008; 99:3975–3981.

[196] Liu Y, Tan H, Zhang X, Yan Y, Hameed B H. Effect of monohydric alcohols on enzymatic transesterification for biodiesel production. Chemical Engineering Journal 2010;157:223–229.

[197] Jeong G T,Park D H. Synthesis of rapeseed biodiesel using short-chained alkyl acetates as acyl acceptor. Appl Biochem Biotechnol 2010; 161:195–208. DOI 10.1007/s12010-009-8777-7.

[198] Su E Z, Zhang M J, Zhang J G, Gaoa J F, Wei D Z. Lipase-catalyzed irreversible transesterification of vegetable oils for fatty acid methyl esters production with dimethyl carbonate as the acyl acceptor. Biochemical Engineering Journal 2007; 36: 167–173.

[199] Du W, Xu Y, Liu D, Zeng J. Comparative study on lipase-catalyzed transformation of soybean oil for biodiesel production with different acyl acceptors. Journal of Molecular Catalysis B: Enzymatic 2004; 30: 125–129.

[200] Xu Y, Du W, Liu D, Zeng J.A novel enzymatic route for biodiesel production from renewable oils in a solvent-free medium.Biotechnology Letters2003; 25:1239–1241.

[201]   Nelson LA, Foglia A, Marmer WN. Lipase-catalyzed production of biodiesel. J Am Oil Chem Soc 1996;73:1191–5.

[202]   Shieh C J, Liao H F, Lee C C. Optimization of lipase-catalyzed biodiesel by response surface methodology. Bioresource Technology 2003; 88: 103-106.

[203]   Dossat V, Combes D, Marty A. Continuous enzymatic transesterication of high oleic sunfower oil in a packed bed reactor:Infuence of the glycerol production. Enzyme Microb. Technol. 1999 ; 25: 194–200.

[204]   Soumanou MM, Bornscheuer UT. Improvement in lipase-catalyzed synthesis of fatty acid methyl esters from sunflower oil. Enz Microb Technol 2003;33(1):97–103.

[205]   Chattopadhyay S, Karemore A, Das S, Deysarkar A, Sen R.Biocatalytic production of biodiesel from cottonseed oil: Standardization of process parameters and comparison of fuel characteristics.Applied Energy 2011; 88: 1251–1256.

[206]   Li L, Du W, Liu D, Wang L, Li Z. Lipase-catalyzed transesterification of rapeseed oils for biodiesel production with a novel organic solvent as the reaction medium. J. Mol. Catal. B:Enzym. 2006 ;43 : 58–62.

[207]   Wang L, Du W, Liu D, Li L, Dai N. Lipase-catalyzed biodiesel production from soybean oil deodorizer distillate with absorbent present in tert-butanol system. Journal of Molecular Catalysis B: Enzymatic 2006;43: 29–32.

[208]   Halim S F Al, Kamaruddin A H. Catalytic studies of lipase on FAME production from waste cooking palm oil in a tert-butanol system. Process Biochemistry 2008; 43: 1436–1439.

[209]   Du W, Liu D, Li L, Dai L. Mechanism exploration during lipase-mediated methanolysis of renewable oils for biodieselproduction in a tert-butanol system. Biotechnol Prog. 2007 ;23: 1087–1090.

[210]   Li Q, Zheng J, Yan Y. Biodiesel preparation catalyzed by compound-lipase in co-solvent. Fuel Processing Technology 2010; 91:1229–1234.

[211]   Ganesan D, Rajendran A, Thangavelu V. An overview on the recent advances in the transesterification of vegetable oils for biodiesel production using chemical and biocatalysts. Rev Environ Sci Biotechnol 2009;8:367–394.

[212]   De Vieira A P A, Da Sılva MAP, Langone MAP, Biodiesel production via esterification reactions catalyzed by lipase. Latin American Applied Research 2006;36:283-288.

[213]   Watanabe Y, Pinsirodom P, Nagao T, Asao Y, Kobayashi T , Nishida Y, Takagi Y, Shimada Y, Conversion of acid oil by-produced in vegetable oil refining to biodiesel fuel by immobilized Candida antarctica lipase, Journal of Molecular Catalysis B: Enzymatic 2007; 44 : 99–105.

[214]   Zaidi A, Gainer J L, Carta G, Mrani A, Kadiri T, Belarbi Y, Mir A, Esterification of fatty acids using nylon-immobilized lipase in n-hexane: kinetic parameters and chain-length effects. Journal of Biotechnology 2002; 93: 209–216.

[215] Chen Y H, Huang Y H, Lin R H, Shang N C. A continous-flow biodiesel production process using a rotating packed bed. Bioresource Technology 2010; 101: 668–673.

[216] Wang X, Liu X, Zhao C, Ding Y, Xu P. Biodiesel production in packed-bed reactors using lipase–nanoparticle biocomposite.Bioresource Technology 2011; 102: 6352–6355.

[217] Hama S, Tamalampudi S, Yoshida A, Tamadani N, Kuratani N, Nodaa H, Fukuda H, Kondo A. Enzymatic packed-bed reactor integrated with glycerol-separating system for solvent-free production of biodiesel fuel.Biochemical Engineering Journal 2011;55: 66–71.

[218] Xu Y, Nordblad M,Woodley JM. A two-stage enzymatic ethanol-based biodiesel production in a packed bed reactor. Journal of Biotechnology2010; doi:10.1016/j.jbiotec.2012.05.017

[219] Yoshida A, Hama S, Tamadani N, Fukuda H, Kondo A. Improved performance of a packed-bed reactor for biodiesel production through whole-cell biocatalysis employing a high-lipase-expression system.Biochemical Engineering Journal 2012; 63: 76–80.

[220] Oliveira D, Oliveira J V. Enzymatic alcoholysis of palm kernel oil in n-hexane and SC CO_2. Journal of Supercritical Fluids 2001; 19 (2): 141–148.

[221] Madras G, Kolluru C, Kumar R. Synthesis of biodiesel in supercritical fluids. Fuel 2004;83 :2029–2033.

[222] Taher H, Al-Zuhair S, AlMarzouqui A, Hashim I. Extracted fat from lamb meat by supercritical CO2 as feedstock for biodiesel production. Biochemical Engineering Journal 2011; 55: 23–31.

Permissions

The contributors of this book come from diverse backgrounds, making this book a truly international effort. This book will bring forth new frontiers with its revolutionizing research information and detailed analysis of the nascent developments around the world.

We would like to thank Prof. Dr. Zhen Fang, for lending his expertise to make the book truly unique. He has played a crucial role in the development of this book. Without his invaluable contribution this book wouldn't have been possible. He has made vital efforts to compile up to date information on the varied aspects of this subject to make this book a valuable addition to the collection of many professionals and students.

This book was conceptualized with the vision of imparting up-to-date information and advanced data in this field. To ensure the same, a matchless editorial board was set up. Every individual on the board went through rigorous rounds of assessment to prove their worth. After which they invested a large part of their time researching and compiling the most relevant data for our readers. Conferences and sessions were held from time to time between the editorial board and the contributing authors to present the data in the most comprehensible form. The editorial team has worked tirelessly to provide valuable and valid information to help people across the globe.

Every chapter published in this book has been scrutinized by our experts. Their significance has been extensively debated. The topics covered herein carry significant findings which will fuel the growth of the discipline. They may even be implemented as practical applications or may be referred to as a beginning point for another development. Chapters in this book were first published by InTech; hereby published with permission under the Creative Commons Attribution License or equivalent.

The editorial board has been involved in producing this book since its inception. They have spent rigorous hours researching and exploring the diverse topics which have resulted in the successful publishing of this book. They have passed on their knowledge of decades through this book. To expedite this challenging task, the publisher supported the team at every step. A small team of assistant editors was also appointed to further simplify the editing procedure and attain best results for the readers.

Our editorial team has been hand-picked from every corner of the world. Their multi-ethnicity adds dynamic inputs to the discussions which result in innovative outcomes. These outcomes are then further discussed with the researchers and contributors who give their valuable feedback and opinion regarding the same. The feedback is then collaborated with the researches and they are edited in a comprehensive manner to aid the understanding of the subject.

Apart from the editorial board, the designing team has also invested a significant amount of their time in understanding the subject and creating the most relevant covers. They scrutinized every image to scout for the most suitable representation of the subject and create an appropriate cover for the book.

The publishing team has been involved in this book since its early stages. They were actively engaged in every process, be it collecting the data, connecting with the contributors or procuring relevant information. The team has been an ardent support to the editorial, designing and production team. Their endless efforts to recruit the best for this project, has resulted in the accomplishment of this book. They are a veteran in the field of academics and their pool of knowledge is as vast as their experience in printing. Their expertise and guidance has proved useful at every step. Their uncompromising quality standards have made this book an exceptional effort. Their encouragement from time to time has been an inspiration for everyone.

The publisher and the editorial board hope that this book will prove to be a valuable piece of knowledge for researchers, students, practitioners and scholars across the globe.

List of Contributors

Didem Özçimen, M. Ömer Gülyurt and Benan İnan
YıldızTechnical University, Faculty of Chemical and Metallurgical Engineering, Bioengineering Department, Istanbul, Turkey

Y.M. Sani, W.M.A.W. Daud and A.R. Abdul Aziz
Department of Chemical Engineering, Faculty of Engineering, University Malaya, Kuala Lumpur, Malaysia
Department of Chemical Engineering, Ahmadu Bello University, Zaria, Nigeria

Mushtaq Ahmad, Muhammad Zafar, Shazia Sultana, Haleema Sadia and Mir Ajab Khan
Biofuel Lab., Department of Plant Sciences, Quaid-i-Azam University Islamabad, Pakistan
School of Chemical Engineering, University of Sains Malaysia, Malaysia

Rosana C. S. Schneider, Thiago R. Bjerk, Pablo D. Gressler, Maiara P. Souza, Valeriano A. Corbellini and Eduardo A. Lobo
Environmental Technology Post-Graduation Program, University of Santa Cruz do Sul, UNISC, Brazil

Alexandre Reis Machado and Olinto Liparini Pereira
Departamento de Fitopatologia, Universidade Federal de Viçosa, Viçosa, MG, Brazil

Rodrigo A. A. Munoz, David M. Fernandes, Tatielli G. G. Barbosa and Raquel M. F. Sousa
Institute of Chemistry, Federal University of Uberlândia (UFU), Uberlândia-MG, Brazil

Douglas Q. Santos
Technical School of Health Sciences, Federal University of Uberlândia (UFU), Uberlândia-MG, Brazil

Camila da Silva and Lúcio Cardozo Filho
Program of Post-Graduation in Chemical Engineering, State University of Maringa, Maringá, Brazil

Ignácio Vieitez and Ivan Jachmanián
Departamento de Ciencia y Tecnología de los Alimentos, Facultad de Química, UDELAR, Montevideo, Uruguay

Fernanda de Castilhos
Department of Chemical Engineering, Paraná Federal University, Curitiba, Brazil

José Vladimir de Oliveira
Department of Chemical and Food Engineering, Federal University of Santa Catarina, Florianópolis, Brazil

Mario Nieves-Soto
Facultad de Ciencias del Mar, Universidad Autónoma de Sinaloa, Mazatlán, México

Oscar M. Hernández-Calderón
Facultad de Ciencias Químico Biológicas, Universidad Autónoma de Sinaloa, Culiacán, México

Carlos Alberto Guerrero-Fajardo
Facultad de Química, Universidad Nacional de Colombia, Bogotá, México

Marco Antonio Sánchez-Castillo
Universidad Autónoma de San Luis Potosí, San Luis Potosí, México

Tomás Viveros-García
Departamento de Ingeniería de Procesos e Hidráulica, Universidad Autónoma Metropolitana-Iztapalapa, México

Ignacio Contreras-Andrade
Facultad de Ciencias Químico Biológicas, Universidad Autónoma de Sinaloa, Culiacán, México

Sevil Yücel and Didem Özçimen
Yıldız Technical University, Faculty of Chemical and Metallurgical Engineering, Bioengineering, Department, Istanbul, Turkey
Mugla Sıtkı Koçman University, Faculty of Sciences, Chemistry Department, Mugla, Turkey

Pınar Terzioğlu
Yıldız Technical University, Faculty of Chemical and Metallurgical Engineering, Bioengineering, Department, Istanbul, Turkey

Printed in the USA
CPSIA information can be obtained
at www.ICGtesting.com
JSHW011455221024
72173JS00005B/1081

9 781632 402554